*The Best American Science
and Nature Writing 2021*

GUEST EDITORS OF
THE BEST AMERICAN SCIENCE
AND NATURE WRITING

2000 DAVID QUAMMEN

2001 EDWARD O. WILSON

2002 NATALIE ANGIER

2003 RICHARD DAWKINS

2004 STEVEN PINKER

2005 JONATHAN WEINER

2006 BRIAN GREENE

2007 RICHARD PRESTON

2008 JEROME GROOPMAN

2009 ELIZABETH KOLBERT

2010 FREEMAN DYSON

2011 MARY ROACH

2012 DAN ARIELY

2013 SIDDHARTHA MUKHERJEE

2014 DEBORAH BLUM

2015 REBECCA SKLOOT

2016 AMY STEWART

2017 HOPE JAHREN

2018 SAM KEAN

2019 SY MONTGOMERY

2020 MICHIO KAKU

2021 ED YONG

The Best American Science and Nature Writing 2021™

Edited and with an Introduction
by Ed Yong

Jaime Green, Series Editor

MARINER BOOKS
An Imprint of HarperCollins*Publishers*
Boston   New York

"Long May They Reign" by Nora Caplan-Bricker. First published in *The Atavist*, June 1, 2020. Copyright © 2020 by Nora Caplan-Bricker. Reprinted by permission of Nora Caplan-Bricker.

"It's Not Too Late to Save Black Lives" by Julia Craven. First published in *Slate*, May 21, 2020. Copyright © 2020 by *Slate*. Reprinted by permission of *Slate*.

"What the Coronavirus Means for Climate Change" by Meehan Crist. First published in *The New York Times*, March 27, 2020. Copyright © 2020 by The New York Times Company. All rights reserved. Used under license.

"The Empty Space Where Normal Once Lived" by Bathsheba Demuth. First published in *The Atlantic*, August 28, 2020. Copyright © 2020 by Bathsheba Demuth. Reprinted by permission of the author and *The Atlantic*.

"The Covid Drug Wars That Pitted Doctor vs. Doctor" by Susan Dominus. First published in *The New York Times Magazine*, August 5, 2020. Copyright © 2020 by The New York Times Company. All rights reserved. Used under license.

"What Happened in Room 10?" by Katie Engelhart. First published in *California Sunday Magazine*, August 19, 2020. Copyright © 2020 by Katie Engelhart. Reprinted by permission of Katie Engelhart.

"The Friendship and Love Hospital" by Jiayang Fan. First published in *The New Yorker*, April 6, 2020. Copyright © 2020 by Jiayang Fan. Reprinted by permission of The Wylie Agency, LLC.

"Out There, Nobody Can Hear You Scream" by Latria Graham. First published

Contents

Connections

Consequences

Foreword

ONE OF THE things I love about helping to edit this series is the breadth of stories I read and the breadth of stories we publish. But this year, of course, feels different. Alongside the essays about the cosmos and earthworms and our relationships with the outdoors are the essays about aerosol spread and exhausted physicians and insufficient recovery, trouble breathing and inequality and illness and death.

I can imagine futures where an unusual confluence of subject in this anthology reflects some thrilling advance or discovery—science and nature writing from the year we discover intelligent life beyond Earth?—and past years have had their own ripples, clusters of essays about animal extinctions, a yearly series of harrowing reporting on wildfires. But for 2020 this book is a portrait of a world upended, drawn in the work of writers who tried to make sense of it for us all.

This was a year when we desperately needed science writing, and also saw how science writing alone wasn't enough. As Ed Yong writes in his introduction, the Covid-19 pandemic was a crisis of science and nature and of so much more—politics, social tensions, education, inequality. It was an "omnicrisis," as Ed puts it, both because it touched on everything and because it was all-consuming. Any time I've been aware that I'm living through history has almost always been awful; 2020 was a whole year of that. But I'm writing this with one dose of a vaccine in my body, and having last week hugged my mom for the first time in over a year. Those moments were historic too. Last year in my foreword I looked out onto the rest of 2020 with thoughts of worlds ending; today there is, if I'm brave enough, some hope (and God I hope I haven't jinxed it).

While Ed spent 2020 as a reporter, I spent it as a reader, at least when it came to the coronavirus. Many science writers, like Ed and like my local paper's sports reporter, were wrangled into writing about the pandemic ("just for a few weeks," I think many of them were assured), but my work was elsewhere. So while I wrote about extraterrestrial life and edited essays about science and technology, my experience with regards to Covid was as a reader. And this year I found science and nature writing to be more vital than ever.

The pandemic revealed to us, over and over, the messy, fitful work of science. Hopefully anyone who once satisfiedly intoned, "I believe science," now sees that science is not a monolith but a process. And this year we watched that process with unprecedented scrutiny—not "we" the science writers, but "we" the public, "we" the people desperate for news and information and, most of all, guidance. We were told to wash our hands, and then told that surface transmission was minimal. We were told that masks were unnecessary, and then that they were our most essential defense, and then that to wear them outside was more deference to politics than public health. None of these changes and reconsiderations meant that science had failed us. Science, to the extent that it's a cohesive entity, was simply doing its job—gathering evidence, testing theories, refining our understanding of the world.

Through this morass, we turned to science writers to help us make sense of the sausage we were watching being made. I want to take a moment for one writer whose reporting you won't read in this anthology: Ed Yong. I've joked to friends that we asked Ed to edit this year's anthology because otherwise the book would have to be half his writing. In truth—well, in addition to that reasoning—I wanted Ed to edit this edition because I've trusted his writing and insight more than anyone else's this year. Prior to the pandemic, Ed did write a brilliant article titled "When the Next Plague Hits" (anthologized in the 2019 edition of this series), but he also wrote about the microbiome and hippo poop and duck penises. I always thought of him as one of the best and funniest science writers out there. This year he turned out to be one of the most vital as well.

Ed's writing on the pandemic offered synthesis and sense-making of a senseless year. He illuminated and assuaged our fears while explaining the frightening realities of the moment. As hard as his writing was sometimes to read—especially for those of us who may have made it through the first months of the pandemic by dissociating just a tiny bit—I know that it was even harder to write. To not only make sense of the chaos but be immersed in it. This goes for all of the writers of the pandemic, in this anthology and elsewhere. So much gratitude.

In this year of survival, writers not only brought us news but made beauty and meaning. Some, like Ed, made sense of the omnicrisis, illuminating the invisible parallels and connections that united seemingly disparate forces and events and revealed the larger scales of significance. In "The Scramble to Pluck 24 Billion Cherries in Eight Weeks," Brooke Jarvis wrote about a seemingly simple, often invisible task—harvesting cherries—to uncover the economic, social, and scientific tensions woven into labor during a pandemic. Meehan Crist, in "What the Coronavirus Means for Climate Change," showed that compounding crises are never separate; instead, they all highlight the need for action and the hopeful possibility of another world. Julia Craven's "It's Not Too Late to Save Black Lives" emphasized the fact that for all that a virus cannot see race, inequality is so entrenched in our society that illness becomes a vector of racism too.

Other writers dove deep into the minutiae. In "They Say Coronavirus Isn't Airborne—but It's Definitely Borne by Air," Roxanne Khamsi addressed the scientific infighting that threatened the communication of some of the most important precautions against Covid. Heather Hogan delved deep into her personal experience with long Covid in "The Soft Butch That Couldn't (Or: I Got Covid-19 in March and Never Got Better)," writing with clarity and searing honesty.

As much as the pandemic was an omnicrisis, it was not all there was to write about this year. Less than half of this book is about it. There is also Shannon Stirone's sweeping story of attempts to map the cosmos and the human desire to understand our place in it; Katy Kelleher's beautiful meditation on the many meanings of a shade of violet; and Sarah Zhang's astonishing reporting, with depth and empathy, on how prenatal testing is changing what it means to be born with Down syndrome.

That's just a glimpse, but you already have the book in your hands. (You can find information for submitting work for consid-

eration for future editions of this anthology at jaimegreen.net/
BASN.) I hope you enjoy the beauty of these writers' work, learn
more about the world, and perhaps reconnect with a tumultuous
and traumatizing year. This is just one snapshot—or twenty-six of
them—of history.

I ended last year's foreword with the wish, "I hope you're doing
okay." I want to end this year's with a moment for the people who
aren't, those who've lost loved ones to Covid, those whose lives
have been upended. To the extent that this book is mine to dedicate (it's not, but I can dedicate the foreword at least), I'd like it
to honor the memory of Rana Zoe Mungin, who died early in the
pandemic, and horribly early in her life. Zoe was a brilliant fiction
writer and a beloved soul. She was young, she had asthma, and she
was Black; the first two times she sought emergency care for Covid,
her concerns were diminished and dismissed. Zoe died on April
27, 2020, and her death will always be an injustice and a great loss.
With love to Zoe's family and friends, and all of you.

JAIME GREEN

Introduction

I ENTERED 2020 THINKING of myself as a science writer. I ended the year less sure.

While the first sparks of the Covid-19 pandemic ignited at the end of 2019, I was traipsing through a hillside in search of radio-tagged rattlesnakes, allowing myself to get electrocuted by an electric catfish, and cradling loggerhead turtle hatchlings in the palm of my hand. As 2020 began and the new coronavirus commenced its ruinous sweep of the world, I was marveling at migratory moths and getting punched in the pinky by a very small and yet surprisingly powerful mantis shrimp. We share a reality with these creatures, but we experience it in profoundly different ways. The rattlesnake can sense — perhaps see — the body heat of its mammalian prey. The catfish can detect the electric fields that other animals involuntarily produce. The moths and the turtles can both sense the magnetic field of the planet and use it to guide their long navigations. The mantis shrimp sees forms of light that we cannot, and it processes colors in a way that no one fully understands. Each species has its own unique coterie of senses. Each is privy to its own narrow slice of the total sights, smells, sounds, and other stimuli that pervade the planet.

My plan was to write a book about those sensory experiences — a travelogue that would take people through the mind of a bat, a bird, or a spider. Such a journey, "not to visit strange lands but to possess other eyes," as Marcel Proust once said, is "the only true voyage."

It quickly became the only voyage I could make. As the pandemic spread, the possibility of international travel disappeared. Commuting turned from daily reality to fading memory. Restaurants, bars, and public spaces closed. Social gatherings became smaller, infrequent, and subject to barriers of cloth and distance. My world contracted to the radius of a few blocks, but the sensory worlds of other animals stayed open, magical and Narnia-like, accessible through the act of writing.

When I had to pause my book leave to report full-time on the pandemic, those worlds closed too.

In theory, 2020 should have been a banner year for science and nature writers. A virus upended the world and gripped its attention. Arcana of epidemiology and immunology—super-spreading, herd immunity, cytokine storms, mRNA vaccines—became dinner-table fodder. Public health experts (and pseudo-experts) gained massive followings on social media. Tony Fauci became a household name. The biggest story of the year—perhaps of the decade—was a science story, and science writers seemed ideally placed to tell it.

When done properly, covering science trains a writer to bring clarity to complexity, to embrace nuance, to run toward uncertainty instead of seeking easy answers, to understand that everything new is built upon old foundations, and to probe the unknown while delimiting the bounds of their own ignorance. The best science writers learn that science is not a procession of facts and breakthroughs, but an erratic stumble toward gradually diminished uncertainty; that peer-reviewed publications are not gospel and even prestigious journals are polluted by nonsense; and that the scientific endeavor is plagued by all-too-human failings like hubris. All of these qualities should have been invaluable in the midst of a global calamity, where clear explanations were needed, misinformation was rife, and answers were in high demand but short supply.

But the pandemic was not just a science story. It was an omni-crisis that warped and upended every aspect of our lives. While the virus assaulted our cells, it also besieged our societies, seeping into every crack and exploiting every weakness it could find. It found many. To understand why the United States fared so badly against Covid-19, despite its enormous wealth and biomedical savvy, one had to understand not just matters of virology but also the nation's history of racism and genocide, its carceral state, its nursing homes, its historical attitudes toward medicine and health, its national idiosyncrasies, the algorithms that govern social media, and the grossly deficient character of its forty-fifth president. I barely covered any of these issues in an eight-thousand-word piece about whether the United States was ready for the next pandemic that I wrote for the *Atlantic* in 2018 (reprinted in the 2019 edition of this anthology). When this pandemic started, my background as a science writer, and one who had specifically reported on pandemics, was undoubtedly useful, but to a limited degree—it gave me a half-mile head start, with a full marathon left to run. Throughout the year, many of my peers caviled about journalists from other beats who wrote about the pandemic without a foundation of expertise. But does anyone truly have the expertise to cover an omnicrisis that, by extension, is also an omnistory?

The all-encompassing nature of epidemics was clear to the German physician Rudolf Virchow, who investigated a typhus outbreak in 1848. Virchow knew nothing about the pathogen responsible for typhus, but he correctly realized that the outbreak was only possible because of poverty, malnutrition, poor sanitation, dangerous working conditions, and inequities perpetuated by incompetent politicians and negligent aristocrats. "Medicine is a social science and politics is nothing but medicine in larger scale," Virchow wrote.

This viewpoint was championed by many of his contemporaries, but it waned as germ theory waxed. In a bid to be objective and politically neutral, scientists focused their attention on pathogens that cause disease and ignored the societal factors that make disease possible. The social and biomedical sciences were cleaved apart, separated into different disciplines, departments, and scholars. Medicine and public health treated diseases as battles between individuals and germs, while sociologists and anthropologists dealt with the wider context that Virchow had identified. This rift began to narrow in the 1980s, but it still remains wide. Covid-19 landed in the middle of it. Throughout much of 2020, the United States (and the White House, specifically) looked to drugs and vaccines for salvation, while furiously debating about masks and social distancing. The latter were the only measures that controlled the pandemic for much of the year; billed as "non-pharmaceutical interventions," they were characterized in opposition to the more highly prized biomedical panaceas. Meanwhile, social interventions like paid sick leave and universal health care, which could have helped so-called essential workers protect their livelihoods without risking their health, were barely considered.

To the extent that the pandemic was a science story, it was also a story about the limitations of what science has become. Perverse academic incentives that reward researchers primarily for publishing papers in high-impact journals have long pushed entire fields toward sloppy, irreproducible work; during the pandemic, scientists flooded the literature with similarly half-baked and misleading research. Pundits urged people to "listen to the science," as if "the science" is a tome of facts and not an amorphous, dynamic entity, born from the collective minds of thousands of individual people who argue and disagree about data that can be interpreted in a range of ways. The long-standing disregard for chronic illnesses like dysautonomia and myalgic encephalomyelitis meant that when thousands of Covid-19 "long-haulers" kept on experiencing symptoms for months, science had almost nothing to offer them. The naive desire for science to remain above politics meant that many researchers were unprepared to cope with a global crisis that was both scientific and political to its core. "There's an ongoing conversation about whether we should do advocacy work or 'stick to the science,'" Whitney Robinson Rivers, a social epidemiologist, told me. "We always talk about how these magic people will take our findings and implement them. We send those findings out and knowledge has increased! But with Covid, that's a lie!"

Virchow's experiences with epidemics radicalized him, pushing the man who would later become known as the "father of pathology" to advocate for social and political reforms. Covid-19 has done the same for many scientists. Many of the issues it brought up were miserably familiar to climate scientists, who drolly welcomed newly traumatized epidemiologists into their ranks. In the light of the pandemic, old debates about whether science (and science writing) is political—many of which have been captured in the introductions of this anthology series—now seem small and antiquated. Science *is undoubtedly political* whether scientists want it to be or not, because it is an inextricably human enterprise. It belongs to society. It is interleaved with society. It is of society.

This is true even of areas of science that seem to be sheltered within some protected corner of intellectual space. My first book was about the microbiome, a bustling area of research that went unnoticed for centuries because it had the misfortune to arise amid the ascent of Darwinism and germ theory. With nature red in tooth and claw, and germs as the root of disease, the idea of animals benefiting from cooperative microbes was anathema. My next book will show that our understanding of animal senses has been influenced by the sociology of science—whether scientists believe one another, whether they successfully communicate their ideas, whether they publish in a prestigious English journal or an obscure foreign-language one. That understanding has also been repeatedly swayed by the trappings of our own senses. Science is often caricatured as a purely empirical and objective pursuit. But in reality, a scientist's interpretation of the world is influenced by the data she collects, which are influenced by the experiments she designs, which are influenced by the questions she thinks to asks, which are influenced by her identity, her values, her predecessors, and her imagination.

When I began to cover Covid-19 in 2020, it became clear that the usual mode of science writing would be grossly insufficient. Much of journalism is fragmentary: big stories are broken down into small components that can be quickly turned into content. For science writing, that means treating individual papers as a sacrosanct atomic unit and writing about them one at a time. But for an omnicrisis, this approach only leads to a messy, confusing, and ever-shifting mound of jigsaw pieces. What I tried to do instead was to unite those pieces. I wrote a series of long features about big issues, attempting to synthesize vast amounts of information and give readers a steady rock upon which they could observe the torrent of information rushing past them without drowning in it. I treated the pandemic as more than a science story, interviewing sociologists, anthropologists, historians, linguists, patients, and more. And I found that the writing I gravitated toward myself did the same. The pandemic clarified that science is inseparable from the rest of society, and that connection works both ways. Science touches on everything; everything touches on science. The walls between beats seemed to crumble. What, I found myself asking, even counts as science writing?

Which is an interesting question to be asking yourself just as you're asked to edit an anthology of science and nature writing.

This is not an anthology about Covid-19, although the pandemic is central to eleven of the twenty-six pieces. This is very much an anthology, however, that reflects a tumultuous, pandemic-suffused year. The stories I have chosen reflect where I feel the field of science and nature writing has landed, and where it could go. They are often full of tragedy, sometimes laced with wonder, but always deeply aware that science does not exist in a social vacuum. They are beautiful, whether in their clarity of ideas, the elegance of their prose, or often both. They extend laterally, into areas that might not traditionally fall within the bucket of science writing. They stretch temporally, drawing on history for context and sending imaginative tendrils into the future. They synthesize, evaluate, dig, unveil, and challenge.

I've loosely organized the pieces into three sections. The first, "Contagion," is entirely about Covid-19. These pieces are not just about the pandemic, but about what the pandemic has revealed about the world in which we live. Zeynep Tufekci leads the set by explaining the crucial idea of overdispersion—the burstiness of the virus's spread. This concept not only explains why some areas were pummeled by the coronavirus while others escaped unscathed, but also why it has been so hard to absorb lessons from the pandemic. Overdispersion, Tufekci writes, "interferes with how we ordinarily think about cause and effect" and our desire to draw patterns from randomness and sense from tragedy.

Next up, Roxanne Khamsi questions the official pronouncements that the coronavirus was not airborne, in a remarkably prescient piece that preceded debates about aerosol transmission by many months. A seemingly simple matter—airborne or not? —boils down to long-standing academic debates about how that word is even defined, Khamsi shows. Amanda Mull then explores why so many Americans seemed bent on taking undue risks in a generation-defining crisis. Drawing on sociology and psychology, she punctures the all-too-common idea that people will simply change their minds if provided with the right information, and she shows how tribal identities, mixed messages, and irresponsible institutions trapped the United States in a cycle of bad decisions.

Drawing on her own experiences and those of Italian colleagues,

Helen Ouyang vividly describes the horrors of working in a hospital that was being overwhelmed by a new disease. Susan Dominus reports on the civil wars that arose between frontline clinicians, who were torn between the need to try something, anything, right now, and the need to accumulate evidence about what treatments actually worked. Heather Hogan writes beautifully about her own experiences as one of the first people in the United States to deal with the symptoms of "long Covid," at a time when the phenomenon was still unknown—"I was the science," she says. In these pieces, the pandemic reveals the edges and weaknesses of modern medicine. In the next, it exposes the flaws in broader society.

Covid-19 laid bare the grievous neglect that we have allowed to befall our elderly, as Katie Engelhart details in her story about the Life Care Center of Kirkland, Washington—a nursing home that was the first Covid-19 hotspot in the United States. The pandemic showed the discriminatory care that Black people have long received, as Julia Craven reveals by juxtaposing the stories of two women against a sweeping look at America's centuries-old legacy of racism. It also showed that people who are billed as "essential" are often treated as disposable, as Brooke Jarvis demonstrates in her piece about the largely immigrant workforce compelled to pick 24 billion cherries in eight weeks. Taking in agricultural science, immigration politics, and the pandemic itself, Jarvis exposes what has been invisible to us: the people behind the fragile system that brings food to our fridges. These stories show that the products of science and technology—longer lives, better health, and readier food—do not exist in a social vacuum but are instead distributed according to whom society values, and whom it does not.

The second section, "Connections," takes a deeper dive into the intimate links between science and humanity at large. Susan Orlean writes about a different pandemic—rabbit hemorrhagic disease. Its recent invasion of the United States can best be understood in the context of humanity's relationship with rabbits, animals that we uniquely treat as both pets and food. Shannon Stirone contrasts a grand plan to create the most detailed 3-D map of the universe against our ancient desire to understand ourselves in the context of where we are and what lies beyond. As part of a series on colors, Katy Kelleher illuminates our cultural connection with periwinkles and purples in a whirlwind essay that takes in botany, oncology, color theory, and art history.

Though science often concerns itself with literally universal mysteries, it is also an acutely personal endeavor, molded by the identities and stories of scientists themselves. Sabrina Imbler tells the story of Elke Mackenzie, a transgender scientist who was one of lichenology's unsung heroes, but whose work and legacy is largely credited to her deadname. In their piece, Imbler portrays a woman who studied organisms that are famously hard to classify and who, "against ease and tradition, did not wish to separate her identity from her research." Jennifer Senior ponders on the life and suicide of psychologist Philip Brickman, who studied the nature of happiness while simultaneously struggling to find it amid the unrelenting pressures of academia.

Latria Graham writes a poignant letter about the challenges of being Black in the outdoors and exploring wild spaces in which she is not always welcome. Bathsheba Demuth, who got Covid-19 in the midst of yet another year of alarming climate change, ruminates on our ability to acclimatize to tragedy: "On the first day of summer, Siberia and I were the same temperature," she writes. Finally, Emily Raboteau catalogs a year of conversations about climate; part diary and part poetry collection, her wonderfully creative piece shows just how immediate and far-reaching climate change truly is.

Climate change looms large over the third and final section, "Consequences," which examines the costs of past and present sins. In a sweeping piece of evidence-based imagination, Meehan Crist considers the effect that the coronavirus might have on our climate; in the overlap between two planetary problems, Crist sees both the unsustainability of modern life and "a rare opportunity, even in the midst of great suffering, for rewiring our sense of what is possible in American society." Climate change is also exacerbating the downfall of the Kariba Dam at the border between Zambia and Zimbabwe—an imminent catastrophe that Namwali Serpell connects to the arrogance and evils of colonialism. Back in the United States, the combination of colonialism and human-caused climate change is also threatening the endangered Yaqui catfish. In the fish's looming extinction, Maya L. Kapoor finds a deeper message about our tendency to destroy nature while asking everything from it.

From global warming to global worming: Julia Rosen explains that earthworms are not native to the eastern United States but were introduced by people who held the Eurocentric idea that worms are good. They aren't good universally, and in their new ranges these ecosystem engineers have reengineered ecosystems to their detriment. Native animals, meanwhile, are disappearing. The monarch butterfly, once "the most ordinary of extraordinary things," in the words of Nora Caplan-Bricker, is now in decline. But by following the people who are toiling to preserve this iconic insect, Caplan-Bricker finds desperation and hope, "the joy of living on this damaged planet, and a will to witness whatever comes next." Rosanna Xia discovers a similar blend of emotions among the scientists who uncovered up to half a million barrels of DDT that were dumped off Santa Catalina Island and are now leaking into the ocean. DDT was once billed as one of science's greatest achievements, but it is now, as Xia shows, a toxic legacy for which we don't have a plan.

The march of science and technology is still leaving a trail of unintended and treacherous potholes. In Boca Chica, Texas, Marina Koren meets the people who became unwitting neighbors to the rocket company SpaceX, their tranquil paradise punctured by Elon Musk's Martian ambitions. In Yangquan, China, Jiayang Fan profiles the Friendship and Love Hospital—a rare hospice in a country where prosperity and taboos around death have left an aging population with little in the way of end-of-life care. And finally, in what is perhaps my favorite piece in this high-caliber collection, Sarah Zhang travels to Denmark, where nigh-universal screening for Down syndrome has dramatically reduced the number of children born with the condition. "The forces of scientific progress are now marching toward ever more testing to detect ever more genetic conditions," Zhang writes, and the route of that march will be defined by our attitude to disability and parenthood. "Recent advances in genetics provoke anxieties about a future where parents choose what kind of child to have, or not have. But that hypothetical future is already here. It's been here for an entire generation."

Some of the writers in this anthology would bill themselves as science or nature writers, but many would not. I consider this a strength. There has long been a view of science writing that imagines it's about opening up the ivory tower and making its obscure contents accessible to the masses. But this is a strange model, laden with troubling corollaries. It implicitly assumes that science is beleaguered and unappreciated, and that unwilling audiences must be convinced of its importance and value. It equates science with journals, universities, and other grand institutions that are indeed opaque and cloistered. And treating science as a special entity that normies are finally being invited to take part in is also somewhat patronizing.

Such invitations are not anyone's to extend. Science is so much more than a library of publications, or the opinions of doctorate-holders and professors. Science writing should be equally expansive. Earlier, I asked: *What even counts as science writing?* Now, here's my reply: *We shouldn't be able to answer that question.* A woman's account of her own illness. A cultural history of a color. An investigation into sunken toxic barrels. A portrait of a town with a rocket company for a neighbor. To me, these pieces show that science and nature are intricately woven into the fabric of our lives—so intricately that science and nature writing *should* be difficult to categorize.

There is an obvious risk here. Of the typical journalistic beats, science is perhaps the only one that draws us out of our human trappings. Culture, politics, business, sport, food: these are all about one species. Science covers the other billions, and the entirety of the universe besides. I feel its expansive nature keenly. I have devoted most of my career to writing about microbes and lichens, hagfish and giraffes, duck penises and hippo poop. I am writing this introduction having resumed my book leave, to finish my travelogue of animal senses. But I do so with a renewed understanding that even as we step away from ourselves, we cannot fully escape. Our understanding of nature has been profoundly shaped by our culture, our social norms, and our collective decisions about who gets to be a scientist at all. And our relationship with nature—whether we succumb to it, whether we learn from it, whether we can save it—depends on our collective decisions too.

I hope this anthology acts as a guide for making better decisions. It is an unusually melancholy medley, and while I didn't deliberately craft it that way, it feels like a fitting reflection of the state of the world at the start of 2021. The pandemic showed us how much we need to fix, and fortunately, science is famous for its capacity to self-correct. The pandemic also revealed the need for unity and connection, to save one another and to feel alive. Good science writing—the best science writing—illuminates those connections, between us and the rest of the world. Even when it is melancholy, I find it beautiful. And I believe it can lead us toward the kind of radical introspection that we so sorely need.

ED YONG

Contagion

ZEYNEP TUFEKCI

This Overlooked Variable Is the Key to the Pandemic

FROM *The Atlantic*

THERE'S SOMETHING STRANGE about this coronavirus pandemic. Even after months of extensive research by the global scientific community, many questions remain open.

Why, for instance, was there such an enormous death toll in northern Italy, but not the rest of the country? Just three contiguous regions in northern Italy have 25,000 of the country's nearly 36,000 total deaths; just one region, Lombardy, has about 17,000 deaths. Almost all of these were concentrated in the first few months of the outbreak. What happened in Guayaquil, Ecuador, in April, when so many died so quickly that bodies were abandoned in the sidewalks and streets? Why, in the spring of 2020, did so few cities account for a substantial portion of global deaths, while many others with similar density, weather, age distribution, and travel patterns were spared? What can we really learn from Sweden, hailed as a great success by some because of its low case counts and deaths as the rest of Europe experiences a second wave, and as a big failure by others because it did not lock down and suffered excessive death rates earlier in the pandemic? Why did widespread predictions of catastrophe in Japan not bear out? The baffling examples go on.

I've heard many explanations for these widely differing trajectories over the past nine months—weather, elderly populations, vitamin D, prior immunity, herd immunity—but none of them explains the timing or the scale of these drastic variations. But there *is* a potential, overlooked way of understanding this pandemic that would help answer these questions, reshuffle many of the current heated arguments, and, crucially, help us get the spread of Covid-19 under control.

By now many people have heard about R0—the basic reproductive number of a pathogen, a measure of its contagiousness on average. But unless you've been reading scientific journals, you're less likely to have encountered k, the measure of its dispersion. The definition of k is a mouthful, but it's simply a way of asking whether a virus spreads in a steady manner or in big bursts, whereby one person infects many, all at once. After nine months of collecting epidemiological data, we know that this is an *overdispersed* pathogen, meaning that it tends to spread in clusters, but this knowledge has not yet fully entered our way of thinking about the pandemic—or our preventive practices.

The now-famed R0 (pronounced as "r-naught") is an *average* measure of a pathogen's contagiousness, or the mean number of susceptible people expected to become infected after being exposed to a person with the disease. If one ill person infects three others on average, the R0 is three. This parameter has been widely touted as a key factor in understanding how the pandemic operates. News media have produced multiple explainers and visualizations for it. Movies praised for their scientific accuracy on pandemics are lauded for having characters explain the "all-important" R0. Dashboards track its real-time evolution, often referred to as R or Rt, in response to our interventions. (If people are masking and isolating or immunity is rising, a disease can't spread the same way anymore, hence the difference between R0 and R.)

Unfortunately, averages aren't always useful for understanding the distribution of a phenomenon, especially if it has widely varying behavior. If Amazon's CEO, Jeff Bezos, walks into a bar with 100 regular people in it, the average wealth in that bar suddenly exceeds $1 billion. If I also walk into that bar, not much will change. Clearly, the average is not that useful a number to understand the distribution of wealth in that bar, or how to change it. Sometimes, the mean is not the message. Meanwhile, if the bar has a person infected with Covid-19, and if it is also poorly ventilated and loud, causing people to speak loudly at close range, almost everyone in the room could potentially be infected—a pattern that's been observed many times since the pandemic begin, and that is similarly not captured by R. That's where the dispersion comes in.

There are Covid-19 incidents in which a single person likely infected 80 percent or more of the people in the room in just a few hours. But, at other times, Covid-19 can be surprisingly much less contagious. Overdispersion and super-spreading of this virus are found in research across the globe. A growing number of studies estimate that a majority of infected people may not infect a single other person. A recent paper found that in Hong Kong, which had extensive testing and contact tracing, about 19 percent of cases were responsible for 80 percent of transmission, while 69 percent of cases did not infect another person. This finding is not rare: multiple studies from the beginning have suggested that as few as 10 to 20 percent of infected people may be responsible for as much as 80 to 90 percent of transmission, and that many people barely transmit it.

This highly skewed, imbalanced distribution means that an early run of bad luck with a few super-spreading events, or clusters, can produce dramatically different outcomes even for otherwise similar countries. Scientists looked globally at known early-introduction events, in which an infected person comes into a country, and found that in some places, such imported cases led to no deaths or known infections, while in others, they sparked sizable outbreaks. Using genomic analysis, researchers in New Zealand looked at more than half the confirmed cases in the country and found a staggering 277 *separate* introductions in the early months, but also that only 19 percent of introductions led to more than one additional case. A recent review shows that this may even be true in congregate living spaces, such as nursing homes, and that multiple introductions may be necessary before an outbreak takes off. Meanwhile, in Daegu, South Korea, just one woman, dubbed Patient 31, generated more than 5,000 known cases in a megachurch cluster.

Unsurprisingly, SARS-CoV, the previous incarnation of SARS-CoV-2 that caused the 2003 SARS outbreak, was also overdispersed in this way: the majority of infected people did not transmit it, but a few super-spreading events caused most of the outbreaks. MERS, another coronavirus cousin of SARS, also appears overdispersed, but luckily, it does not—yet—transmit well among humans.

This kind of behavior, alternating between being super-infectious and fairly non-infectious, is exactly what k captures, and what focusing solely on R hides. Samuel Scarpino, an assistant professor of epidemiology and complex systems at Northeastern, told me that this has been a huge challenge, especially for health authorities in Western societies, where the pandemic playbook was geared toward the flu—and not without reason, because pandemic flu *is* a genuine threat. However, influenza does not have the same level of clustering behavior.

We can think of disease patterns as leaning deterministic or stochastic: in the former, an outbreak's distribution is more linear and predictable; in the latter, randomness plays a much larger role and predictions are hard, if not impossible, to make. In deterministic trajectories, we expect what happened yesterday to give us a good sense of what to expect tomorrow. Stochastic phenomena, however, don't operate like that—the same inputs don't always produce the same outputs, and things can tip over quickly from one state to the other. As Scarpino told me, "Diseases like the flu are pretty nearly deterministic and R0 (while flawed) paints about the right picture (nearly impossible to stop until there's a vaccine)." That's not necessarily the case with super-spreading diseases.

Nature and society are replete with such imbalanced phenomena, some of which are said to work according to the Pareto principle, named after the sociologist Vilfredo Pareto. Pareto's insight is sometimes called the 80 / 20 principle—80 percent of outcomes of interest are caused by 20 percent of inputs—though the numbers don't have to be that strict. Rather, the Pareto principle means that a small number of events or people are responsible for the majority of consequences. This will come as no surprise to anyone who has worked in the service sector, for example, where a small group of problem customers can create almost all the extra work. In cases like those, booting just those customers from the business or giving them a hefty discount may solve the problem, but if the complaints are evenly distributed, different strategies will be necessary. Similarly, focusing on the R alone, or using a flu-pandemic playbook, won't necessarily work well for an overdispersed pandemic.

Hitoshi Oshitani, a member of the National Covid-19 Cluster Taskforce at Japan's Ministry of Health, Labour, and Welfare and a professor at Tohoku University who told me that Japan focused on the overdispersion impact from early on, likens his country's approach to looking at a forest and trying to find the clusters, not the trees. Meanwhile, he believes, the Western world was getting distracted by the trees, and got lost among them. To fight a super-spreading disease effectively, policymakers need to figure out why super-spreading happens, and they need to understand how it affects everything, including our contact-tracing methods and our testing regimes.

There may be many different reasons a pathogen super-spreads. Yellow fever spreads mainly via the mosquito *Aedes aegypti,* but until the insect's role was discovered, its transmission pattern bedeviled many scientists. Tuberculosis was thought to be spread by close-range droplets until an ingenious set of experiments proved that it was airborne. Much is still unknown about the super-spreading of SARS-CoV-2. It might be that some people are super-emitters of the virus, in that they spread it a lot more than other people. Like other diseases, contact patterns surely play a part: a politician on the campaign trail or a student in a college dorm is very different in how many people they could potentially expose compared with, say, an elderly person living in a small household. However, looking at nine months of epidemiological data, we have important clues to some of the factors.

In study after study, we see that super-spreading clusters of Covid-19 almost overwhelmingly occur in poorly ventilated, indoor environments where many people congregate over time—weddings, churches, choirs, gyms, funerals, restaurants, and such—especially when there is loud talking or singing without masks. For super-spreading events to occur, multiple things have to be happening at the same time, and the risk is not equal in every setting and activity, Müge Çevik, a clinical lecturer in infectious diseases and medical virology at the University of St Andrews and a co-author of a recent extensive review of transmission conditions for Covid-19, told me.

Çevik identifies "prolonged contact, poor ventilation, [a] highly infectious person, [and] crowding" as the key elements for a super-spreader event. Super-spreading can also occur indoors beyond the six-feet guideline, because SARS-CoV-2, the pathogen causing Covid-19, can travel through the air and accumulate, especially if ventilation is poor. Given that some people infect others before they show symptoms, or when they have very mild or even no symptoms, it's not always possible to know if we are highly infectious ourselves. We don't even know if there are more factors yet to be discovered that influence super-spreading. But we don't need to know all the *sufficient* factors that go into a super-spreading event to avoid what seems to be a *necessary* condition most of the time: many people, especially in a poorly ventilated indoor setting, and especially not wearing masks. As Natalie Dean, a biostatistician at the University of Florida, told me, given the huge numbers associated with these clusters, targeting them would be very effective in getting our transmission numbers down.

Overdispersion should also inform our contact-tracing efforts. In fact, we may need to turn them upside down. Right now, many states and nations engage in what is called forward or prospective contact tracing. Once an infected person is identified, we try to find out with whom they interacted afterward so that we can warn, test, isolate, and quarantine these potential exposures. But that's not the only way to trace contacts. And, because of overdispersion, it's not necessarily where the most bang for the buck lies. Instead, in many cases, we should try to work *backwards* to see who first infected the subject.

Because of overdispersion, most people will have been infected by someone who also infected other people, because only a small percentage of people infect many at a time, whereas most infect zero or maybe one person. As Adam Kucharski, an epidemiologist and the author of the book *The Rules of Contagion,* explained to me, if we can use retrospective contact tracing to find the person who infected our patient, and *then* trace the forward contacts of the infecting person, we are generally going to find a lot more cases compared with forward-tracing contacts of the infected patient, which will merely identify *potential* exposures, many of which will not happen anyway, because most transmission chains die out on their own.

The reason for backward tracing's importance is similar to what the sociologist Scott L. Feld called the friendship paradox: your friends are, on average, going to have more friends than you. (Sorry!) It's straightforward once you take the network-level view. Friendships are not distributed equally; some people have a lot of friends, and your friend circle is more likely to include those social butterflies, because how could it not? They friended you and others. And those social butterflies will drive up the average number of friends that your friends have compared with you, a regular person. (Of course, this will not hold for the social butterflies themselves, but overdispersion means that there are much fewer of them.) Similarly, the infectious person who is transmitting the disease is like the pandemic social butterfly: the average number of people they infect will be much higher than most of the population, who will transmit the disease much less frequently. Indeed, as Kucharski and his co-authors show mathematically, overdispersion means that "forward tracing alone can, on average, identify at most the mean number of secondary infections (i.e. R)"; in contrast, "backward tracing increases this maximum number of traceable individuals by a factor of 2–3, as index cases are more likely to come from clusters than a case is to generate a cluster."

Even in an overdispersed pandemic, it's not pointless to do forward tracing to be able to warn and test people, *if* there are extra resources and testing capacity. But it doesn't make sense to do forward tracing while not devoting enough resources to backward tracing and finding clusters, which cause so much damage.

Another significant consequence of overdispersion is that it highlights the importance of certain kinds of rapid, cheap tests. Consider the current dominant model of test and trace. In many places, health authorities try to trace and find forward contacts of an infected person: everyone they were in touch with since getting infected. They then try to test all of them with expensive, slow, but highly accurate PCR (polymerase chain reaction) tests. But that's not necessarily the best way when clusters are so important in spreading the disease.

PCR tests identify RNA segments of the coronavirus in samples from nasal swabs—like looking for its signature. Such diagnostic tests are measured on two different dimensions: Are they good at identifying people who are not infected (specificity), and are they good at identifying people who are infected (sensitivity)? PCR tests are highly accurate for both dimensions. However, PCR tests are also slow and expensive, and they require a long, uncomfortable swab up the nose at a medical facility. The slow processing times means that people don't get timely information when they need it. Worse, PCR tests are so responsive that they can find tiny remnants of coronavirus signatures long after someone has stopped being contagious, which can cause unnecessary quarantines.

Meanwhile, researchers have shown that rapid tests that are very accurate for identifying people who do *not* have the disease, but not as good at identifying infected individuals, can help us contain this pandemic. As Dylan Morris, a doctoral candidate in ecology and evolutionary biology at Princeton, told me, cheap, low-sensitivity tests can help mitigate a pandemic even if it is not overdispersed, but they are particularly valuable for cluster identi- fication during an overdispersed one. This is especially helpful be- cause some of these tests can be administered via saliva and other less-invasive methods, and be distributed outside medical facilities.

In an overdispersed regime, identifying *transmission events* (someone infected someone else) is more important than iden- tifying *infected individuals*. Consider an infected person and their twenty forward contacts—people they met since they got infected. Let's say we test ten of them with a cheap, rapid test and get our results back in an hour or two. This isn't a great way to determine exactly who is sick out of that ten, because our test will miss some positives, but that's fine for our purposes. If everyone is negative, we can act as if nobody is infected, because the test is pretty good at finding negatives. However, the moment we find a few transmis- sions, we know we may have a super-spreader event, and we can tell all twenty people to assume they are positive and to self-isolate —if there are one or two transmissions, there are likely more, ex- actly because of the clustering behavior. Depending on age and other factors, we can test those people individually using PCR tests, which can pinpoint who is infected, or ask them all to wait it out.

Scarpino told me that overdispersion also enhances the utility of other aggregate methods, such as wastewater testing, especially in congregate settings like dorms or nursing homes, allowing us to detect clusters without testing everyone. Wastewater testing also has low sensitivity; it may miss positives if too few people are in- fected, but that's fine for population-screening purposes. If the wastewater testing is signaling that there are *likely* no infections, we do not need to test everyone to find every last potential case. How- ever, the moment we see signs of a cluster, we can rapidly isolate everyone, again while awaiting further individualized testing via PCR tests, depending on the situation.

Unfortunately, until recently, many such cheap tests had been held up by regulatory agencies in the United States, partly because they were concerned with their relative lack of accuracy in identify- ing positive cases compared with PCR tests—a worry that missed their population-level usefulness for this particular overdispersed pathogen.

*

To return to the mysteries of this pandemic, what *did* happen early on to cause such drastically different trajectories in otherwise similar places? Why haven't our usual analytic tools—case studies, multi-country comparisons—given us better answers? It's not intellectually satisfying, but because of the overdispersion and its stochasticity, there may not be an explanation beyond that the worst-hit regions, at least initially, simply had a few unlucky early super-spreading events. It wasn't just pure luck: dense populations, older citizens, and congregate living, for example, made cities around the world more susceptible to outbreaks compared with rural, less dense places and those with younger populations, less mass transit, or healthier citizenry. But why Daegu in February and not Seoul, despite the two cities being in the same country, under the same government, people, weather, and more? As frustrating as it may be, sometimes, the answer is merely where Patient 31 and the megachurch she attended happened to be.

Overdispersion makes it harder for us to absorb lessons from the world, because it interferes with how we ordinarily think about cause and effect. For example, it means that events that result in spreading and nonspreading of the virus are asymmetric in their ability to inform us. Take the highly publicized case in Springfield, Missouri, in which two infected hairstylists, both of whom wore masks, continued to work with clients while symptomatic. It turns out that no apparent infections were found among the 139 exposed clients (67 were directly tested; the rest did not report getting sick). While there is a lot of evidence that masks are crucial in dampening transmission, that event *alone* wouldn't tell us if masks work. In contrast, studying transmission, the rarer event, can be quite informative. Had those two hairstylists transmitted the virus to large numbers of people despite everyone wearing masks, it would be important evidence that, perhaps, masks aren't useful in preventing super-spreading.

Comparisons, too, give us less information compared with phenomena for which input and output are more tightly coupled. When that's the case, we can check for the presence of a factor (say, sunshine or vitamin D) and see if it correlates with a consequence (infection rate). But that's much harder when the consequence can vary widely depending on a few strokes of luck, the way that the wrong person was in the wrong place sometime in mid-February in South Korea. That's one reason multi-country comparisons have struggled to identify dynamics that sufficiently explain the trajectories of different places.

Once we recognize super-spreading as a key lever, countries that look as if they were too relaxed in some aspects appear very different, and our usual polarized debates about the pandemic are scrambled too. Take Sweden, an alleged example of the great success or the terrible failure of herd immunity without lockdowns, depending on whom you ask. In reality, although Sweden joins many other countries in failing to protect elderly populations in congregate-living facilities, its measures that target super-spreading have been stricter than many other European countries. Although it did not have a complete lockdown, as Kucharski pointed out to me, Sweden imposed a fifty-person limit on indoor gatherings in March, and did not remove the cap even as many other European countries eased such restrictions after beating back the first wave. (Many are once again restricting gathering sizes after seeing a resurgence.) Plus, the country has a small household size and fewer multigenerational households compared with most of Europe, which further limits transmission and cluster possibilities. It kept schools fully open without distancing or masks, but only for children under sixteen, who are unlikely to be super-spreaders of this disease. Both transmission and illness risks go up with age, and Sweden went all online for higher-risk high school and university students—the opposite of what we did in the United States. It also encouraged social distancing, and closed down indoor places that failed to observe the rules. From an overdispersion and super-spreading point of view, Sweden would not necessarily be classified as among the most lax countries, but nor is it the most strict. It simply doesn't deserve this oversize place in our debates assessing different strategies.

Although overdispersion makes some usual methods of studying causal connections harder, we can study failures to understand which conditions turn bad luck into catastrophes. We can also study sustained success, because bad luck will eventually hit everyone, and the response matters.

The most informative case studies may well be those who had terrible luck initially, like South Korea, and yet managed to bring about significant suppression. In contrast, Europe was widely praised for its opening early on, but that was premature; many countries there are now experiencing widespread rises in cases and look similar to the United States in some measures. In fact, Europe's achieving a measure of success this summer and relaxing, including opening up indoor events with larger numbers, is instructive in another important aspect of managing an overdispersed pathogen: compared with a steadier regime, success in a stochastic scenario can be more fragile than it looks.

Once a country has too many outbreaks, it's almost as if the pandemic switches into "flu mode," as Scarpino put it, meaning high, sustained levels of community spread even though a majority of infected people may not be transmitting onward. Scarpino explained that barring truly drastic measures, once in that widespread and elevated mode, Covid-19 can keep spreading because of the sheer number of chains already out there. Plus, the overwhelming numbers may eventually spark more clusters, further worsening the situation.

As Kucharski put it, a relatively quiet period can hide how quickly things can tip over into large outbreaks and how a few chained amplification events can rapidly turn a seemingly undercontrol situation into a disaster. We're often told that if Rt, the real-time measure of the average spread, is above one, the pandemic is growing, and that below one, it's dying out. That may be true for an epidemic that is not overdispersed, and while an Rt below one is certainly good, it's misleading to take too much comfort from a low Rt when just a few events can reignite massive numbers. No country should forget South Korea's Patient 31.

That said, overdispersion is also a cause for hope, as South Korea's aggressive and successful response to that outbreak—with a massive testing, tracing, and isolating regime—shows. Since then, South Korea has also been practicing sustained vigilance, and has demonstrated the importance of backward tracing. When a series of clusters linked to nightclubs broke out in Seoul recently, health authorities aggressively traced and tested tens of thousands of people *linked to the venues,* regardless of their interactions with the index case, six feet apart or not—a sensible response, given that we know the pathogen is airborne.

Perhaps one of the most interesting cases has been Japan, a country with middling luck that got hit early on and followed what appeared to be an unconventional model, not deploying mass testing and never fully shutting down. By the end of March, influential economists were publishing reports with dire warnings, predicting overloads in the hospital system and huge spikes in deaths. The predicted catastrophe never came to be, however, and although the country faced some future waves, there was never a large spike in deaths despite its aging population, uninterrupted use of mass transportation, dense cities, and lack of a formal lockdown.

It's not that Japan was better situated than the United States in the beginning. Similar to the United States and Europe, Oshitani told me, Japan did not initially have the PCR capacity to do widespread testing. Nor could it impose a full lockdown or strict stay-at-home orders; even if that had been desirable, it would not have been legally possible in Japan.

Oshitani told me that in Japan, they had noticed the overdispersion characteristics of Covid-19 as early as February, and thus created a strategy focusing mostly on cluster-busting, which tries to prevent one cluster from igniting another. Oshitani said he believes that "the chain of transmission cannot be sustained without a chain of clusters or a megacluster." Japan thus carried out a cluster-busting approach, including undertaking aggressive backward tracing to uncover clusters. Japan also focused on ventilation, counseling its population to avoid places where the three C's come together—crowds in closed spaces in close contact, especially if there's talking or singing—bringing together the science of overdispersion with the recognition of airborne aerosol transmission, as well as presymptomatic and asymptomatic transmission.

Oshitani contrasts the Japanese strategy, nailing almost every important feature of the pandemic early on, with the Western response, trying to eliminate the disease "one by one" when that's not necessarily the main way it spreads. Indeed, Japan got its cases down, but kept up its vigilance: when the government started noticing an uptick in community cases, it initiated a state of emergency in April and tried hard to incentivize the kinds of businesses that could lead to super-spreading events, such as theaters, music venues, and sports stadiums, to close down temporarily. Now schools are back in session in person, and even stadiums are open —but without chanting.

It's not always the restrictiveness of the rules, but whether they target the right dangers. As Morris put it, "Japan's commitment to 'cluster-busting' allowed it to achieve impressive mitigation with judiciously chosen restrictions. Countries that have ignored super-spreading have risked getting the worst of both worlds: burden-some restrictions that fail to achieve substantial mitigation. The United Kingdom's recent decision to limit outdoor gatherings to six people while allowing pubs and bars to remain open is just one of many such examples."

Could we get back to a much more normal life by focusing on limiting the conditions for super-spreading events, aggressively en-gaging in cluster-busting, and deploying cheap, rapid mass tests —that is, once we get our case numbers down to low enough num-bers to carry out such a strategy? (Many places with low commu-nity transmission could start immediately.) Once we look for and see the forest, it becomes easier to find our way out.

ROXANNE KHAMSI

They Say Coronavirus Isn't Airborne — but It's Definitely Borne by Air

FROM *Wired*

AMID THE HOURLY updates on the new coronavirus, a single, calming fact stands out: a particle of happy news, hanging in a cloud of dread. The germ that causes Covid-19 may be responsible for a terrifying public health disaster, but hallelujah, thank the lord, *at least it isn't airborne.*

This message is now dogma for news outlets and public health officials. They impress on us that droplets laced with the new coronavirus don't remain aloft for long — that they only sail for six feet at the most before they fall onto the ground. That's why we're told that soap and water are the best protections one can find: twenty seconds' worth of hand-related hygiene, repeated many times throughout the day. *The virus isn't airborne;* so keep on washing when you can. *The virus isn't airborne;* so you'd be wise to trade your grubby handshake for an elbow bump. *The virus isn't airborne;* so don't forget to keep your fingers off your face.

But I'm afraid this standard line — this single, calming fact about the new coronavirus — may not be as simple as it seems. When health officials say the pathogen isn't "airborne," they're relying on a narrow definition of the term, and one that's been disputed by some leading scholars of viral transmission through the air. If these scholars' fears bear out — if the new coronavirus does, in fact, have the potential to travel farther through the air than officials have been saying — then we might need to reevaluate our standards for protecting health care workers at the front lines of fighting Covid-19. In fact, we might need to make some tweaks to all our public health advice.

From early on, any spread of the new virus through the air has been downplayed from the top. World Health Organization director-general Tedros Adhanom Ghebreyesus assured people on Twitter last week that "actually it's not airborne." He went on to clarify that "[i]t spreads from person to person through small droplets from the nose or mouth which are spread when a person with #COVID19 coughs or exhales." According to this way of thinking, the blobs of viral particles that get expelled from coughs and exhales are too big to float around; so they mainly cause infection by landing onto someone close, or by dropping on a surface from which they're later transferred to someone's body via touch.

For public health officials such as Tedros (who goes by his first name), a *truly* airborne virus is one that floats around for extended periods—like measles, which is known to be infectious in the air for at least half an hour. A pathogen like this can create a nightmare scenario. A sick person might ride an elevator, for instance, and shed some virus along the way. Later on, someone else who got into the same elevator might breathe in those germs and develop the disease.

There are very good reasons to believe—and good reasons for public health officials to assure the public—that the new coronavirus virus isn't "airborne" in that specific and apocalyptic sense. But the definition used by these officials may also be obscuring vital details of transmission. In particular, it papers over all the nuances in how someone's virus-laden cough or sneeze or breath really travels through the air. The authorities employ a rule of thumb for distinguishing what they call "droplets" from "aerosols." Droplets are often defined as being larger than five microns in diameter, and forming a direct spray that is propelled by cough or sneeze up to two meters away from the source patient. Aerosols, in this scenario, are smaller gobs of potentially biohazardous material that may remain afloat for longer distances.

This black-and-white division between droplets and aerosols doesn't sit well with researchers who spend their lives studying the intricate patterns of airborne viral transmission. The five-micron cutoff is arbitrary and ill advised, according to Lydia Bourouiba, whose lab at the Massachusetts Institute of Technology focuses on how fluid dynamics influence the spread of pathogens. "This creates confusion," she says. First of all, it garbles terminology. Strictly speaking, the aerosols are droplets too. When you breathe out or cough, you release bits of watery mucus from inside your body in a wide array of sizes, ranging from bigger, wetter ones to finer ones. All of these are *droplets*. The smallest droplets are commonly described as *aerosols*. Whatever you call them, though, any of these bits of mucus may be laced with viral pathogens. To make matters more complicated, when the water component of droplets dries up in the air, the remaining bits of floating virus are called "droplet nuclei," which are even lighter and more apt to travel long distances. Aside from size, other factors, such as local humidity and any drafts of air, will also affect how far a droplet flies.

Even the fattest droplets may not always fall right to the ground within a few feet. When you go to the ocean on a windy day and feel the sea spray on your face, you've just encountered droplets of a size that might be described as "not airborne" in a public health briefing. Even breezes that are far more subtle than the ones coming off the ocean can lift and push a droplet. Oddly, though, many traditional studies of droplet trajectories have made use of simplified models that don't account for the gust of air released when a person coughs or sneezes, which gives those droplets an extra push. Bourouiba calls this a mistake. Her lab has found that coughs and sneezes, which they call "violent expiratory events," force out a cloud of air that carries droplets of various sizes much farther than they would go otherwise. Whereas previous modeling might have suggested that five-micron droplets can travel only a meter or two—as we've heard about the new coronavirus—her work suggests these same droplets can travel up to eight meters when taking into account the gaseous form of a cough.

For researchers like Bourouiba, who study the physics of pathogens' paths, any virus traveling in the air might as well be described as "airborne." But there is no consensus among scientists as to which pathogens should get that label and which shouldn't. Julian Tang, a virologist at the University of Leicester in England, co-authored a review article on this very topic last year. The paper noted that for some researchers, "airborne transmission" involves only fine aerosols. For others, it can involve both aerosols and larger droplets. Ultimately, in their paper, Tang and his colleagues settled on using the phrase to mean transmission by particles of fewer than ten microns in diameter—a cutoff twice as large as what the WHO has used.

The debate over whether something is "airborne" is particularly sensitive around pathogens that cause the most acute, deadliest outbreaks. But there's not even agreement among experts as to how regular old influenza transmits through the air. Those who say the flu does this well point to a curious incident from the 1970s in which an airplane with fifty-four passengers was grounded on the tarmac for three hours because of engine issues during a take-off attempt. There was one person who had been ill onboard; and within three days, three-quarters of the other people who had been on the plane showed symptoms of flu such as cough, fever, and fatigue. The majority of those tested were positive for the virus. Donald Milton, whose research at the University of Maryland School of Public Health includes studies of infectious bioaerosols, says that all these years later he and his peers are still trying to convince other scientists that influenza is substantially airborne. He published a paper in 2018 asserting that, contrary to what some might think, sneezing and coughing are not required for influenza virus to be released in an aerosol form that can float around.

Meanwhile, the aerodynamics of more exotic pathogens have stirred controversy. One infectious-disease expert warned, in 2014, that Ebola might become highly transmissible by air. This proved to be a false alarm. There is some evidence that coronaviruses such as SARS and MERS can travel in hospital air. Some researchers still dispute these data: the MERS research, for example, did not use a hospital room without infectious patients as a control. But others take it as a given that these coronaviruses were floating in their infectious form around parts of hospitals.

As for the airborne behavior of the new coronavirus, scientists are racing to obtain data. A study published in the *Journal of the American Medical Association* on March 4 looked at the hospital isolation rooms of three patients in Singapore with Covid-19. The study offered some solace because it didn't find evidence of the virus in air samples. However, the air vent blades in one patient's room did test positive. A second study, described in a preprint paper published on March 10, examined the hospital environments of Covid-19 patients in Wuhan, China. Although the levels of the microbe that causes Covid-19 in most rooms were undetectable or low, the study did find the presence of the virus in aerosol form.

That there would be non-negligible amounts of virus in the air does not surprise Linsey Marr, a researcher at Virginia Tech who studies the dynamics of viruses in the air. "This is exactly what I suspected," she says. Even before that paper came out, she'd told me it's "unfortunate" that the WHO insists on saying that the new coronavirus "is not airborne."

Crucially, the hospital studies only looked for the genetic signature of the virus, as opposed to mixing the viral material with animal cells to see whether it would wreak havoc. As such they could not know whether the viral material present in the ventilation system or the air was infectious. This is a critical point— virologists emphasize that the presence of residual RNA or DNA left by pathogens in no way guarantees that people might get sick from it. However, the question of whether the new coronavirus is infectious as an aerosol was explored in another paper posted as a preprint this week. In that study, scientists used a laboratory machine to force the virus into aerosolized form and then tracked it for three hours. They found the pathogen was still able to infect animal cells at the end of that time frame, although there was substantially less virus suspended in the air from one hour to the next.

These three new papers should not be overinterpreted. Only one of them has been vetted by peer review at this point. It also remains unclear, and undemonstrated, whether the Covid-19 virus released from patients' lungs comes out in aerosol form; whether aerosolized particles of this virus travel significant distances; and, if so, whether they do so in sufficient number to cause infection. Notably, while the joint WHO-China mission report published in late February said that although airborne particles were "not believed to be a major driver of transmission," it noted that such a mode "can be envisaged if certain aerosol-generating procedures are conducted in health care facilities."

Given that much research on airborne transmission in outbreaks is focused on medical settings, it's also less than clear how even the most common viruses might pass from person to person under everyday circumstances. Julian Tang and his colleagues have created a visualization of the breaths exchanged by two people in conversation standing three feet apart. Most of the time, the puffs of air they let out remain separate; but portions of their exhalations do sneak from each person's breathing space into the other's. Given all this uncertainty, some experts say there needs to be better public messaging on the spread of the new coronavirus. "Crowded public transport where people can breathe on each other may also lead to transmission of infection," Tang says, echoing public health advice that, while widespread, may not be getting as much emphasis as hand-washing. Milton agrees, adding that it might be wise to shut off air-recirculation systems in cars, which could potentially spread the pathogen among passengers.

Even if it turns out that the new coronavirus is meaningfully airborne, at least in rare circumstances, you shouldn't rush out to buy masks, including N95 respirators. Don't do that. We've already witnessed grave shortages of masks for health workers and people who are immunocompromised. To buy one now is to put those people's lives in danger.

The scientists I spoke with for this story do not want people to shutter themselves inside in fear of toxic vapors. They point out that being outdoors, in fresh air exposed to UV light, is healthy. They do not want to encourage anyone to cower from all social interaction. This article is not meant to induce panic among the worried well, who clog health systems needed for people who are actually ill. But there needs to be a more nuanced understanding of this issue.

When public health officials say a pathogen is or isn't "airborne," they create a false dichotomy that doesn't keep people safe. In this particular case, the folks who are most at risk for airborne transmission are medical workers. Just this week, amidst concerns about insufficient supplies of respirators, the U.S. Centers for Disease Control and Prevention updated its guidance for health care personnel dealing with the Covid-19 pandemic. Based on its assertion that "airborne transmission from person-to-person over long distances is unlikely," the agency said that "facemasks" —presumably the floppy surgical masks that do not do as much to protect against floating pathogens—constitute an acceptable alternative for health care workers. (It does note that N95s should be prioritized for procedures that are especially likely to release virus into the air.) But if the *JAMA* study and preprint articles from this week prove correct, and the new coronavirus falls somewhere on the spectrum of airborne-ness besides *not at all*, then this advice might be counterproductive.

When it comes to this virus's ability to travel in air—in hospitals or elsewhere—it's hard to know where things will ultimately land. Until then, describing it in absolute terms seems risky.

AMANDA MULL

The Difference Between Feeling Safe and Being Safe

FROM *The Atlantic*

ON A NORMAL DAY, the White House is one of the safest build-ings in the world. Secret Service snipers stand guard on the roof, their aim tested monthly to ensure their accuracy up to 1,000 feet. Their heavily armed colleagues patrol the ground below and staff security checkpoints. Belgian Malinois guard dogs lie in wait for anyone who manages to jump the property's massive iron fence.

But safety means something different in a pandemic. Over the past few days, several aides to Vice President Mike Pence, includ-ing his chief of staff, have tested positive for the coronavirus. The outbreak is the second in the White House in a month, after doz-ens of people, including President Donald Trump himself, tested positive following the apparent super-spreader event hosted by the administration to celebrate the Supreme Court nominee Amy Co-ney Barrett.

The outbreaks have been both utterly predictable and totally shocking. The Trump administration has consistently downplayed the severity of the coronavirus, encouraged Americans to resist safety measures, and promised that the pandemic is nearing its end. But the people orchestrating the country's disastrous corona-virus response had no plausible deniability: the very best experts, information, and precautions were all available to them, even if they refused to pass that help on to others.

People will write books on everything Donald Trump did wrong during the pandemic, with explanations both personal and ideological for his administration's often willful failures. But for a group of people for whom self-preservation has long been an obvious goal, their willingness to put themselves in optional danger, given all the resources at their disposal, can't be completely explained by Trump's lack of empathy or his advisers' policy goals. It suggests that on top of everything else, the administration fell prey to an error of intuition: presumably, Trump and his coterie felt safe, despite the mortal danger nipping at their heels for all to see.

Trumpworld's infection fiasco is an especially bizarre case study of one of the pandemic's defining features: how different *feeling* safe and *being* safe actually are. This misperception has played out in millions of homes and workplaces across the country as regular people make good-faith efforts to grapple with the swiftly changing circumstances of American life, absent the resources available to the federal government. Things that used to be safe, such as visiting grandparents and attending a friend's wedding, are now potentially deadly. Things that used to be foreboding, such as the sight of many masked strangers in public, are now a source of comfort.

This new sort of safety is difficult to adapt to, both practically and emotionally. Over the summer, previously innocuous private social gatherings, such as dinner parties and birthday celebrations, were cited as a primary driver of new infections all over the United States. In some instances, the people involved perhaps didn't care about the risk or thought the pandemic was fake. But in others, they likely couldn't imagine why they should be scared of time with loved ones. Many of these same people were wearing masks to the grocery store, using hand sanitizer, and otherwise doing what they understood to be asked of them.

Safety is among the most powerful motivators of human behavior, which also makes the drive to feel safe a potent accelerant for confusion, disinformation, and panic. Staying safe requires an accurate, mutually agreed-upon understanding of reality on which to assess threats and base decisions. Since the pandemic arrived in the United States, however, politicians have sparred over basic safety precautions and aggressive reopenings. The federal government and many of its allies at the state and local levels have actively undermined efforts to get people on the same page. These contradictions have sown confusion, even among those who disagree po-

litically with the leaders encouraging people to flout masking and
social distancing. When everyone is left to write their own version
of *Choose Your Own Pandemic Adventure,* no one is safe.

To understand how humans think about safety, you have to understand how they think about fear. To be safe, people need to be
free from the threat of physical or mental harm. But to *feel* safe,
people need to be free from the *perception* of potential harm, confident that they understand what the likeliest threats are and that
they are capable of avoiding them. Whether their perception is
accurate is often incidental, at best, to the feeling itself. "Fear reactions are very primitive," Arash Javanbakht, a psychiatry researcher
at Wayne State University, told me. "We don't react so well or so
accurately to conceptual threats."

People learn what or whom to fear in a few different ways, according to Javanbakht. The things we have experienced or observed ourselves, such as car accidents or the kinds of violence
frequently depicted on the news, have a significant impact. So do
the warnings of peers and authority figures. This assemblage of
influences—family members, friends, co-workers, religious or cultural leaders—is as much a tribe now as it was when these instincts
evolved, and the security and support that it can provide create
a profound psychological incentive to remain a member in good
standing of one's group.

People's dependence on group affiliation for safety and support can be so strong, in fact, that it sometimes overrides more
logical assessments of fear and safety, Javanbakht said. Even in situations where the actions of the tribe's leaders contribute to the
group's collective misery, many members will find it difficult to
reject that leadership. Instead, studies have shown, people dig in
their heels when confronted with evidence that challenges their
beliefs or identity: they redouble their support for trusted authority figures and reject outside criticism, which they'll often paint as
proof that the group is under threat. Javanbakht compared this dynamic to softer forms of American tribalism, such as being a fan of
the Cleveland Browns. The team's leadership has been antagoniz-
ing its fan base for decades, but some people cannot be mistreated
into retracting their emotional and monetary support.

Many Americans have come to understand their political affiliations in much the same way they do their affinity for particular

sports teams or movie franchises, but with much darker implications than, say, getting your hopes up about an Atlanta Braves postseason run. On a basic level, you could see tribalist fear in how people scrambled for clarity early in the pandemic. The methods humans use to understand and pursue safety aren't really built for quick, competent responses to novel threats. Fear abounds as people realize they don't have a script to follow. (Buy six cases of bottled water.) Panic sets in. (Lysol the groceries.) They wait for guidance from existing leaders and search for previously overlooked ones. (Nice to meet you, Dr. Fauci.) They monitor the behavior of their peers. (Wear a mask.)

In a country with a dwindling consensus on the basic experience of reality, though, tribal affiliation can be especially fraught. Trump and other leaders have conscripted supporters into cultural warfare, recasting safety measures as political attacks from the opposition and tools of social control. Some people have duly followed, rejecting simple precautions recommended by scientific experts, such as masks and open windows, and promoting some of the more extreme reactions to the pandemic: the virus is a hoax; it's a bioweapon meant to hurt Trump's reelection chances; it really is just the flu.

Tribal affiliations have been exploited in American politics for centuries; influential people stoke fear against those outside their constituencies, such as ethnic and religious minorities, immigrants, and the poor, which intensifies supporters' belief that loyalty to the group itself is their best bet for safety. But Trump, from whom this narrative radiates, has been especially adept at fomenting and wielding fear as a source of power in his brief political career. People who have built a significant portion of their identity around Trump fandom by attending rallies, joining Facebook groups, and buying merchandise likely have a psychological investment in his version of reality that's too high to consider abandoning; for some of them, losing those beliefs might feel like a fate worse than the coronavirus. When the spell is that strong, according to Javanbakht, it generally requires a more immediate measure of danger to break it, such as the Covid-19 death of a loved one.

Intuitive failures of safety extend beyond Trump and his acolytes. Ultra-strict adherence to pandemic precautions can itself be a display of tribal identity, especially as scientific understanding of the virus evolves and some safety measures, such as wiping down groceries with sanitizing wipes, can be dispensed with. No matter where their beliefs fall on the country's political spectrum, people whose identity is more weakly tied to a political group have an advantage when it comes to adapting in response to new information. A relatively small proportion of conservatives are virulently anti-mask, and most liberals are not fully sequestered in their homes, Lysol-soaking their mail-in ballots.

While these sets of behaviors share an underlying psychological mechanism, they are not equivalent. To belong to one tribe, people accept an outsize and sometimes irrational portion of responsibility for their own safety and that of others through self-abnegation and personal fastidiousness. To belong to the second tribe, people must refuse to care about others at all.

Over time, theoretically, the country's collective understanding of safety should both improve and normalize, as the definition of safety is expanded to account for the pandemic, and the group —in this case, Americans as a whole—forms new norms to achieve it. In most of America, that hasn't necessarily been the case so far, and any progress has been hard-won over sometimes-violent opposition. But the tribalist fear that causes those big variations often doesn't account for smaller ones among families and social circles, which tend to be politically similar. Anyone who's been holed up in their home for months, watching friends and family proceed with indoor weddings or spend their weekends inside bars—or anyone who's invited a buddy to a day at the beach only to be rejected over amorphous safety concerns—knows that something more complicated is going on.

Even among people with similar beliefs, a smooth, uniform pivot to a new understanding of safety requires diligent, competent, well-intentioned leadership. According to Eric Scott Geller, a psychologist at Virginia Tech who has studied safety for more than forty years, the best way to get lots of people to adopt new safety precautions is to be explicit and consistent about what they are and why they're important, and then demonstrate examples of people adhering to them repeatedly over time. As a whole, American leadership continues to do the opposite. "I've been in the business for a long time, and I've never seen anything quite like this," Geller told me. "The biggest problem we have right now is mixed messages."

Left to their own devices, people chart their paths based on their personality, how they see the world, and how they relate to risk. According to Geller, many people presented with a barrage of contradictory instructions just grow tired and give up. Others become hypervigilant, their behavior calcifying against new information that might let them ease up and enjoy life a little more. Still others simply choose optimism, no matter how dangerously misguided—such as the belief that "herd immunity" is near, or the assumption that catching the virus will have no long-term consequences for them. "People will gravitate to the positive message because it's convenient, and it's not scary, it's not fearful," Geller said.

And so the chaos of a country becomes the chaos within its families and communities. People spar over their assumptions and hastily made decisions based on half-understandings of scientific evidence. They're forced to conduct their own awkward, fraught behavioral micro-negotiations before visiting relatives, celebrating a birthday, or going out for a beer on a bar's patio. Americans have no common conception of the pandemic, which means you can't assume that someone you've trusted for years isn't about to expose you to a deadly disease, or even that you live on the same plane of reality. People feel bad about enforcing their boundaries, or they simply grow tired of constant vigilance. Occasionally, they just forget.

In some ways, these tragic errors in intuition are convenient for leaders, both within and outside the government. Birthday parties and vacations and nights out on the town are easily framed as personal choices free from government influence, even though other countries have gone much further in giving their citizens tools to keep themselves safe and make good decisions—nationally coordinated testing programs, extensive government aid for businesses, a clear and consistent message about safety. You can see the same finger-pointing dynamic in how some college presidents set their charges up to fail, and then punished them for becoming infected. This approach reflects how, in America, blame for large-scale destruction and death is often shunted onto those with the least power to change policy or protect themselves. "Think about all the accidents on oil platforms, or drilling rigs," Susan Silbey, a sociologist who studies safety at MIT, told me. "It's the same few companies over and over again. They always blame the workers."

Even with mixed messages from above, pandemic-era safety would be a little bit clearer to negotiate if talking about your behavior and asking questions about others' weren't so excruciatingly awkward. Ensuring pandemic safety requires interrogating loved ones about whom they've been with and what they've been doing, and if they've been tested recently. If that sounds familiar, it's because health educators ask people to go over the same topics anytime they have a new sexual partner. As it turns out, it wasn't just the sex that made those conversations notoriously difficult to have, but the same strategies that public health experts have developed to talk about sex might be just as useful in these new types of negotiations.

As masks became both more common and more controversial in the United States during the spring, Logan Levkoff, a sex-and-relationship educator, found herself having the same conversation over and over again with her peers. The dynamics of pandemic health and sexual health make people anxious and cagey in very similar ways, about very similar types of precautions. Should you refuse to attend your cousin's wedding unless she moves it outside? Should you tell your maskless friend to mask up? These situations are "still fraught with shame and stigma and assumptions about people's politics," Levkoff told me. "Often, if we don't want to have someone make assumptions about us, and we don't want them to feel we're making assumptions about them, we just don't say anything at all."

That avoidance can put people in riskier situations than they'd choose for themselves if they felt free to be picky. Even so, the desire not to anger loved ones in an already anxious and fearful time can be strong. "That's where we get tripped up a bit," Levkoff said. "We think that in those intimate relationships, it would be a violation of trust to ask." She said that the politicization of public health measures has likely exacerbated this dynamic. Conversations about safety can quickly devolve into arguments about political differences; less than two weeks out from the election, many people are probably happy to avoid poking that bear if they can.

These conversations can even be hard among political peers,

which in theory seems silly. We're in a pandemic, after all. Isn't transparency just reasonable? "We have this belief that our health is a measure of our character," Levkoff explained. In America, being disease- and disability-free is often assumed to be an indicator of a person's moral righteousness and good priorities. To you, asking a friend if she's been to a house party recently might feel like checking a box on a to-do list. For her, it might feel like being told she's some kind of degenerate.

These dynamics rear their ugly head in less intimate interactions too. After the University of Notre Dame's president, John I. Jenkins, became one of the many people who tested positive for the coronavirus after attending the apparent White House superspreader event celebrating Barrett, an unnamed member of the Notre Dame delegation told the *New York Times* that the group's decision to go maskless was politeness, not politics—an attempt to blend in and adhere to the conventions set by the event's powerful hosts. There are many ways in which people are expected not to rock the boat in American social culture. Those niceties can set people up to spread a deadly disease in an environment where the circumstances of safety have changed swiftly and confusingly.

Looking for ways to have these conversations—or even just thinking about their implications at all—is half the battle. Broaching an uncomfortable issue or figuring out a new way to understand our friends and loved ones is awkward, but it's certainly doable. Tread lightly. Reassure friends and family that you're just trying to help everyone have fun. Volunteer information about your own behavior. Starting these conversations is often the hardest part, especially among people whose goals are the same, even if their methods differ. They've already made the most difficult concession: banishing the idea that the way we understood one another seven months ago is enough to get us through the months ahead.

"We think we share the same values and norms," Sibley, the sociologist, told me. "We think we know each other."

HELEN OUYANG

I'm an ER Doctor in New York. None of Us Will Ever Be the Same.

FROM *The New York Times Magazine*

First Week of March
NYC Covid-19 cases, March 1: 1

I AM IN KARACHI, Pakistan, on March 2, when I read the news: New York City has its first patient hospitalized with the coronavirus. Though I am more than 7,000 miles away—reporting on a different disease outbreak—I am already worried about what I will face when I return home in two days to my job as an emergency-room doctor in the city. Even in the best of circumstances, the ER can be swamped, with patients doubled up in rooms and too few monitors and beds to go around. Doctors and nurses are always multitasking at the edge of their limits. "Damage control," we call it.

I know the situation with Covid-19 is already dire in different parts of the world, Italy especially. Could our hospitals also be overtaken that quickly? What would that look like? I need to know what might come, what decisions I might be confronted with. I want to hear about them directly from health care workers in Italy.

A few days from now, I will come across the name of Guido Bertolini, a clinical epidemiologist who studies intensive care. Through a colleague of his, I reach out to him over WhatsApp, and we begin corresponding. He had been high up in the Italian Alps through the last day of February, when the distressing messages started to come in from colleagues asking him to join a new Coronavirus Crisis Unit for Lombardy, a region in northern

Italy. Some of the pleas had an Excel file attached. When Bertolini opened it, he tells me, he couldn't believe the numbers. He had to see the situation for himself.

With an ER doctor from Milan, he drove to the Lombardy city of Lodi the next day. He was horrified by what he witnessed. "So many patients, in every corner," he says. "They were attached to oxygen in all possible ways." Individual oxygen dispensers, meant for single patients, were being split among four people at a time. "When we came out, we were silent for all the journey home," he says. "We could not speak." He knows the hospital has already passed its maximum capacity.

"From my position in the crisis unit, I see the whole picture," he says. "Which is dramatic." Lombardy is one of Italy's richest areas, where there is "almost no limit on resources," he explains. Yet the region has only half the number of ICU beds it needs to care for all the critically ill patients infected with Covid-19. He knows doctors are soon going to have to decide who lives and who doesn't. How could he help them do that?

He begins rounding up—virtually, over Skype—a group of bioethicists and ICU physicians. They include Marco Vergano, a forty-five-year-old ICU doctor in Turin, in the neighboring province of Piedmont, who is also the chairman of the bioethics group of Italy's Society of Intensivists (SIAARTI). He's working back-to-back shifts in the ICU, but he jumps online with the six other members of the task force that Bertolini has set up. That night, he begins drafting a document. The first version includes strict criteria. If you are over eighty or one of your organs isn't functioning well or your dementia has advanced past a certain point, you are unlikely to get a breathing tube or a spot in the ICU.

Soon after, the group decides to delete the specific cutoffs, so that hospitals can adapt their responses to circumstances, which are changing hourly. They want doctors to have flexibility but use these principles to guide and justify their decision-making. The document's fundamental thrust, though, is that those with the highest chances of survival—the young and the healthy—get priority. They strongly advise against allocating precious resources, like ventilators and beds, on the traditional basis of first-come, first-served, which would reduce the number of lives a hospital could save. "Soft utilitarian" is how Vergano characterizes the approach.

Swift and fierce denunciation of the group and its recommendations follows the document's release. "You cannot imagine to what extent we have to face harsh criticism," Vergano says. "From colleagues to journalists to bioethicists—we are in firing lines these days," Bertolini adds. "They say we are God-playing."

Vergano notes that most of the criticism has come from regions in Italy that have yet to be hit as hard as Lombardy. They are "completely living in another world," Bertolini says, because "unless you are inside this situation, you cannot understand fully." People are, it seems, woefully bad at grasping how future events will unfold, whether in the city next door tomorrow or across the Atlantic a couple of weeks later.

Back in New York, I work a couple of shifts in the ER. Though we've been given updated instructions for screening patients for coronavirus—which say that a person need not have a history of recent travel to qualify for testing—the hospital feels mostly like its normal, hectic self, as it did before I left the country.

Second Week of March
NYC Covid-19 cases, March 8: 14

We have started to hold regular Covid-focused meetings over Zoom. Participants ask questions about the availability of tests and how we should protect ourselves, but no one seems very worried by what's unfolding in Italy.

Bergamo, a city of 120,000, with about a million more in the surrounding province, sits at the foothills of the Alps, 25 miles northeast of Milan. Travel guides describe how the upper part of the city, perched high on a hill and encircled by walls, is connected to the lower part by walking paths and a funicular. The city is known for its spectacular medieval architecture. The area, home to San Pellegrino sparkling water and a manufacturer of brakes for Formula One cars, is also a busy transit hub, with an airport that serves over 12 million passengers a year. Doctors tell me the province of Bergamo has been hit the hardest by this pandemic.

Papa Giovanni XXIII Hospital, which provides advanced, state-of-the-art medical care, is one of the biggest hospitals in the region, housing more than 900 beds. It probably has the highest number of Covid infections in the country. Andrea Duca, an ER doctor there, has been treating these patients for a couple of weeks now, since the first case was detected. They had only sore throats and mild coughs to start, but after a few days, patients were showing up with more severe symptoms. They had significant lung infections and low oxygen levels, even when they didn't look that ill. Some of them had diarrhea instead of respiratory complaints, which made diagnosis confusing. The clinical picture was different from what Duca and his colleagues expected. "The virus is as free as the wind," Pietro Brambillasca, an anesthesiologist who works with Duca, tells me over the phone. "It does whatever it wants."

The patients keep coming. Beds fill up. Ventilators get parceled out. Quickly, there are many more patients than equipment and space. Doctors can be recruited, or take on more patients than they are usually comfortable with, but what to do about the lack of resources? Who gets the precious few ventilators?

Those deemed too old or too sick don't get ventilators or have them taken away so that they can be used for patients who are more likely to survive. Duca recalls for me one of the first patients he subjected to this calculation. The man, sixty-eight, had transplanted lungs. His oxygen level had dropped; his breathing rate increased. "I knew that he was not doing well," Duca says. But there were no spots in the ICU, because they were filled with younger and healthier patients whose prospects of recovery were greater. Duca made the difficult decision not to give the patient a breathing tube, to save the ventilator for someone more likely to live.

Family members weren't allowed into the hospital because they, too, could get infected or spread the virus to others if they themselves were sick. But Duca asked for permission from his supervisor to let the man's wife and daughter in, just for a few minutes. "I saw his face when he looked at his wife coming inside this room," Duca recalls. "He smiled at her. It was a fraction of a second. He had this wonderful smile." He continues: "Then I saw that he was looking at me. He realized that there was something wrong if only *his* relatives were coming inside." The man knew in that instant that he was going to die, Duca says. As the man's breathing worsened, morphine was started. He died twelve hours later.

"Which one is the lucky man of the day?" Brambillasca asks. He normally cares for very sick children who have had organ transplants, but since the outbreak, he has been called to float between the ER and the ICU. When we speak by phone one morning, on

one of my days off, he sounds defeated. His wife, an otolaryngolo-
gist, has also been recruited to the effort: she is now working in a
Covid unit in a neighboring hospital. I can hear their one-year-old
daughter in the background. Every day, Brambillasca feels inadequate. "I ask myself if I'm more useful if I go outside my home,
take paper and alcohol and disinfect the doorknobs of my neighbors instead of going to work as a doctor," he says.

Brambillasca tells me about how he had two patients side by
side one day. One man was around sixty-five and had been on
a ventilator for ten days. He had heart problems, and he wasn't
improving. To his left was another man, about the same age but
healthy. His breathing was becoming faster and shallower. Over
the course of two minutes, Brambillasca decided to take the ventilator away from the first man and give it to the second one. "If you
think of it as saving the most number of lives, that's it, you have to
do it," he says. "But I'll become an ice-cream maker instead of a
doctor if I have to go on this way."

Will I, too, feel that way soon? We are starting to see some cases
in our hospitals, but it's nothing like what doctors in Italy are describing. They warn me that we are about two weeks behind them.
Could we really get to where they are in such a short time?

Third Week of March
NYC Covid-19 cases, March 15: 330

One of our ER doctors, who also works in the ICU, proposes an
extreme case during a Zoom meeting: we know from China's experience that once a patient is in cardiac arrest from Covid-19, the
chance of survival is essentially zero. His words hang in the air,
but the question is clear: Should we try to resuscitate this patient,
despite our equipment shortages and the risks to ourselves? As this
hypothetical situation plays out in my head, I immediately want
to know the age of the patient. In practice, this decision comes
sooner for me than I expect. For now, it's only a shadow of what
my Italian counterparts are facing, but it forces the very real question of how to allocate resources, whether ventilators or beds—or
those of us who work in the ER.

A man in his late eighties is sent in from a nursing home with
a fever, cough, and diarrhea. He is my first patient who is most

likely Covid-positive; I can't know for sure, because tests are taking up to twenty-four hours to come back in our internal lab. Although the man is designated DNR/DNI—"do not resuscitate" and "do not intubate," which instruct us not to pursue aggressive interventions like electric shocks and breathing tubes—his family, with death now looming, reverses his no-resuscitation order and decides, instead, that he should receive even the most extraordinary lifesaving maneuvers. The man hasn't walked in years; he has advanced dementia and was unable to talk even before this most recent illness. He can't tell me what he wants, so under normal circumstances we are to follow the family's orders. They are in the waiting room, unable to come in because of our new, strictly enforced no-visitor policy, to prevent virus transmission. With the man's breathing rapidly worsening, I don't have time to call them. I am supposed to obey their wishes, which the doctor from the nursing home had, in his spare cursive handwriting, documented in a statement.

We are weeks away from the full impact of this outbreak, but we are already trying to conserve masks, gowns, and face shields. Because of how infectious the virus is and the country's lack of preparation and equipment, the decision to intervene is a question not only of how to apportion tangible supplies, but also of how to best distribute risk among health care workers. I want to do everything for my patients, as much as they and their families want, just as we have always done. But what do I owe future patients? What do I owe my colleagues? As I look at my team of doctors and nurses and consider our next steps, I think of a recent Facebook post from one of my supervising physicians, who trained me during residency, William Binder, who is now in his sixties. "As an emergency physician, I understand anything can happen to anyone at any time, but I have never felt exposed nor susceptible," he wrote. "The coronavirus has stripped away my veneer of invincibility."

Fourth Week of March
NYC Covid-19 cases, March 22: 9,045

It has been only a week since my colleague first posed the hypothetical case about resuscitating a Covid-infected patient whose heart has stopped. I feel like that was a different world back then, one in which we all held onto a thread of optimism that we would not have to face Italy's choices. Early on, I joined several private Facebook groups for doctors and browsed health care workers' feeds on Twitter. But these posts soon feel unbearable; it's suddenly too much to see clinical scenarios discussed hypothetically. In New York City, the hypothetical is here. A bunch of us in the ER have started communicating through a WhatsApp group chat so that we can openly discuss how we're feeling about the pandemic response. Three New York City hospitals are rumored to be out of ventilators. Thirteen Covid patients died in one hospital in twenty-four hours. A refrigerated truck is sheltering dead bodies there because the morgue is already full.

We are starting trial runs of putting two patients on one ventilator at my hospital. I can't believe we are coming up to this point already. So many patients are overflowing into the hallways, relying on oxygen tanks instead of the dispensers on the walls. Do we have enough tanks in the ER? No one in the chat group knows. Should someone take control of their supply? Yes, good idea. Someone suggests medical students, but the school wants to protect them from exposure to the virus in the ER.

It seems impossible to avoid getting infected. You would have to be perfect, and in the mayhem of the ER, it's nearly impossible to be even good. I make mental calculations to keep all protective equipment on for my eight-hour shifts; during my twelve-hour shifts, I'll remove it only twice, to eat or drink. Two Italian colleagues—a doctor and a nurse—have already warned me about the physical toll of wearing this equipment on their aching faces, their noses rubbed raw, the tracing of their masks etched into their skin.

On days off, I try to learn what I can about this virus and its many tricks. I watch videos on how to best manage patients on their ventilators. Their respiratory needs are different from what I'm used to. Go high on the oxygen and the post-exhalation pressure. Keep the breaths small, though, because Covid lungs are thought to be stiff and might overstretch. It's a delicate balance between trying to protect the healthy parts of the lung while giving injured areas time to rest. We're still learning about the disease, though, and it seems some patients' lungs might have different needs than this.

Okay, onto the heart. Someone sends me ultrasound images of profound heart failure in a Covid patient he cared for. There's the gut too—patients can experience a lot of diarrhea. But we can't give them too much in the way of IV fluids or we could flood the respiratory system. I recall a mantra from my days in residency: "A dry lung is a happy lung." An Italian doctor tells me that she's learning that the kidneys could also take a hit, compromising their ability to filter waste from the blood. Every part of the body comes under attack, it seems.

We're temporarily out of the proper disinfectant wipes at the ER in one of our hospitals. Someone intubates two patients—a procedure that risks exposing the medical worker to discharge from a patient's nose and mouth—without a face shield because none were immediately available. A co-worker is collaborating with others to 3-D-print face shields. Emails come through from hospital leadership and the city's health department telling us to be "appropriate" and conserve our N95 masks. A physician assistant is baking her masks in the oven to sterilize them. She shares her recipe: 170 degrees for thirty minutes. Others spray theirs down with Lysol after every shift. I was shocked when they told us to use these single-use masks for the whole day; now we are told they must last multiple shifts. We can discard them only if they become visibly soiled. Otherwise, wear the same one—"for multiple patients, for multiple shifts." How am I supposed to know when a mask should be thrown out? What does a virus particle look like anyway? I start telling my residents that it's better to be lucky than to be good.

The hospitals I work at are nearing maximum occupancy, even as new quarters are constantly being opened to accommodate more patients infected with Covid. In the meantime, updated clinical recommendations are given to us to follow: if patients' oxygen levels are slightly below normal, send them home anyway if they look okay. *Let's hope they know when to come back,* I think. Brambillasca, the Italian anesthesiologist, tells me that his patients often look well, but if their oxygen reading is slightly low, they can "crump"—medical slang for getting very ill very quickly. When I share this with colleagues, a couple of them start to counter: "The current evidence says . . ." What evidence? The novel coronavirus has been around for only a few months.

Everyone in medicine knows that one of the most heart-dropping phrases you can hear is: "You know that patient you saw the other day? Well, he came back and . . ." I think of all the doctors who sent their patients home because they looked well or are young or didn't have medical problems, and they came back to the ER needing a breathing tube. I'm sure these patients all looked okay a few days ago.

Don't worry, we hear, Andrew Cuomo, New York's governor, is protecting us from lawsuits. He had issued an executive order stating that physicians "shall be immune from civil liability for any injury or death" while caring for patients during the Covid outbreak, unless it's a case of "gross negligence." I ask my co-workers if anyone is still concerned about getting sued. I think we're much more anxious about having to live with people dying—and possibly getting sick ourselves. Not only do we have to think about patients not getting ventilators, but now we have to worry about sending infected people home, where they will likely worsen and may become critically sick, unable to make it back to the hospital in time. Paramedics say they are seeing three hundred "dead on arrival" cases in one day, citywide, instead of the usual fifty or so.

As soon as I hear this, I venture out that night to buy two pulse oximeters, small devices that go on a person's finger to monitor his or her respiratory status. I can keep track of friends and neighbors who fall ill. Even when I'm at home, I can help triage. I bring one to work, to test patients' oxygen levels, to see how much they drop when they walk. I'm told we will give them to patients soon, so they can monitor themselves—and maybe to-go oxygen containers as well, if they're needed. I hope this will be effective. But it seems a lot to ask of someone who's very sick.

One colleague, who is over sixty, already has a plan if she feels ill. She'll check her oxygen measurement, and if it's less than normal, she'll consult an outbreak map online and survey the surrounding states. She'll pick the closest city with the smallest number of cases. Then she'll drive there and hope that her age won't be considered when it comes to the care she gets. I find out that more doctors are hospitalized with the virus. One ER colleague across town is intubated. I get texts from colleagues across the country about doctors who are infected and hospitalized, some in the ICU, some intubated. I look at my reused mask. It doesn't seem soiled yet. I put it back on my face. *Better to be lucky than to be good,* I tell myself.

On Twitter, I see a photo of resident doctors at Massachusetts General Hospital, where I trained, holding up official documents explicitly designating who should make decisions for their care if they become critically ill. The thought is overwhelming, but I know, as a doctor, I want my patients to do the same. I decide to do it unofficially, texting a close doctor friend I work with and telling him what I want in writing. Please try, for as long as possible, if there's a chance I can make a decent recovery. If not, well, you know what to do. After all, someone else could probably use that ventilator.

Fourth Week of March
NYC Covid-19 cases, March 24: 14,905

Though it has been only two weeks, I desperately ache for that time when a patient testing positive for Covid was a surprise. I think back to the man from the nursing home. I had made the decision that day to intubate him, which would necessitate giving him a ventilator and an ICU bed. It was early on in New York's outbreak, and we were still in patient-centered mode, as the doctors in Italy put it. They are deep into community-centered care now. "As physicians, we normally choose the best option for the patient," Giovanna Colombo, an ICU doctor at Papa Giovanni XXIII Hospital, tells me. "We don't have to think of the community implications of what we're doing. But the epidemic setting is completely different."

I call up Mirco Nacoti, another ICU doctor there. "Without guidelines," he tells me, "it's impossible to work. And there is no space for imagination during humanitarian crisis. If you use a lot for the first patient, then you have no treatment for the next patient. You have to reorganize everything. You have to reorganize your mind; you have to reorganize your work; you have to reorganize your personnel and health care people."

Marco Vergano, a co-author of the controversial SIAARTI guidelines, had removed the criteria from the document because he wanted to give doctors flexibility—and because he knew the criticism would be overwhelming. Yet clear criteria are what physicians want most. The doctors I speak to in Italy all want a specific formula and decisions from a third-party team, one whose members aren't directly treating the patients. "You're caring for patients who are complex enough," Colombo tells me. "If you keep thinking of this problem, you can never do this job."

To help with this task in Bergamo, a few weeks into the outbreak, a doctor at the hospital comes up with a scoring system. It's
not meant to be a strict make-or-break guideline, but it functions
as a tool to help in decision-making. It has specifications that the
SIAARTI document lacks. In this tiered system, patients are scored
for age, medical problems, and the severity of their current respiratory status. The higher a patient's final tally, the higher the
priority for intubation.

In New York City this week, the conversation shifts. The question of who gets a ventilator and who doesn't comes up in every
single Zoom meeting among ER physicians that I participate in. A
hospital committee is discussing that, we're told. We want guidelines; nobody wants to exclusively treat people first-come, first-
served. I've thought and written about what makes a meaningful
life, and I generally agree that means autonomy for patients and
families; they should get to make decisions about their treatment.
But I do believe that when resources are scarce, doctors can and
should make judgments about who should get more care. A colleague, feeling similarly, announces during a meeting: soon I'm
just not going to intubate the eighty-something-year-old patient
who doesn't talk or walk so that there will be a ventilator available
for the thirty-year-old who comes in later. It sounds heartless, but
we agree with her. Future patients like the thirty-year-old are not
yet here, but they are definitely on their way.

One of my residents asks me, "Will there be ventilators for us if
we need them?" As with many questions I've been receiving lately,
I don't know the answer to that one.

Fourth Week of March
NYC Covid-19 cases, March 26: 23,112

I am scheduled to be off from work for several days. The evening
before I'm due to return to the hospital, a colleague messages our
group to say that a forty-nine-year-old Covid patient of hers, who
was waiting in the ER for an inpatient bed, was found blue and
dead in a chair. Nobody even knows if he gasped before he died.
On my way to work, I hear on the radio that a forty-eight-year-old
nurse from another New York City hospital has died from coronavirus. Someone else tells me that an anesthesiologist at our hospi-

tal is on a ventilator. A surgeon and an ER doctor across town are in similar states.

When I walk through the hospital doors, the ER is a place I no longer recognize. Intubated patients, of every age, are on ventilators everywhere. It feels simultaneously electrifying and oppressive. But it's also eerily quiet. Family members and friends haven't been allowed into the ER for more than a week; most of the patients are too sick to talk; the few without breathing tubes who are able to cough are muffled by their masks. Oxygen hisses in the background. A couple of hours into my shift, one of the nurses comes to me. She falls apart, tears streaming down her inflamed, marked cheeks. She sobs out words of anger and frustration and sadness. The morning, on top of the last several days, has crushed her. I want to hug her, but I can't.

Soon after that, someone asks, "Doctor, is it okay to take the patient to the morgue?" The other physician on duty and I look at each other. The morgue? Who just died? Apparently, a patient who was waiting for an inpatient bed, whose family had decided against extreme resuscitative measures, had died, without us even knowing.

Several days ago, only a few patients had Covid, but suddenly it seems we have become, like facilities in Italy, a Covid hospital. Every patient seems to test positive for it. I am shocked by the one or two negative results I receive during a shift. We have to function as if everyone is infected.

A co-worker tells me he used three masks during the course of his shift. Three masks?! I respond. That's crazy! Then I realize I am the absurd one. The masks are meant for single use, one per patient encounter; my colleague had used three masks over a twelve-hour shift, most likely having seen upward of thirty patients who potentially have Covid. It's idiotic that I was shocked by his using three masks, especially when many of our co-workers in the city have fallen ill.

Patients who test positive for the virus are unintentionally roomed with those who test negative or whose tests are still pending, because the ER is bursting. Even if we are exposed to a patient without proper personal protective equipment, we are expected to return to work if we don't have symptoms. In Italy, where sixty-one doctors have already died from Covid (a number that will grow past one hundred in the next couple of weeks), health care workers believe that they themselves expedited the spread of the virus. There, the doctors are routinely tested for any exposures, even if they are asymptomatic.

I have to shut down thoughts about my own risks and mortality. I recall the words of my old mentor, but I don't think I can do this job unless I force myself to believe in my own invincibility. Otherwise, with every violation of the protective barrier, every instance of less-than-ideal protection, which is almost every time, I would be paralyzed by thoughts of having infected myself. I see a patient around my age intubated, hear about a hospital colleague getting critically ill. A co-worker texts that her classmate from residency is now intubated. I read an article about how health care workers seem to suffer more from serious Covid infections, even if they're young, possibly as a result of being exposed to higher initial doses of the virus. I'm not even sure this is true anymore—I've seen plenty of critically ill patients in their thirties and forties. I push these thoughts away, immediately. *Better to be lucky than to be good,* I remind myself. It's the only thing that provides some reassurance. If I feel like it's not totally in my control, then I won't completely lose my mind over every mistake I make donning and removing my PPE and recycling single-use equipment.

I look in the mirror for the first time when I get home one night. My eyes are bloodshot. Deep horizontal creases run across my cheeks. A faint abrasion has already settled into the bridge of my nose. I just want to fall into my bed, but I force myself to shower. When I turn my phone back on, a nurse in Bergamo, Stefania Cornelli, has texted me that she crashed her car. The vehicle was totaled, but she wasn't seriously hurt. It had been about one month into this crisis for her. "We are so tired, tired of a tiredness that no sleep can relieve," she writes. "I think I really need to ask help to a psychologist."

Fourth Week of March
NYC Covid-19 cases, March 28: 30,766

When I get to work the next day, a patient who had a breathing tube inserted overnight had woken up enough to pull it out. She was delirious, lacking oxygen to her brain, and had also yanked out her IV lines. Sputum and blood and sweat are flying everywhere in the room. My instinct is to run in to help, but I force myself to pause, put on all the equipment. I place the N95 respirator on my face—and a surgical mask over the N95 to keep it clean and reusable, as we're instructed—as well as a gown, goggles, gloves, and a face shield, 3-D-printed by my colleague. It's so hot. I start sweating immediately. We manage to reinsert her breathing tube and replace her IV lines; she safely makes it to the ICU. After an hour working like this, I feel lightheaded, but it is too early to remove the mask and drink water. How do I make it through the next twelve hours?

Later in the day, I start getting chills underneath all my equipment. I briefly wonder if I'm getting sick, then I decide that it will become obvious if I am, that I should just go on for the day. Even if I develop symptoms, I'm not able to get a test from employee-health services at my hospital anyway. Whenever I have patients come in telling me that they tested positive at their doctor's office or at urgent care, I immediately take note of where they got that done.

I get statistics from my hospital indicating that over 80 percent of the admitted patients from the previous day have tested positive for Covid-19. I feel at odds with myself, conflicted between my emotional response and my intellectual curiosity about this virus, which seems, as Brambillasca said, to be mercurial—reckless in what it inflicts on its victim. I scroll through the lab tests of patients. I excitedly exclaim out loud that one patient's lymphocytes, a type of white blood cell, are very low, something I had read about. Then I pause, realizing that this is a sign that the patient probably won't do well. These observations happen repeatedly; I pendulum back and forth between my fascination with the disease and my despair for my patients.

Six hours into my shift, I go to the bathroom for the first time. I stand in the unvented bathroom for a minute and pull my mask away from my face. The air is stale, but the rush of oxygen into my lungs feels wonderful. I take big gulps of it through my nostrils before letting the mask compress my face again. When I'm not in the hospital, I feel a phantom mask on my face. I wake from sleep trying to adjust it, thinking it's still on.

I try to preserve the equipment that I do have, but the steps seem futile. In the ER, I sanitize, glove, remove glove, sanitize again. I have to touch a door handle to go into the workroom to type my clinical notes. I'm unable to sanitize again because there are no more portable hand sanitizers left. I get flustered when I accidentally touch my face, wondering how I forgot and slipped. Sometimes, I can't remember if the gloves on my hands are clean or dirty.

At so many points I ask myself, *Does it even matter anymore?* It feels like the virus is everywhere, breathing on all the surfaces, exhaling itself into the atmosphere. It feels exhausting wearing one mask all day and covering it with another to keep it clean, having to think so much about not getting it soiled and wondering if I accidentally contaminate the inside of it when I hold it away from my face to breathe for a minute or take it off to chug water. Sometimes I see the individual virus particles—round with red, protruding crown-shaped spikes, like the CDC's rendering—everywhere in the hospital, on beds and monitors and phones and computers. I shudder, forcing myself to erase the image from my mind.

I've taken part in humanitarian relief missions in more than twenty countries, in settings as resource-poor as mobile clinics in South Sudan immediately after its secession, refugee camps in Kenya, an abandoned war hospital in Liberia, medical facilities in Somalia. Never have I personally felt unsafe, like I didn't have enough protection for myself. People are now referring to ours as "a third-world country," but in terms of PPE in this pandemic, it's actually worse than those overseas hospitals. While most of the specialists have been unflinchingly generous, offering extra hands in the ER and imperiling their own lives, a few doctors who are consulted for their expertise on certain medical conditions have balked at having to see patients here at all. They feel unsafe, they say. Deep down, I know they're probably right.

By the end of my shift, every patient begins to blend into a single patient. "Fever and cough," "fever and cough and shortness of breath," "cough and trouble breathing," "sent by doctor's office for Covid rule-out," "sent from urgent care for Covid test." I can't even keep track of them anymore. Usually I remember patients by their faces, but they all have masks on too, so all I see is their eyes, which more often than not are closed.

I become obsessed with oxygen levels, which seem to be the only reliable indication of how patients are doing. Is 92 percent much better than 90 percent? Should 93 be the cutoff to send someone home, or should I make it 94? I used to be able to rely on my gut and clinical judgment when I walked into a room and looked at the patient, but coronavirus is lawless. It obeys no rules. What is unusual, in this illness, is that many people come in talking to you, even as their breathing worsens. They can speak, but their oxygen readings are frighteningly low. As the hours tick by, they rapidly get sicker, to the point where they need a breathing tube. In most other situations, people who require breathing tubes in the ER arrive at the hospital too ill to interact with me, needing mechanical ventilation right away. That makes it a little easier.

Patients' oxygen tanks run out. (It's impossible to know unless you bend over, look behind the stretcher, and glimpse the thin black needle ticked over to the red zone on the gauge.) Or whatever oxygen you did give them becomes suddenly insufficient, as their lungs grasp for ever more. Maybe an alarm bell sounds because their oxygen level has dropped. Or more likely, they've become disconnected from the monitor, a far-too-frequent occurrence, and you see them frantically trying to breathe. Or most likely, the oxygen, even if it's blowing, is of no use, because they're unable to take it in, barely inhaling at all, silently dying, alone.

What may have been unimaginable even a week ago seems completely possible, even likely, now. A colleague informs me that she had to push aside a dead body to plug in a ventilator for a new patient who was recently intubated. Is this how the dead leave the world now?

Before, I would check in with the Italian doctors, concerned for their and their patients' well-being, but our roles have now reversed. I am now at the receiving end of their grief and sympathy. "How are you?" one texts me. "We hear it's so bad there." Yes, it really is. "Stay strong," another says.

We're unable to reliably predict who does well and who doesn't. Old or young, all seem wholly vulnerable. Politicians, epidemiologists, even doctors have been saying that people in their twenties and thirties who get sick already have medical problems or are obese, but then, right after I hear that, I need to put a young and fit patient on a ventilator. The virus is impulsive, attacking one person more ferociously than another. I feel the compression from all sides—the ICU is full, the ER is full—I just don't see the end of this in sight. When I think about that, I feel submerged, and my instinct is to rip off my mask and leave the hospital. Then I try to convince myself that it's like running. When you start off, your lungs burn and your legs ache, but as your stride hits a rhythm, you start to feel good, and you know you can go on for miles. I hope intensely for that moment to come soon.

I know many New York hospitals are working on their own resource-allocation guidelines and designating a third-party team of in-house doctors to decide which patients get to have their care escalated. Now that I'm already involved in helping to make those decisions, I'm less worried about getting the criteria in my hands. I'm also hopeful that external relief will come. I used to travel to others to provide humanitarian assistance, and now people and materials are coming here to help. Makeshift hospitals are opening around the city and will take some of the load off. When we hear that the Javits Center and the Navy hospital ship *Comfort* will care only for non-Covid patients, my colleagues and I find this laughable, because everyone has the virus. They'll realize it soon enough, we say to one another. (A few days after it opens, the Javits temporary hospital changes its admission policy to take in Covid patients; the *Comfort* does so the following week.)

Health care workers and equipment are coming in from other states. I am optimistic that for those who have a chance of surviving, we will be able to do everything for them. Of course, hard choices will still have to be made—it's never easy withholding care from a patient—but I believe they will be rational decisions that most doctors would agree on. We are not playing God, as those who made the SIAARTI guidelines were accused of, but we have been doing this long enough to know which patients will have a possibility of recovery and which ones will needlessly suffer. Even in Italy, Vergano tells me, his critics have all backed off.

Still, mental health professionals, especially those who treat combat veterans, worry that doctors will sustain moral injury from having to allocate medical equipment and care. The truth is, when treatment is rationed or withheld, the decisions are almost always reasonable, and hopefully the family will be involved. I've already had a few of those conversations on the phone with family members, guiding them through what would happen to their loved ones, explaining the extensive medical procedures involved and the thin likelihood of survival, assuring them that they should feel no guilt, that I would do the same for my mother. I can't say with 100 percent certainty that they would not have survived, but I can say that I didn't prolong their suffering. There is a bit of solace in that.

I think back again to the elderly man I intubated, when we were still at the foothills of this pandemic. If I were given a do-over, I would not do it. I would save the ventilator for a future someone else. I would override what the family wanted and hope that afterward, they would understand. I believe it will be fairly obvious that in most of the cases where we don't push forward with extreme medical interventions, we would not have been able to save the patients anyway.

What I think will actually cause moral injury is seeing people die after getting the most advanced care available. People who come in talking, with stories to share. They get care—the best that modern medicine has to offer—with life-prolonging machines and IV drips of all sorts of critical-care drugs. We put our full minds and whole hearts into trying to save them. Then I see their bodies shut down anyway. They are alone. I'll see that over and over again, and it will reach a point when it is numbing. What will affect me the most is not remembering them as individual people, no particular detail that separates a person from the one before and the one after, because they all come in sick with the same symptoms, the same history, until they morph together, become breathless bodies. That I am the last person they see before they die—not their families—and that I won't remember them at all because there will be hundreds more just like them. That it will become routine.

Last Days of March
NYC Covid-19 cases, March 30: 38,087

"By now, I think it's very hard to stay human," says Duca, the ER doctor in Bergamo. "You go on, you forget you have a person, a human in front of you. You forget the patient has a life. I think that we do this to protect ourselves. Otherwise, it would be impossible to work every day." Colombo, his ICU colleague, tells me she goes through an "emotional shutdown," as she calls it, when she arrives at the hospital. "The first few days, I was crying when I was home," Brambillasca says. "Then you transform, because you have to do it. You become tough in a few days."

The first patient hospitalized in New York is finally discharged, nearly a month after his diagnosis. Many have died in the meantime, and many more are uncounted in the Covid-19 death toll because they succumbed at home or weren't tested. Some others wait days in the ER for an inpatient bed, languishing in hallways. Their fates remain unknown. I pass by them when I first arrive at the ER and when I leave at the end of the day. Sometimes they are still there the next day. If they are awake, I'm hesitant to make eye contact. I'm too ashamed that after nearly fifteen years as a doctor, I can't do much more for them except put an oxygen mask over their nose and mouth.

One day I see a grandfatherly man, who speaks softly and smiles sweetly, come in with oxygen numbers dipping as low as 75 percent. He feels good, he says, and his breathing is fine. Just a little tired, don't worry, he says. Despite everything I know so far, I think he will do okay because he looks so well. The next day, when I return to the ER, I see he is now confused. Even wearing an oxygen mask, he could not sustain levels above 90 percent overnight. He had previously decided that he did not want extraordinary measures taken to save his life; he did not want to be on a breathing machine. His family, over the phone, is clear about his wishes, so we make him comfortable with morphine.

I want to spend time with him, but more patients, much younger patients, keep arriving, struggling to breathe. I have to tend to them instead. The disease has won against him; the new patients have a chance. I don't want to think that way, but it is the dismal truth of our new situation. I hope the morphine is enough to blur the reality that he's all alone. I move on, forcing myself not to think about him again. Too concerned about the new patients, I never take the time to check on him again. Too exhausted at the end of my shift, I don't say goodbye to him either. He dies later that night.

Before the pandemic, I would typically see a fair number of nonwhite patients. Yet Hispanic and Black patients appear to be arriving at our ER at higher rates now—and they seem sicker than patients of other ethnicities. A paramedic points out a similar pattern in what he's seeing. (Data that comes out later confirms as much: Black and Hispanic patients are dying at twice the rates of their white and Asian counterparts.)

I keep hearing about this "apex," that we're still weeks away from it. I can't bear this word anymore. Apex. When is it coming? How will we know when we've reached it? What if cases start to slow down, then increase again? But mostly, I think, how can I think that far ahead, when I have to coach myself just to get through the next hour?

On the last day of March, I get several texts from Duca and his colleagues. For the first time, they are seeing some light: the number of new patients seems to be finally decreasing. Of all the messages I've received from friends and strangers all over the world, these are the ones that keep me going. To know that as bad as this is now, it will end someday. But when?

First Days of April
NYC Covid-19 cases, April 1: 47,440

In the early evening, toward the end of one shift, a woman with ash-blond hair in her fifties walks into the ER. She converses with the nurse about her week of fever and cough, but while an EKG is being done, she suddenly becomes unresponsive. She loses her pulse. We shock her out of the irregular, rapid rhythm her heart is in, put a breathing tube down her throat, and start drips of multiple IV medications to stimulate her heart and constrict her blood vessels. Later that night, I get a text from a colleague in her sixties, who had walked by during the resuscitation. "I have the sense that the world is ending," she writes. "The person you were coding was six years younger than me."

The next morning, as I'm getting ready for work, I panic: I might not have showered last night when I got home from the hospital. I try to retrace my actions but fail. I simply cannot recall. Did I just fall asleep? Am I infected? Should I change my sheets, scrub my apartment? But I have to get to the hospital for my shift. There's nothing I can do about this now. Better to be lucky.

When I arrive in the ER, I look up the woman's electronic medical record from yesterday's shift. I badly want to be able to text back to my colleague that the patient is doing okay, that we'll all be okay. When I open her chart, a warning flashes across the computer screen: "You are entering the medical record of a deceased patient. Are you sure you want to proceed?" She barely made it to daybreak.

A few days ago, palliative-care doctors started helping us with some of the life-or-death conversations. They call families and talk to them about procedures that patients might have to undergo if they want to escalate the interventions; these doctors help figure out where the limits should be drawn. They also explain how patients could otherwise be made comfortable, if they don't want to continue with more aggressive treatments. But the doctors are soon overloaded, unable to tend to all the consultations. I try to do what I can. An eighty-nine-year-old patient is brought in by ambulance, with an oxygen mask covering most of her small face. I don't think she'll be able to talk, but she is actually able to express herself and tell me: "I don't want a breathing tube. I'm almost ninety years old. I've lived." She's originally from North Carolina, she says. I call her niece, who is her health care proxy. She conferences in other family members. "Well, can't we overrule what she wants?" one of them asks me.

I'm not formally trained in this, as our palliative-care doctors are, but I've had many of these discussions over the years. I tell them that she has clearly expressed what she wants, and I promise to make her comfortable. Think about what you know of her, I say. What does she value? Would she want to be hooked up to a machine? How can we stay true not only to her wishes but also to who she is as a person? They agree that dying peacefully would be what she would want.

The patient is still awake, interacting with me. I call the patient's family through FaceTime on my cell phone. Her niece comes on, her smooth cheeks shiny with tears. She tries an upbeat hello. My patient isn't fooled. "Everyone's got to stop crying," she says. "They're taking good care of me here." We all laugh a little through our tears. I order some morphine for the patient. Her breathing gets easier.

I run around, trying to care for more patients. I'm not sure if anything I do makes a difference. I can't run away from Brambillasca's words about the virus: "It does whatever it wants."

I wonder if I'm more useful FaceTiming patients' families rather than applying my skills as a doctor. Three hours later, I pull out my phone again and call my patient's niece. "I love you," she says to her aunt. My patient flutters her eyes half-open. "I love you too," she slowly replies, her voice noticeably weaker now. I put my hand on her hand. She grabs my fingers, tells me she feels cared for. She doesn't want to let go. I don't want to either. I look down at my purple-gloved hand holding hers, delicate and bony. I hate that she has to feel synthetic rubber, that she doesn't get an actual human touch before she departs from the living.

The next morning, a much-needed message comes through from Italy. "Please, don't give up," writes Cornelli, the nurse in Lombardy. "Our jobs are difficult but are the most beautiful ones."

We try putting a few patients prone on their stomachs. I first heard about this weeks ago, from one of the private Facebook groups devoted to caring for critically ill Covid patients. It's said to help intubated patients—why not give this a try with those who don't have breathing tubes but aren't oxygenating well? I see a patient's oxygen level shoot up. It works, I yell out, elatedly, prematurely. Something actually works! We need massage tables with the cutout face holes for our patients, I joke to my resident.

A couple days later, I see on Twitter that a Detroit-area oral-surgery resident has died. His name and photo are in the tweet. "Christopher Firlit." I say his name out loud; I look at his photo. I want to honor his death. I want people to know; I don't want doctors to die in anonymity. Eventually, I put my phone away. Then I think back to my own resident's question: What would happen if they need to be put on ventilators?

Second Week of April
NYC Covid-19 cases, April 5: 67,552

"Messaging with you helps," I text Brambillasca. "To hear it will end." (I punctuate using a period, but in my mind it's a giant question mark.)

"And it will," he immediately replies. "We are seeing it here. So it will come to NYC as well."

I read his words three times. *This will end,* I tell myself. *This will end.* I am hardly responding to family and friends anymore. It feels impossible to explain to them what's going on. I can think of nothing else, but the last thing I want to do is describe to each person what's happening in the hospital. I rely on my co-workers—they grasp everything I'm feeling with just one glance or a three-word

text. Even doctor friends—in Philadelphia, Boston, Los Angeles—
seem like separate species now. No one from another region could
understand what was happening in Lombardy. Can someone from
another city understand what's happening in New York?

My phone vibrates again. Brambillasca just got his first non-
Covid patient in the "clean" ICU, intended for patients not infected with coronavirus. The patient's heart had stopped twice
during a rescue abdominal surgery—a terribly sick person with
severe complications whose outlook remained poor. But Brambillasca was still grateful, still happy: "What a soft lung to inflate."

I happened to have been assigned to work at one hospital for a
chain of shifts, so I hadn't been inside one of our other hospitals
in over a week. As soon as I open the ER doors there, I shrink from
the sights and smells. Patients are now triple-bunked into single-
person spaces, curtains pushed aside. In one room, three men,
who appear to be in their eighties or so, are side by side in their
stretchers, each one pulling at his oxygen mask, confused, their
frail limbs swinging in the air. Some have sat in their own feces for
a day. Puddles of urine have pooled around the wheels of some
patients' stretchers. Nurses are out sick; the remaining ones are
coping the best they can. I have gotten texts from colleagues about
the chaos here, but I thought that those were just about one bad
day, that they had already gone through the worst.

Another doctor notices the bewilderment on my face and
comes over. "There are people literally dying of hypoxia in the
hallways," he says, "and there's empty space with oxygen dispensers on the wall and no one using them." What is he talking about?
Isn't the hospital full? He suggests that I take a walk down the hall
and make a right, less than 100 yards away.

I swipe open the unit, which usually serves as a postoperative
area, with my ID. I see a room about half the size of the ER. It's a
Sunday, a slow day usually, but still, there's only one patient, who's
being tended to by a nursing assistant. A nurse hovers nearby. I
track the green oxygen dispensers on the walls, these fountains of
life that my patients gravely need. I go upstairs to one of the regular floors. It's calm and quiet. Unlike in the ER, where I dodge
patients, colleagues, and stretchers to get around—forget six feet
of separation; we're not able to maintain six inches—here the hallways are free and unobstructed. It's just a regular hospital floor,
but the space feels glorious, luxurious.

It's the first day of our pulse-ox to-go program. Until this point, I have been opposed to the idea of sending hypoxic patients home with pulse oximeters, especially after learning from the Italian doctors that their oxygen numbers often drop quickly to life-threatening levels—sometimes before the patients feel it. These guidelines seem too unsafe to me. A colleague begs me to rethink this, telling me they will get better care at home with their family members than here in the ER, at least in its current state. "And I'm saying this as someone who doesn't believe in these guidelines," he adds. After witnessing how many patients are suffering in the ER, I immediately discharge two to self-monitor. I know I'll probably soon hear the dreaded words—"You know that patient you sent home the other day? . . ."—but I have to do what's best for them right now, with what I have in front of me. I'm hopeful that the field hospital being built at Columbia University's soccer facility, to be staffed largely by former military personnel, will open soon with a capacity for nearly 300 patients.

This week, our employee-health services is at last starting to routinely test medical workers who develop symptoms that could be Covid-related. Still, I wish we could regularly get swabbed and checked when we know we have been exposed, even those of us without symptoms, so that we don't inadvertently pass it on to our patients. Some of us are also eager for antibody testing, seeking a sense of security if we end up having antibodies, though it's probably too early to say whether or for how long that could actually provide immunity.

In the ER, I run into two co-workers who have recovered from the virus and are back at work. Our ER colleague across town is out of the ICU. I look at a photo of her eating and smiling on Facebook. The next day, I see on Twitter that James Pruden, a seventy-year-old doctor in New Jersey, is leaving the hospital after spending nearly a month in the ICU. He was one of the first doctors hospitalized for coronavirus infection in the United States. I didn't think he would make it, because of his age and how sick he seemed. In a video clip, Pruden, in a blue dress shirt, is wheeled out on a stretcher and points energetically at the surrounding crowd. I've never met him, but I'm immediately tearful. I replay the recording four more times. Then I send the tweet to a colleague who works with him. "Something going our way for a change," he responds. "If he can do this, we sure can."

Later that same day, though, I get a text that several more of our ER staff members are hospitalized, requiring oxygen. I learn that another died a few days earlier. More co-workers are ill at home with symptoms. At night, I open an email that a doctor in Brooklyn forwarded to me with the names of more health care workers in New York City who have died. I hadn't even heard of their deaths.

Over the next several days, I notice the tone changing during my shifts. Conversations about dying and death are all around me now, the only kind I hear. Either I am having one or the physician next to me is. We spend our days talking to patients and families about the limits of medicine and what doctors can do; we call people to tell them their loved ones have passed away. Then we make another call. And another.

Those of us who work in the ER are accustomed to pushing our patients' mortality to the edge. My promise to them has always been that when they come through those ER doors, I will do everything I can to help them live. This is how we approached every shift. In a way, that job was easy. Do everything possible, unless the patient or family has explicitly expressed otherwise.

This is no longer the sole operating principle of emergency medicine in New York City. It has been less than six weeks, but I've never felt less useful as a doctor. The one thing I can do—what I think will matter most, in the end—is just to be a person first, for these patients and their families.

"Staying human is painful, but it is what I need to keep working," Duca says. "I realize now that keeping the emotions outside of me can help to manage the shift and the stress, but I need to be human to keep working." I know exactly what he means. It's no longer getting through this day or this week; we are in the deep now, the interminable. For doctors to survive this pandemic, we have to feel each moment—even if it makes each moment more difficult to endure.

SUSAN DOMINUS

The Covid Drug Wars That Pitted Doctor vs. Doctor

FROM *The New York Times Magazine*

MANGALA NARASIMHAN, AN intensive-care-unit doctor, started feeling impatient soon after the start of a meeting she attended at Long Island Jewish Medical Center on May 13. She wanted to get back to the unit, but instead she was sitting in a conference room with about a dozen colleagues. By then, the surge of Covid-19 cases, the waves of suffering that had crashed down on her hospital for months, was beginning, miraculously, to recede. The throngs of out-of-town health care workers who had come to New York City to help were also diminishing, heading home to regions whose own times would come. Narasimhan and her team now had fewer hands to oversee new patients coming in and the long-suffering ones on ventilators who were still in need of meticulous care. Long Island Jewish, in Queens, had, at the time, treated more Covid-19 patients than any other hospital in the country; the doctors there were still weary, still battered, their energy and time in need of careful rationing.

Narasimhan, who was in charge of more than twenty ICUs across the Northwell Health system, knew heading into the meeting that it might be tense. Adey Tsegaye, a pulmonary critical-care doctor who was calling in remotely, shared some of Narasimhan's concerns. The meeting's agenda included time for remarks from Alex Spyropoulos, a lead researcher at the Feinstein Institutes for Medical Research—the research arm of Northwell—who was running a clinical trial. The research was trying to determine whether a standard dose of an anticoagulant or a higher dose yielded bet-

ter outcomes for Covid-19 patients who were already on oxygen
or a ventilator and were at high risk of organ failure and clotting.

A doctor on Narasimhan's unit had recently been at odds with
a member of Spyropoulos's research team. Stella Hahn, a pulmonary critical-care doctor, arrived at work the day before the meeting to find that a Covid-19 patient had gone into cardiac arrest.
She knew that the patient was enrolled in the clinical trial and
had been randomly assigned to receive either the standard dose
of the anticoagulant or the higher one. As is always the case in the
most rigorous trials, neither the patient nor Hahn was supposed
to know to which group this woman belonged. Double-blind, randomized, controlled trials—RCTs—are considered the gold standard in research because they do not allow findings to be mud-
died by any individual doctor's biases or assumptions. But Hahn
believed that the patient's condition now called for the higher
dose, which could potentially require the patient's removal from
the trial.

Word made it back to a doctor working with Spyropoulos, and
that doctor called Hahn to urge her to reconsider, or at least to
get more tests before acting. They exchanged heated words, as
the colleague implored her to stay the course. Hahn pushed back:
She had to rely on her clinical judgment and believed that it was
unethical to wait for more information. How could researchers
dictate care to a doctor right there at the bedside, especially when
a patient's condition was so dire?

The point of contention would be discussed at the May 13 meeting. Dozens of doctors from the Northwell system videoconfer-
enced in, including Spyropoulos, who was seated in his home in
Westchester. Hahn's colleagues, a tight-knit unit who had seen one
another through so much, sat together in the conference room,
occasionally checking their phones or exchanging glances as the
meeting went on. As Spyropoulos recalls, he talked to the group
about the importance of high-quality, randomized trials in making scientific progress, and the risks of trying experimental treatments without them. "I stressed to the group that we should not
abandon this principle, even in the very stressful environment of
a pandemic that was overwhelming our hospitals at Northwell," he
said. Relying on gut instinct rather than evidence, he told them,
was essentially "witchcraft."

For Tsegaye, the word landed like a blow. "There was a chill in the air," said Tsegaye, who registered it even by videoconference. "Followed by rapid backpedaling." Spyropoulos quickly explained that he had so much respect for what those doctors had done—he had not been in those critical-care units, in the emergency room, which he knew were unlike any other he had ever experienced. "But it was like a retraction sent to the newspaper the next day," Tsegaye said. "The headline says it all. The retraction the next day? It doesn't have the same impact."

In the days to come, whenever Tsegaye thought about what Spyropoulos said in that meeting, she felt appalled all over again. She knew that she had never extended herself on behalf of her patients the way she had since March. She kept flashing back to a day when she was told that a ventilated patient's endotracheal tube had fallen out, a situation that can be fatal for the patient and is also dangerous for the physician: replacing it requires the doctor to come into close contact with the patient's breath. Tsegaye was putting on her N95 mask to enter the patient's room when its elastic snapped in two. There was no time to go to the supply area to get a new mask. What was the right thing to do? With a sense of dread, she found her feet and moved toward the patient's room. As she prepared to enter, one of her fellows, whose mask was intact, told her to leave—she could manage it on her own.

Looking back, Tsegaye felt that the agony of making those kinds of decisions all day long compounded the grief she felt while treating so many patients she could not help. "These are the decisions we have had to face," Tsegaye said. "For someone like me, who had been in that situation, to have someone tell you that you have been practicing witchcraft is kind of giving no value to the sacrifice that I have made—that my colleagues have made."

As doctors face new spikes of Covid-19 cases around the country, they are also confronting a harsh reality: the virus's deadly secrets remain largely intact. The medical community now has some research-backed drug treatments—remdesivir, an antiviral drug found to shorten hospital stays, and dexamethasone, a cheap, readily available steroid that seems to cut deaths of patients on ventilators by a third. But six months after the first patient tested positive on the West Coast, there is still no treatment that reliably slows progression of the illness, much less a cure. In July, the number of patients dying in this country topped 1,000 five days in a row, according to the Covid Tracking Project.

In these early months, doctors have faced two unknowns in trying to fight the devastation. The first is the virus itself: deadly, contagious, and entirely novel. The standard of care for most intractable illnesses develops over years, as doctors build a body of research that tests various theories, compares and contrasts dosages, measures one drug's power against another. Here doctors were starting from scratch: any treatment protocol beyond supportive care—oxygen, hydration, antibiotics, and ventilation—was conjecture. The second, equally novel challenge has been the sheer scale of the outbreak. Few doctors in this country had encountered the overwhelming volume of patients, the sense of helplessness, the exhaustion, and the desperation to save lives. Hospital administrators found themselves plunging headlong into making difficult decisions in the absence of strong, unifying federal guidance. Most did so without the benefit of perfectly parallel case studies or personal experience in hospitals so overrun by suffering.

When there is no precedent, when there is an information vacuum, decisions are inevitably subject to challenge. In an already heated environment, some of the worst of the tensions played out between research-oriented doctors and those who saw themselves primarily as clinicians. Many treating patients on the floor considered it axiomatic that, with so many dying so fast and so little to go on, they would rely on their experience to make judgment calls about treatment options. They would try using medications that had been approved for other illnesses but not yet for this one— what the medical community calls off-label uses—if they felt they had good reasons to do so. They would take into consideration any information that was available: the observations of doctors in Milan and China, conversations among doctors in WhatsApp group texts and in Covid-19 physician Facebook groups, tidbits of research that made medical sense but had not yet been peer-reviewed.

Other clinicians, and especially doctors more heavily involved in research, were frustrated that many of their colleagues were not sufficiently invested in the importance of empirical research to figure out which treatments worked best and were safest. Kevin Tracey, president of the Feinstein Institutes, tried to emphasize to the doctors affiliated with the Northwell hospital system that if they were going to try drugs off-label, they should always be doing so in the context of a clinical research trial: the drug might help some patients but could hurt even more of them. If that was the case, it was better to know than to operate out of a mix of hope and conviction. He understood, he said, the impulse for doctors to try drugs off-label out of compassion—and the "raw emotion of humans trying to help each other survive and not knowing what to do." But he did not approve of it. "Emotions cannot carry the day," he said. "You need evidence-based medicine, and you need clinical trials. You don't make an exception in the middle of a pandemic."

Ethan Weiss, a cardiologist at the University of California, San Francisco, who specializes in metabolic research, spent two weeks treating patients at a hospital in New York and was also distressed by how quickly doctors were trying untested therapies outside clinical trials. "I mean, it felt like it wasn't even World War I medicine," he said. "It was almost like Civil War–level medicine." He asked that the name of that New York hospital be withheld out of respect for his colleagues, whom he knows were not only risking their lives but were also overwhelmed by their clinical demands and had no research to rely on. He nonetheless was surprised to see many of them making decisions "based on the sort of opinion or written protocol of one or a couple of people that was based on kind of nothing that I could see, other than just, 'This seems like a good idea.'"

Many clinicians on the ground felt the urgency of treating the hundreds of patients dying in front of them; researchers, with their literal and intellectual distance from the ICU, were pressing them to think about the thousands of patients who were sure to follow —to slow down long enough to build a body of evidence that they knew with more certainty could help. The tensions between these two ways of thinking about medicine have always existed. But during the early months of the pandemic, the disagreements—what one critical-care doctor called, on his well-read blog, the profession's "intellectual food fight"—provided another layer of painful stress to some doctors already near their limits. "It became like Republicans and Democrats," said Pierre Kory, a critical-care doctor who faced that tension himself at the University of Wisconsin Hospital and Clinics. "The two sides can't talk to each other."

*

As they prepare to evaluate a given medication or procedure, researchers are expected to approach their task with a certain neutral mindset. The official term for that stance sounds both scientific and strangely poetic: "clinical equipoise." It's a point at which a doctor's curiosity is greater than her conviction that any one result is the most likely one. Clinical equipoise is an elegant characterization of a humble admission: *I have no idea which of these two choices is better.*

Equipoise gave way to unbridled enthusiasm among some physicians at Lenox Hill Hospital on the Upper East Side of New York in April when the city was in the thick of the surge. Many doctors there believed they were seeing great results by providing tocilizumab, an anti-inflammatory drug that tamps down the autoimmune response and is used for rheumatoid arthritis. The doctors were prescribing the drug, sometimes in conjunction with a steroid, to Covid-19 patients, particularly those who were not yet on ventilators but whose blood tests suggested that they were about to take a turn for the worse. In using it, doctors hoped to stave off what's known as a cytokine storm, a potentially deadly immune-system overreaction in which a torrent of cytokines—proteins that can trigger infection-fighting forces—is released. Tocilizumab, which blocks the pathway of a cytokine called IL-6, might prevent that deadly storm from gathering force. But any anti-inflammatory carries risk, because in fighting inflammation, it can also hamper the body's ability to clear the primary infection or others that follow; tocilizumab is also thought to carry some elevated risk of anaphylactic shock and lower-intestinal perforation.

Patients with extreme flu in the ICU sometimes receive tocilizumab; it is also used to treat cytokine storms that some cancer patients experience as a side effect of treatment. Doctors at Lenox Hill did not believe it was a leap to think that the drug could address cytokine storms in Covid-19 patients. They knew that doctors in Milan were leaning heavily on the drug; they were in conversation with doctors at Yale New Haven Health, considered a fortress of research-heavy medicine, which also incorporated tocilizumab into their protocol. In addition, some small studies showed support for the drug's effectiveness, though none were randomized, controlled trials.

"I understand that it has never been trialed," John Boockvar, a neurosurgeon at Lenox Hill who is affiliated with the Feinstein

Institutes, told me in late April. (Boockvar is one of the doctors featured on the documentary series *Lenox Hill.*) "But there is clearly enough data to support its use." The doctors at Lenox Hill had also briefly participated in a randomized, controlled trial for another drug with a similar mechanism, called sarilumab. But to Boockvar, enrolling a patient in that trial, which might result in a patient receiving a placebo, posed an ethical challenge when he could simply prescribe tocilizumab—doctors refer to it as "toci"— instead. In April, he learned that Massachusetts General Hospital was starting a randomized, controlled trial for tocilizumab. "If that was *my* loved one," he said, imagining a family member who might receive a placebo in that trial, "I'd be upset. I'd think, *Why am I doing this?* If it's an off-label use with an approved drug—give the damn drug to everybody."

At Long Island Jewish, some doctors who were hearing about the drug from colleagues at Lenox Hill, a part of the Northwell Health consortium, started clamoring for liberal access to it. And yet sometimes, when doctors placed orders with the hospital pharmacist, their prescriptions were declined; those patients didn't meet criteria Northwell had established for administering tocilizumab, which was in short supply. Physicians were frustrated that patients who they believed would benefit from the drug could not receive it. Northwell wanted to be conservative about the off-label use of drugs outside clinical trials. "There's no proof that *anything* works!" Tsegaye thought at the time. "*Everything* is experimental!" As for enrolling patients in a trial, as overwhelmed as she was, she hardly felt she was in a position to take that on.

Tsegaye's supervisor, Narasimhan, also knew researchers were concerned that in prescribing tocilizumab so readily, physicians were possibly hampering enrollment in the trial under way at her hospital for sarilumab—a patient who received tocilizumab could not also receive sarilumab. She and her team did not prioritize the trials, she said; they wanted to provide the drugs they thought were needed. "We've always been allowed to choose treatment, right or wrong, based on what we thought was best," Narasimhan said in May. "And that was gone. It was hard."

In addition to fighting resistance from their administrators, the doctors were sometimes also at odds with their colleagues, especially infectious-disease doctors, many of whom believed that anti-inflammatories like tocilizumab and steroids could do more harm than good. "You're killing these patients," one infectious-disease doctor told Hahn at Long Island Jewish.

In the Mount Sinai Health System, tocilizumab was also in demand. Administrators felt the stress of making decisions in the absence of clear data. Judith Aberg, the chief of the division of infectious diseases for Mount Sinai, fielded demands from doctors working on wards who wanted to use tocilizumab, early and often. "I have to give her credit; she was single-handedly fighting off a lot of pressure from hematologists," said Keren Osman, a Mount Sinai oncologist and hematologist who was on some of those calls. As experts in blood cancers and diseases, hematologists had experience working with tocilizumab to treat cytokine storms that were a side effect of some cancer treatments. "She was saying, 'I'm not comfortable just giving patients willy-nilly anything we have—we don't know.'"

Patients and their families, who heard through the news media about the drug, also started to demand it, even for Covid-19 patients whose inflammatory markers were normal. "People were calling for us to give it, just to give it, because there were no other therapies," Aberg said. At first, a medical team that included Aberg agreed to put some patients who were on ventilators on the drug—in those patients, it was obvious that systemic inflammation was already evident; also, the closer the patient was to dying, the more the risk seemed justified. Eventually, the thinking at Aberg's hospitals and at others evolved to favor use of the drug earlier, before systemic inflammation did so much damage that the patient was already clinging to life.

By May, doctors at Long Island Jewish and Mount Sinai had stopped pressing for tocilizumab—if it was effective, it was not such a miracle drug that they could see its effects clearly. Many had started to pin their hopes instead on convalescent plasma, another experimental treatment in which sick patients are given plasma from recovered patients with antibodies, though its effectiveness is still unknown. "We did rush," Aberg says now. "I mean, we were pushed. We were grasping for anything that we could possibly do."

In early July, the drug company studying sarilumab, the drug similar to tocilizumab, announced that it was halting its trial; researchers found, as Aberg put it, "nada." A few weeks later, the pharmaceutical company Roche announced preliminary results of a tocilizumab trial that was run on Covid patients with pneumonia. The drug's effects were no better than a placebo. By then, Narasimhan was also starting to see preliminary reports of other research that showed the drug could, in fact, be dangerous, increasing the risk of fatal secondary and fungal infections.

"My take-home is that I wish we had done more randomized, controlled trials so we could have some real answers, so that we could tell Florida and Texas, 'This works, and this doesn't work,' " Narasimhan, who is now in charge of intensive-care units throughout the Northwell Health system, told me in July. "We could have had so many more answers in a way that was meaningful. We had this fixation that all these drugs were curative. And they weren't."

The story of hydroxychloroquine will most likely be recalled as a classic medical parable of the pandemic. It was a drug that seemed so promising that physicians were desperate to use it, and researchers were equally driven to see if it actually delivered the hoped-for results. In the end, the enthusiasm of the first camp most likely slowed the speed with which the second could study the drug— only to find that the enthusiasm was never really justified in the first place.

In mid-March, Steven Libutti, director of the Rutgers Cancer Institute of New Jersey, read about a small hydroxychloroquine trial in France that was generating attention, having found that the anti-malarial might be effective in the treatment of Covid-19. "It looked interesting, exciting, promising, but it looked very far from convincing," Libutti said. Although his specialty is cancer, he wanted to bring his extensive research knowledge to bear on the pressing question of the drug's effectiveness. He wrote a proposal for a randomized, controlled trial that would measure the effectiveness of hydroxychloroquine on a patient's viral load. (He was comparing the effect of the drug alone with placebo, as well as with the drug when administered with another drug called azithromycin.)

The Food and Drug Administration and the ethical review board at the Rutgers Cancer Institute approved his trial in record time, as has been typical for many proposed drug trials during the pandemic. He enrolled the first patient on April 1, hoping he could easily reach 150, calling on doctors to recruit patients at six hospitals in New Jersey.

By then President Trump had claimed in mid-March that the

drug was a "game changer." Some doctors in New York were quietly taking it prophylactically. That month, the FDA authorized hydroxychloroquine for emergency use, a special dispensation that
facilitated doctors' access to the drug even outside the context of
a trial. Many New York hospitals' standard treatment protocols
encouraged doctors to consider hydroxychloroquine for patients,
even though the evidence that it worked remained slim and reports were emerging that in some patients it was causing heart
problems.

Thousands of patients were pouring through those six New Jersey hospitals, but Libutti waited, for weeks, with great frustration
as only a handful of patients were enrolled in his trial each day.
Typically, in clinical trials, after a patient is admitted to the hospital, a doctor or nurse, often affiliated with the research, talks to
the patient about the possibility of enrolling in a clinical trial. But
Libutti's team was finding that by the time a nurse could begin
the conversation with the patient, that person had already been
administered hydroxychloroquine—which meant the researchers could not get a baseline reading of that patient's viral load.
Patient after patient was disqualified from the study. They had
"been handed hydroxychloroquine along with their toothbrush
and slippers when they got to the emergency room," Libutti told
me. "They were giving it out like dinner mints." The researcher
said he "was shocked by the number of folks whom I thought were
incredibly well-read, knowledgeable physicians but were just panic-
prescribing hydroxychloroquine. I've never seen anything like it.
It just shows how lost in the storm folks were." (Michael Steinberg,
who helps oversee trials as well as clinical care at Robert Wood
Johnson University Hospital, which was involved in Libutti's trial,
said that although physicians use their clinical judgment to make
decisions about treatment, they strongly encourage doctors to use
evidence-based criteria.)

Other doctors shared Libutti's experience. Arthur Caplan,
a bioethicist at New York University's medical school, said he is
aware of three medical centers where researchers trying to study
hydroxychloroquine felt that the early ardor for the drug among
doctors and patients made it difficult for them to recruit subjects
—to determine, essentially, whether the embrace of the drug was
at all justified. Caplan and a colleague argued, in an article published online in April in the *Journal of Clinical Investigation*, that

"panicked rhetoric about right-to-try must be aggressively discouraged in order for scientists to learn what regimens or vaccines actually work." Communicating directly with doctors at various hospitals who were making the drug part of the official protocol, he used more plain language: "This is nuts!"

The Montefiore Health System in New York was one of the many that included hydroxychloroquine as an option in its treatment protocol, starting in late March. Michelle Ng Gong, the director of critical-care research, did not actively fight to have the drug removed from the protocol. But when she was working in her capacity as a critical-care doctor, she does not recall ever prescribing the medication, and she sometimes took patients who had received it in the emergency room off it. "When so many people are dying, you want to do something," she said. But very sick patients are more susceptible to adverse events. "The problem is that we know from critical-care literature, as well as trials in the past, that we can always do more harm."

In the end, the biggest randomized, controlled trial on hydroxychloroquine came out of Britain in June, and preliminary results found that the drug was not an effective treatment for Covid-19. In contrast to American doctors whose access to the use of the drug, even outside trials, had been eased by a federal agency, British physicians were given the opposite message. On April 1, the highest medical officials in England, Wales, Northern Ireland, and Scotland each sent a letter to every hospital in their respective countries, urging doctors not to prescribe medications off-label outside trials. Instead they encouraged doctors to enroll their patients in large, multi-center, randomized, controlled trials, like a study run by the University of Oxford called Recovery, which looked at the efficacy of hydroxychloroquine, tocilizumab, convalescent plasma, dexamethasone, and two other treatments. At some hospitals in Britain, as many as about 60 percent of patients were enrolled in Recovery trials; even the Northwell system, which is committed to research, was able to enroll, at its most trial-driven hospital, North Shore University Hospital, around only 20 percent of its patients in clinical trials.

A flood of patients all with the same illness presents logistical challenges to trials, but also the perfect conditions for them; that the American medical system could not harness more of those patients into randomized, controlled trials, said Peter Horby, one of the two chief investigators for Oxford's Recovery trials, represents a lost opportunity. Whether or not convalescent plasma actually helps patients, for example, has not yet been resolved by a randomized, controlled trial despite the tens of thousands of doses that American patients have received, numbers that dwarf those in Britain. Given those numbers, American researchers "could have nailed it by now," said Horby, whose own trial on convalescent plasma is still under way.

Caplan, the NYU bioethicist, acknowledges that doctors in the United States did manage to enroll more patients in trials more quickly than ever. But even still, he believes that the commitment to long-shot efforts to rescue patients was stronger than the commitment to science, which slowed results and possibly cost more lives. "We did a lot," he said. "But we could have gone faster and resolved questions sooner."

If researchers see hydroxychloroquine's failure as a cautionary tale about the perils of acting without evidence, Pierre Kory, the Wisconsin critical-care doctor, sees a different medical lesson emerging from the pandemic: that the emphasis on randomized, controlled trials can get in the way of doctors' providing common-sense, lifesaving treatments.

In April, supportive care alone was considered the best option for patients with Covid-19, given that there was no evidence yet to back other treatments. Kory, who was then the chief of critical-care service at the University of Wisconsin Hospital and Clinics, believed instead that medications commonly used in critical care would most likely help critically ill Covid-19 patients too. That month, at a well-attended meeting with fellows, residents, and leadership, including Lynn Schnapp, the chair of the department of medicine at the University of Wisconsin medical school, Kory suggested an approach that went beyond supportive care. He had been consulting with senior hematologists at the hospital and had observed alarming blood clotting in Covid-19 patients. He and the hematologists proposed that the hospital consider administering an aggressive dose of anticoagulants to patients whose blood tests showed elevated risks for clotting. (Many medical society guidelines that once called for only supportive care now recommend the use of anticoagulants in Covid-19 patients, but not in doses as aggressive as those that Kory and specialists at the hospital had proposed.)

The meeting among Kory and his colleagues took an adversarial turn. "No one else is doing this," said Lynn Schnapp, as Kory recalls. (She denies saying that, although a former colleague of Kory's who attended the meeting confirmed Kory's account.) "There is no evidence," a fellow ICU doctor said more than once, her voice raised. Kory, who pointed out at the meeting that his suggestion was based on the opinion of the hospital's own experts, says he fired back with equal intensity. "And this is Wisconsin," he told me. "People don't yell here." Other colleagues who were supposed to jump off the call to attend another meeting later confided to Kory that they couldn't bring themselves to leave, for fear of missing out on this unusual hospital drama.

At a subsequent, smaller meeting, Kory brought up with Nizar Jarjour, a division chief, the possibility of giving steroids, commonly used on critical-care patients, to Covid-19 patients in the ICU. "I don't want to talk about it," Jarjour said.

In a lengthy email Jarjour later sent me, he explained that open discussion was welcome during that period of time; he also sympathized with the sentiments of the ICU colleague who was urging caution while facing a novel virus.

Corticosteroids have a complicated and controversial history in critical-care medicine. Numerous trials over the past fifty years have been conducted on their efficacy in patients with acute respiratory distress syndrome, or ARDS, a diagnosis for patients who have reached a stage of perilous respiratory failure. Because many of those patients at that stage of illness have confounding factors, findings are far from definitive. But based largely on some meta-analyses, including those looking at how patients with MERS and SARS fared, the World Health Organization advised, early in the pandemic in this country, against the use of steroids in Covid-19 patients experiencing ARDS, which is to say, most patients on ventilators.

Kory and several colleagues at hospitals around the country noted that the studies that the WHO cited, for example, were largely not randomized and controlled; other relevant institutions like the Society of Critical Care Medicine, whose doctors treat the most ill patients, and the European Society of Intensive Care Medicine did recommend the use of steroids for ventilated Covid-19 patients with ARDS. Also, in Kory's own clinical experience, corticosteroids could be lifesavers. He did not see them as a wild-card drug for this disease, like hydroxychloroquine; he used them for non-Covid-19 patients who were facing cytokine storms or ARDS. He was surprised by the heat with which colleagues challenged him when he made the recommendation, and he believed that his own leadership role in conference calls subsequently diminished. He and Jarjour, he said, had more disagreements in three days than they had in the previous five years.

On April 7, Kory's colleague Ellie Golestanian sent an email to Kory and others, at 1:32 a.m., in response to another colleague's call for the use of corticosteroids and anticoagulants: "In patients with severe Covid-19, we are fumbling in the dark, clutching at anything that might work. But as you are well aware, just because a therapy 'should' work, or we desperately 'want' it to work—it does not follow that it 'will' work."

"When I hear stuff like corticosteroids described as experimental and unproven, I want to jump out a window," Kory told me later that month. "They make it sound like we are experimenting on people. I want to be respectful of my colleagues, but I feel like they are getting it 100 percent wrong. I've never seen smarter people get a problem more wrong. Because they are running hypotheses in a lab and so many of them fail, they think when I approach a patient, I am testing out a hypothesis. It's not like a hypothesis, but more like a problem, and I have to figure out how to fix it with a couple of decades of experience to back me up. It's a stretch to call it a hypothesis. It's just me doctoring."

Kory was so frustrated about the hospital's approach that in May he resigned, taking a job instead at Aurora St. Luke's Medical Center in Milwaukee. "Our differences were so far apart, I felt I couldn't be a part of it," said Kory, who foresaw, in April, a "catastrophe" if doctors at any hospital could apply only supportive care. A colleague of his in New York, an ICU doctor affiliated with a major medical center, confirmed that he, too, resigned from his hospital, in part because of tensions around his decision to try an FDA-approved medication off-label and outside a trial. In May, Kory, following his disagreement in Wisconsin, spent several weeks in New York treating patients, often with steroids.

In June, Oxford posted a preliminary report for its Recovery trial of more than 6,000 patients who received either standard care or dexamethasone, a steroid similar to the ones that Kory and other ICU doctors had been advocating. At least when administered to patients who were already on oxygen or ventilators, the drug saved lives.

Kory sees, in the Oxford results, a story of triumph. He believes that he successfully treated patients with steroids and that the Recovery trial results prove it. And yet if patients did not respond, he would go further, increasing the dose, in a few instances, to a level ten times as strong as that in the trial. Did the higher dosage increase the risk? The answer to that question, a research purist like Kevin Tracey would point out, is still unknown. Despite the enthusiasm for the Recovery trial, Tracey maintains that even one stellar randomized, controlled trial does not settle the question of the use of steroids for patients with Covid-19. "It needs to be replicated," he said. Given the long, complicated history of steroid studies, he predicts that sometime down the road another statistically powerful randomized, controlled trial will yield contradictory findings. In Tracey's reservations, Kory sees not rational evaluation but bias. "That's a 6,000-person trial he's discrediting," Kory said. "That's a person who will never be convinced."

Kory is also part of a group of critical-care doctors who widely disseminated a protocol for treating Covid-19 that includes anticoagulants and steroids but also other treatments—including Pepcid and intravenous vitamin C—whose efficacy is hotly contested among doctors.

Should Kory and his colleagues have been administering steroids when they did? Were they right? Kory thinks so. But Eric Rubin, the editor of the *New England Journal of Medicine*, thinks it's not so clear-cut. "You could also say he was lucky," Rubin said.

At times, over the course of several conversations, Rubin defended the bond between doctors and patients, the need for physician autonomy, the necessity of making judgments in the absence of evidence, especially when mortality rates were so high; at other times he seemed frustrated that doctors were still relying on treatments for which there was no evidence, concerned that a lack of equipoise had possibly muddled the course of research. "I know I seem to be saying opposite things," he admitted. "And I agree with myself."

Rubin is an infectious-disease doctor at Brigham and Women's

Hospital in Boston, and like many doctors in that specialty, he had strong reservations about steroids. He advised colleagues against using them. "And I was wrong," he said. He also acknowledges that he was less opposed to the use of tocilizumab, although that, too, was untested and could also increase the risk of infection. I asked him whether perhaps there was something about tocilizumab's novelty, even its scarcity, its high price, that may have given it a sheen of credibility? In the absence of evidence, we are all suscep-tible to predictable irrational biases. "We know less about toci," is how Rubin said he thought about it at the time: it left open the possibility that it could be helpful, although of course that means it also left open the possibility that it could do more harm than supportive care, more harm even than steroids. Unlike steroids, to-cilizumab was not yet the devil he knew. Rubin could see, months later, his decision-making process with humility. "I can't argue that I was super-rational either," he finally said.

At the peak of the surge, Rubin was far from the only doctor whose usual commitment to evidence faltered. "There would be a physi-cian that would have said, 'No, no, no,' and all of the sudden, it's his mother," Aberg, the Mount Sinai doctor, told me. "All of a sudden, let me tell you, they wanted everything. We had some of our own physicians admitted, and people were just crazed about what they wanted to do to those individuals." Kevin Tracey, the Feinstein researcher, says that overwhelming uncertainty was driv-ing people's reactions: "We just lived through a plague. It was life and death. Fear. Ignorance. You were seeing raw human behavior in survival mode, a classic reaction to threat."

In recent months, a relative calm has set in. Since that tense mid-May meeting between the researcher Spyropoulos and his colleagues on the ground at Long Island Jewish Medical Center, clinicians and researchers have forged more compromise through a series of conversations. Looking back at that time, one critical-care doctor mentioned that Spyropoulos, when he called in by videoconference that day, seemed tired and stressed; perhaps, the critical-care doctor thought, that accounted for why Spyropoulos had spoken so harshly to the group.

Spyropoulos, the director of the anticoagulation program at Northwell, had in fact been working nineteen-hour days to try to get the trial up and running at breakneck speed. He also mentioned to me in passing, during an interview, that both he and his wife had been sick with Covid-19, his wife more so than him. Perhaps, like many doctors, he was laboring under the additional stress of having not just a professional but a deeply personal struggle with the power of the disease. In response to a text I sent asking about the extent of his wife's illness, Spyropoulos wrote back on July 11. "My wife was very sick and bedridden for 3 days (I was mildly symptomatic) but she responded well to hydroxychloroquine," he wrote.

His casual certainty about the cause and effect of her recovery was surprising. His wife had been ill, taken hydroxychloroquine, and recovered. Whether the second event caused the third was at best unknown but statistically unlikely, given the now-significant body of research showing that the drug does not help. As if even Spyropoulos recognized that his comment was less than rational, he went on to rehearse the argument he made to his colleagues at Long Island Jewish about the importance of relying on research. Anything else, he repeated, was witchcraft.

HEATHER HOGAN

The Soft Butch That Couldn't (Or: I Got Covid-19 in March and Never Got Better)

FROM *Autostraddle*

I WAS ALREADY IN a hospital the first time I realized I needed a wheelchair. A sprawling, full city-block of a hospital in midtown Manhattan with a lobby that looks strikingly similar to the cavernous Ministry of Magic atrium. My neurosurgeon had sent me in for an emergency MRI three months after I'd been diagnosed with Covid-19. Death or a two-week flu were the only options for people who contracted Covid, according to the Centers for Disease Control, and I wasn't dead, so surely lingering coronavirus wasn't the thing that was causing my body to go so berserk. Racing heart, palpitations, stroke–high blood pressure, chest pain, weak legs, fatigue that felt like my body was made of lead, nausea, loss of appetite, extreme weight loss, shortness of breath, brain fog that caused me to forget how to form sentences, bladder dysfunction, and creeping numbness and tingling in my feet and legs.

It was those last things that caused my neurosurgeon's alarm. I have what's called "remarkable" cervical stenosis. When my neurosurgeon first showed me my CT scan and diagnosed me last fall, I thought he meant it in a good way. Exceptional. Impressive. Miraculous. He did not. Remarkable, he explained, in the sense that I have the spine of a person who is 300 years old. And with such a spine, when you can't remember the word "carrot" or how to turn on the oven, and your legs are tingling all the way up to your knees, there's a chance your vertebrae are crushing your spinal cord en route to paralysis. It could happen over time, or it could happen *right now.*

As I stood swaying in the hospital lobby while the nurses and security guards debated what to do with me, I realized I should have taken my neurosurgeon up on his offer to send an ambulance to my house to get me. But I'd already racked up one ambulance ride when my feet initially lost their feeling. An Uber was faster and, frankly, less expensive.

The problem was I didn't have a positive Covid test or a negative Covid test, because when I'd gotten sick over ninety days before, no testing was available in New York City, unless you were admitted to the hospital, and the telemedicine doctor told me not to go to the hospital unless I couldn't finish my sentences or my lips were turning blue. But also, I still had Covid symptoms. But also, I wasn't there to be treated for Covid symptoms. I was there because my neurosurgeon told me to meet him there. Did I belong in the Covid triage tent? Did I belong in the regular emergency room? Was I a danger to everyone in the lobby? Were they a danger to me? Who did I think I was, anyway, walking in off the street demanding an MRI in the middle of a global pandemic?

"I'm sorry, I hate to be dramatic," I said, as I waited to see if they were going to let me into the hospital to find out if my spine was squashing the feeling out of my legs permanently, "but is there anywhere I could possibly maybe please sit down?"

There was not. The visitor chairs had been removed for social distancing. I glanced over at the security guard's stool. "Don't," he said.

I decided my only hope was utilizing my gift for defusing tense situations with humor, so I decided to do a Ministry of Magic gag. "Broom Regulatory Control, level six" is what I meant to say, but I couldn't remember the word "broom." I started my joke; "broom" fell out of my head; I made the motion of sweeping, then stepping over the handle of a broom, then lifting up off the ground and whooshing around. I heard myself make a *vrooooooom* sound, like when an airplane spoon is heading toward a child's mouth.

"Ma'am," the security guard said, even more apprehensive than he had been, "have you been drinking?"

I called my neurosurgeon and he sent an intern down to get me.

*

My Covid onset was pretty normal, in terms of what's considered normal for a novel coronavirus that shuts down the entire world. I started displaying symptoms a week after New York City went into lockdown in March. Slight chills and a mild sore throat that progressed to chest congestion and tightness, a cough, and shortness of breath. I was really tired and I didn't have much of an appetite. I got winded just walking down the hallway to the bathroom. Hot showers left me hacky-coughing for hours. It was terrifying—hospitals were overflowing; the only sound outside my bedroom window was the constant scream of ambulance sirens; the death toll skyrocketed every day; the Empire State Building was programmed to flash red like a beating heart as a showcase of solidarity with health care workers, but it looked like some kind of apocalyptic lighthouse—but after two weeks, my body started feeling better. Slowly, the band around my chest seemed to loosen, my cough eased up, my breath came back, and I could walk down the stairs again.

I thought, *I survived! It really was, for me, just a bad flu!*

But I never got all the way better. At some point, I started to feel like I was relapsing—and then: new symptoms. The heart stuff and lung stuff and fatigue stuff that had me woozy-wobbling that night in the hospital waiting for my MRI, and a whole new kind of panic attack where the adrenaline that flooded my body was so extreme it caused tremors in my legs and arms that lasted for hours.

My startle reflex started operating in overdrive. If a loud noise woke me up in the middle of the night, I'd immediately burst into a panic attack. Nothing I did could get them under control. Not meditation, not medication, and exercise was out of the question —the only way I could get down the stairs at that point was to basically fling myself forward from the top and hope for the best.

The weakness seemed to settle into my bones.

I lay in bed, day after day, week after week, too tired to sit up for more than a few minutes. Stacy made my meals and brought them to me. Each time, I ate a few bites and had to lie back down to gather up enough energy to sit up and eat a few more. I tried to work, but couldn't make it through a full day, and then couldn't make it through a full morning. I asked for an extended leave of absence. I moved all of my toiletries from the bathroom to my nightstand. I could hardly manage a two-minute shower.

The telemedicine doctors all told me I'd be fine; of course the virus would knock me out for a couple of weeks, but I'd bounce back in no time. I explained it had actually been quite a lot longer than "no time" and they smiled and thanked me for calling. My primary care physician said it was anxiety. My psychiatrist said it was depression. New York City's newly opened Long-Term Covid Care Center told me I didn't have antibodies so I hadn't had Covid. And anyway, my blood work was excellent.

"Is it possible," I asked the doctor, "that this antibody test might not be a foolproof way to determine who actually had Covid? And that you might not have all the answers for this brand-new global pandemic–causing virus we find out something new about every day?"

"No," she said. "Not possible."

I needed a wheelchair to leave the Long-term Covid Care Center, but I was scared to ask for one. The doctor seemed to think I was overreacting about having a simple flu. I paused and leaned against walls to steady my legs and catch my breath instead. It took me twenty minutes to finally make it out of the building, and as I waited for my Uber, I sat down on the sticky summer sidewalk right in the middle of Union Square.

My neurosurgeon is renowned. In the good way; not in the way that "remarkable" means "yikes" in spinal vernacular. When his team arrived in the emergency room to handle my emergency MRI, everyone stopped treating me like a woman who'd stumbled in and mimed flying a magical broom around in the lobby and started treating me like a quadruple sapphire iridium medallion frequent flyer. Harried nurses who'd been grumbling about demanding doctors behind their backs for an hour were happy to volunteer when my neurosurgeon's intern announced that he needed someone to help him check my anal tone. Doctors suddenly had time to come by and offer me a kind word. Nurses kept bringing me juice.

It was the most attention I'd had from medical professionals in months, despite having spoken to dozens of doctors about my Covid symptoms by then. I asked every nurse and doctor who came to look in on me if they knew anything about people suffering from Covid long-term, about unusual symptoms, about the antibody tests. What I found out over the course of my night in the ER is that doctors and nurses still had no idea what Covid was about, but they all agreed it was *weird* and getting infected in mid-March at the onset of the outbreak meant that I was one of the first people in the United States to be dealing with long Covid symptoms; I was the science.

My neurosurgeon likes me, as a person; I can tell because in my after-visit notes from the first time I saw him, he wrote that I was "bright and amiable" and transcribed what I told him when he asked if my neck cracked when I moved it: "like an undead lich rising from an ancient throne in his tomb in the empire of necromancers." I'd been in bad shape on my initial visit for a pinched nerve; I'd lost 70 percent of the strength in my left arm, shoulder, and hand. He'd told me I could try physical therapy but that surgery was probably inevitable. I went so hard at PT that the next time I saw him and he gave me the strength test, I sent his wheely chair whizzing across the room and smashed him into the wall. I said, "Now, that's remarkable!"

The emergency MRI showed that my spine was a little worse than last time, not by much. Surgery was not urgent. But my neurosurgeon was still worried about me. He said I seemed heavy, faint, deeply exhausted, muted. And there was still the matter of words slipping out of my brain, pins and needles in my jelly legs, a wildly overactive bladder. He had the ER doctors run a battery of other tests; all of them were clear.

I said, "Do you think all this could be Covid?"

He leaned back in his chair, pinched his eyebrow, and said, "I don't know. But you're not yourself. So maybe." He studied me for a long minute and decided: "Yes."

I knew exactly who I was before I got Covid: a woman disposed to rise to every occasion. A bitch who gets stuff done. The person everyone relies on to do the thing no one else has the heart or guts or fortitude to do. A soft butch holding my family and friends and my whole little world together with nothing but love and tenacity.

I had huge plans when lockdown started. I was going to finish my book, remake our outdoor furniture, grow fruits and vegetables in my container garden, learn to cook Stacy's favorite pie (strawberry-rhubarb), run a delivery service for my neighbors for prescriptions and groceries, and connect (and reconnect) with my dearest family and friends.

I didn't get a chance to do any of those things.

Stacy took over doing all the laundry and dishes and vacuuming and toilet cleaning and shopping and cat feeding and grooming—my jobs, the things I love to do, the homemaking projects I'd longed to be in charge of my entire life, the caretaking tasks that made my days feel full and valuable—while working day and night from her makeshift editing suite in the living room. The only thing I saw beyond my bedroom walls was the sky outside my window: gray then blue then purple and gold and cinnamon and orange then black then gray again.

Before Covid, my friends and I spent glorious weekends gathered around a table in my living room, sharing meals and wine and stories from our weeks and hours and hours playing Dungeons & Dragons. The real world so often tried to rob us of our power, but inside our D&D campaign we were unstoppable heroes. We saved towns. We slayed beasts. We made an entire queer universe of inside jokes. We moved our game online during lockdown and I told my friends maybe next week I could play, and maybe next week I would feel better, and maybe next week I'd be back to my old self. When I finally told them they should start a new Dungeons & Dragons game without me, that I wasn't really getting better and I didn't know when I would feel okay again, I lay in Stacy's lap and sobbed with such fierce and broken hoarseness I didn't even recognize the sounds as my own. I told my therapist, over Zoom, in my bed in my pajamas, that I'd only ever heard people cry like that at funerals before.

"I'm losing everything," I told Stacy. "I'm losing me."

My neurosurgeon called me on a Saturday, out of the blue, two weeks after my MRI, and said, almost giddily, that he'd asked all of his colleagues and finally found something he thought explained what was going on with me: dysautonomia. That same day, during a Q&A with my long Covid support group, a different neurologist made the same guess. And so I made an appointment with a cardiologist who specializes in dysautonomia and dragged myself back into Manhattan to be disbelieved by another doctor. I dressed nicely. I had all my paperwork in order in a crisp manila folder. I typed out the main words that kept falling out of my head in the Notes app on my phone, just in case. I'd taken a series of videos on my phone of my heart rate and blood pressure using five different devices total, over the course of two weeks.

I started listing off my symptoms as soon as I sat down in the cardiologist's office, and within thirty seconds, she held up her hand to stop me. *Here we go*, I thought. *She wants a positive Covid swab or a positive antibody test or this is just anxiety or what happens after a cold and buck up and take a nap and you'll be fine.*

Instead, she said, "I know exactly what's wrong with you."

I blinked at her, stunned into total silence.

"You have postural orthostatic tachycardia syndrome. POTS."

POTS is, in fact, a form of dysautonomia. A person's autonomic nervous system controls all the things we don't think about, like heart rate, blood pressure, circulation, digestion, body temperature. When people with POTS sit up or stand up, our autonomic nervous systems can't properly control our circulation, so all the blood rushes out of our heads and down our bodies, ultimately pooling in our feet. This, of course, makes us very dizzy, to the point of passing out, and also causes our hearts to start beating like mad and our blood pressure to go berserk to try to get our blood back up into our brains. It also causes big-time spikes in adrenaline, because our fight-or-flight systems are almost always activated. Our hearts are often in the cardio zone all day long, and so of course we're exhausted.

POTS can be caused by many things, one of which is a viral infection. I was this specialist's first post-Covid case; she said maybe I was the first diagnosed post-Covid POTS case in the entire city of New York.

I jumped up when she diagnosed me; nearly passed out; sat back down, hard; and started to cry so hard my tears soaked through my face mask.

"It's called an 'invisible' illness," the cardiologist explained, "because you look fine and your tests and lab work also look completely normal. But it affects every single system in your body. And you feel absolutely miserable. Doctors almost always write it off as depression or anxiety."

Every morning now when I wake up, I sit up slowly in bed and lean back against my headboard. I drink a full liter of water, eat some salted almonds, put on my compression socks for the day, and take a beta blocker for my heart and an SNRI to keep my adrenaline more in check. After thirty minutes, when my body has adjusted to sitting up, I can stand.

Downstairs, I make the first of three Liquid IV drinks of my day and eat a small breakfast. I drink four liters of water, total, and take more SNRIs and beta blockers as the day progresses. I sit on a stool when I take a shower. I sit down at a little portable table to do all the vegetable chopping and potato peeling for our meals. I wrap an ice scarf around me and sit down near the oven to cook. I use my office chair to wheel around the kitchen when I'm putting away groceries or dishes. I use a cane when I leave the house, which I only do for doctor's appointments; it folds out into a stool so I don't have to sit on the ground.

My cardiologist asked me on my second visit how I was adjusting, emotionally, to having a disability.

I said, "Do I have a disability?"

She said, "Well, yes. I thought you knew."

Before I left for my most recent trip to the hospital, Stacy checked my backpack to make sure I had everything I needed. She tucked my face mask straps behind my ears, and kissed me on the forehead. "If they don't have your wheelchair ready when you get there, ask for it," she said. "Okay?"

I said, "I will, I promise," and smiled with my eyes so she could see it.

Stacy is scared to be overbearing, because I've always hated being told what to do. She's scared to be *under*bearing, because my brain fog has made my cognitive functioning less sharp, especially in the morning and at night. She's scared I'll do too much, because I've always done too much my entire life, that I won't listen to my body, that I don't even know how to listen to my body. She's scared I'll begin to feel angry at her for what she can do that I can no longer do, that becoming a caretaker to me when I've always been a caretaker to everyone else will create an emotional wound that will grow and fester.

She watches me open my pillbox first thing in the morning and slowly work out what I need to take right then and what I need to save for later; she asks if I need a hand; she pretends not to notice that I clench my jaw against her offer and the knowledge that, actually, yes, I could use some help. She leaves my bike, my most prized possession and my lifetime beloved hobby, untouched in the living room on its stand, because maybe one day I'll be able to ride it again.

I'm scared too. Overachieving isn't something I do; it's always been *who I am.* Now that my sympathetic nervous system is misfiring in a way that makes simply getting through the day out of bed an achievement, there's no energy left to *over*do anything.

What will happen to my career now that I can't show up early and stay late to help our community survive? What will happen to mine and Stacy's lease now that I can't keep my landlord extra happy by doing all the yard work and fixing all the leaky, crumbly, broken things in our house? What will happen to my friendships if I never have the energy or brainpower to sit at a table for five hours and role-play zombie battles and villager rescues again? What will happen to my relationships with my family when talking on the phone wears me out, and I don't know how to answer the question about if I'm feeling better? How is it possible that Stacy won't grow to resent me when I can't even walk two blocks to pick up my own prescriptions, when my newly diagnosed illness is already eating into our savings, when I can't stand in the hot kitchen long enough to make our favorite soup, when I can't even really carry on a conversation at night because my brain and body are so spent?

What if I'm not disabled enough to use the word "disabled"? I'm a person with a huge platform; what if I talk about disability in the wrong way, or miss the mark on my advocacy because it's (shamefully, mostly) new to me? What if I hurt people who are already hurting with my naïveté, or accidentally dishonor the work of the queer disabled activists who came before me, some of whom I love and cherish as dear friends? What if I can't find the balance between hope and acceptance? What if I become one of the 30 percent of POTS patients who are too disabled to work at all?

Is a soft butch a soft butch if she can barely hold even herself together? Is a soft butch a soft butch without her swagger?

Yesterday, I called Stacy on a banana-phone; walked from the kitchen into the living room where her office is set up and *ring-ring, ring-ring*-ed. She looked up from her computer monitor and saw me holding a banana to my ear and quirked her eyebrow. "Hi yes," I said. "This is the last banana and I was wondering if you were planning to go to the store for more bananas in the next few days and so I can go ahead and use this in my smoothie or should I save it for a banana emergency?" She wanted to be annoyed with me because I was interrupting her workday and she was on a deadline. She wanted to be miffed because I'd pulled her out of her creative flow. But her mouth twitched into a smile because I was asking her for something I needed and could not do for myself, and I was being ridiculous, and it'd been so long since I'd been ridiculous. She said she'd run out to the grocery store tomorrow. I thanked her for her time, hung up the banana-phone, and turned it into a microphone to interview our cat Socks about the rumors that he's a marshmallow head.

I'm scared, but I'm alive.

I'm scared, but I'm not broken.

In long Covid support groups, we say our name and where we're from and how long we've been sick. I'm Heather Hogan from New York City. Week 19 / Day 133.

KATIE ENGELHART

What Happened in Room 10?

FROM *California Sunday Magazine*

THAT TUESDAY NIGHT, Helen lay awake and listened to her roommate dying. She heard the nurses moving around. Their whispers. She heard the heaving of the oxygen machine. At some point, someone had closed the curtain that divided the room, but it didn't do much to mute the noise. The beds were so close together that each woman could hear the other breathing—and that was true on a normal day, before the coughing.

It was four days into the outbreak. Or, rather, it was four days since the Life Care Center of Kirkland, a nursing home in Washington State, had publicly confirmed the existence of a coronavirus outbreak. From Room 10, where Helen and Twilla had lived for more than a year, the women couldn't see the nurses wheeling sick residents out the front door to meet the ambulances in the parking lot—sometimes holding white bedsheets around the stretchers to shield the patients from the photographers waiting at the side of the road. Room 10 faced inward, toward the courtyard, and it was quiet there. Still, from their beds, the women could hear nurses running down the hallway. The sound was conspicuous because people don't usually run inside nursing homes.

Later, the story of the Life Care outbreak would be flattened by the ubiquitous metaphors of pandemic. People would say that Covid-19 hit like a bomb, or an earthquake, or a tidal wave. They would say it spread like wildfire. But inside the facility, it felt more like a spectral haunting. A nurse named Chelsey Earnest said that fighting Covid was like "chasing the devil."

By that day, March 3, the facility's nearly 120 residents had been told to stay inside their rooms. Now and again, someone with dementia would forget the new rules and wander into the hallway, but she would quickly be redirected back to bed. Sometimes, she would cry because the nurses redirecting her looked alien and strange in their surgical masks.

Since there were so few remaining staff members—by then, more than a third of Life Care's staff had called out sick—some residents had not been showered or helped out of bed in days. Some weren't properly covered up because their sheets had fallen down; their legs stuck out, bare and exposed. These were small indignities, maybe, given everything, but still. A few residents had called their sons and daughters to say how awful it felt to be lying that way: stiff and dirty. Others had wanted to call home but couldn't because, amid the chaos, no one had remembered to replace their hearing-aid batteries.

Many of Life Care's residents had spent that Tuesday watching their wall-mounted TVs. From the news channels, they had learned about the coronavirus and how their nursing home was the first in the country to be infected by it—how Life Care was, in fact, "the epicenter" of the coronavirus in the United States. On that very day, three residents had died of the virus, bringing the facility's death count to seven and the national death count to nine. But other residents had not watched the news or were confused by the keyed-up chatter of the network correspondents. They understood nothing, beyond that their children had stopped visiting.

Later, some staff members would wish that they had done more —that they had done anything, really—to explain things. *You see, there's this new virus* . . . But it was hard to think straight because there were 911 calls to make: for a sixty-five-year-old with multiple sclerosis (Room 14, beside the window) who was running a 103-degree temperature; for a seventy-seven-year-old man (Room 21, beside the door) whose oxygen levels were falling; for a fifty-one-year-old with lung cancer (near the nurses' station, on the ground) who was kicking and screaming and refusing to be touched. Also, the phones just kept ringing. Family members wanted to know how their mothers and fathers were doing. Did they have fevers? Had they eaten dinner? Were they alive? Some were nice about it; they thanked the nurses and told them to keep their chins up. Others were not nice about it. They threatened to sue. They threatened to call CNN. They said the nurses were murderers who should die and go to hell.

Helen was ninety-eight years old, and she understood everything. This was largely because she never allowed the door to her and Twilla's room to be closed—all the better for eavesdropping —and had listened to the administrators talking in the hallway. That evening, Helen's daughter called her at 7:30, and Helen told her that she was feeling okay. A nurse had come by to take her vital signs, and Helen relayed the numbers with precision: her temperature, her blood pressure, her oxygen levels. She said she wasn't scared because there was no point in being scared if there was nothing you could do about it.

It was harder to say what Twilla understood. She was eighty-five and had dementia. In the last few months especially, Twilla had succumbed to the temporal vertigo that can accompany the disease and ravage a person's sense of her own chronology. She was here, and then she was ten years ago, and then she was ten years old. She cried out for her long-dead mother and hissed at her long-dead husband and repeated words over and over: "Come on, come on, come on." Yet, she had moments that approximated clarity. Sometimes, she remembered that she had grandsons and was proud of them. Sometimes, when someone complimented the lipstick that a nursing aide had layered over her pale lips, she would say, "Thank you, and you look very nice yourself." Was Twilla tormented on that night she lay dying? Was she even awake, when nurses came to sit with her so that she wouldn't be alone? Who could say? Helen could only lie still and listen, her permed white curls pressed into the pillow.

Around 2 a.m., it was over. Chelsey, the nurse, came into the room and placed a stethoscope against Twilla's chest. When she looked up, she saw that the curtain separating the room's two beds had fallen open, and she went to close it. Her eyes met Helen's for a moment, but neither woman said a word as Chelsey pulled the curtain tight. Then a mortician came in with a gurney and took the body away.

When a nurse called Twilla's daughter, Debbie de los Angeles, around 2:40 a.m. on Wednesday morning to deliver the news, she sounded weepy. "Your mom died," she told Debbie. "I'll miss her." Debbie thanked the nurse. It was sad, of course, but Twilla's kidneys had been failing, and Debbie had confirmed with the staff that week that her mother was not to be resuscitated if she stopped breathing. But still, Debbie couldn't stop thinking about the day she left Twilla at the nursing home: how it was one of the worst days of her life and her mother's too. Twilla hadn't wanted to go.

A week later, on March 12, Debbie drove to Life Care and stood outside, on the lawn, to listen as a "crisis communications specialist" recently hired by the nursing home's parent company, Life Care Centers of America, addressed a cluster of reporters. He said that the facility's occupancy had fallen from 121 residents to just 47. He said that 26 former residents were dead and that 26 more had tested positive and that 66 staff members were showing symptoms, though it was hard to say anything conclusive about the staff because only a sixth of them had managed to get tested. "We do need more help here," he said. "We've been asking for it."

After the press conference, Debbie spoke with some of the other family members. One told her that the nursing home had waited days to report that residents were getting sick. Another wanted to know why every time she called the facility and asked to speak to the doctor, she was told he wasn't there. A pair of sisters said that they had seen nursing aides go room to room without washing their hands. Even now. Even after everything. Debbie listened to what they had to say and then drove back home. That's when all her floating thoughts began to shape themselves into a question: Was it possible that Life Care had done something wrong that caused her mother to die?

On April 2, federal regulators fined the Life Care Center $611,325 after finding evidence of "serious deficiencies" in its handling of the outbreak, some of which placed residents in "immediate jeopardy." A few days after that, a personal-injury lawyer named Brian Mickelsen called Debbie. He said he was from South Carolina, and he wanted to know if Debbie had thought about suing the nursing home. Well, sort of. Would it make a difference if she didn't have to pay any money up front? Well, sure. (Mickelsen did not respond to several requests for an interview.) A week later, Debbie filed what was likely the first wrongful-death lawsuit related to Covid in an American nursing home.

When Helen's daughter, Carolyn Croshaw, heard about Debbie's lawsuit, she called her mother right away. "Twilla's daughter is talking trash about Life Care," she said. Carolyn had always

loved Life Care and so had Helen and so, she thought, had Twilla
—to the extent that you could love a place when you didn't know
where exactly you were and what exactly you were doing there.
"Oh Lordy," said Helen. She thought that once the virus got inside, nobody could have saved those poor souls.

In the months that followed, the lawsuit against Life Care and
the prospect of more lawsuits to come have divided Life Care residents and their families. In part, this is because the lawsuit speaks
to a fundamental question: To blame or not to blame? To some,
Debbie's case is a greedy assault against the unluckiest nursing
home in America. To others, a successful case against Life Care
would be a fitting comeuppance for a facility that made terrible
errors and whose errors, they argue, killed dozens of people—at
least forty-six people—who otherwise would not have died. And
then there are those who think that, yes, blame should be assigned,
but that it belongs with forces much larger than a single nursing
home: with the county, with the state, with federal regulators, with
the system. This is a debate about perspective: whether Life Care
failed or was failed, or whether this was just the inevitable way of
the virus.

The nursing-home population, of course, was always going to be
vulnerable because nursing homes are full of the oldest and frail-
est. In Canada and Italy and Sweden, too, residents have died in
extraordinary numbers. The debate now is whether all that death
can be explained by biology and demographics—or whether, in
spite of biology and demographics, more nursing-home residents
could have been spared. "That is the essential question," said Toby
Edelman, a senior policy attorney with the Center for Medicare
Advocacy, a nonprofit advocacy group. "Are the facilities totally
blameless? In which case we just need to help them. Or did some of
them make mistakes? In which case something needs to change."

As of mid-August, 177,129 nursing-home residents in the United
States had tested positive for Covid-19, and 45,958 had died of it.
This means that while nursing-home residents represent just a
fraction of 1 percent of the American population, they account
for more than a quarter of total pandemic deaths. Since March,
approximately 3 percent of all nursing-home residents have died
of Covid. Because the Life Care Center of Kirkland was the first
to be infected—and because it had one of the country's deadliest
outbreaks—the facility has acquired outsize significance in the his-

tory of the pandemic and has become a shorthand for all that has gone wrong in the American nursing home.

When I spoke to Tim Killian, Life Care's spokesperson, he was unequivocal on the question of blame. Life Care, he said, was appealing its federal citations. "I will take a nonapologetic view of the ascribed mistakes that we have made. I will say outright: I think it has largely been nonsense . . . I'm not going to take a we're-so-sorry-this-happened-to-us approach. I'm not going to apologize at all."

For a few weeks after Twilla died, Debbie had expected someone from Life Care to call and apologize for her mother's death —or, at least, to say they felt bad about how things had ended. But nobody called, and Debbie gave up waiting. "The nursing home didn't do shit."

In retrospect, almost everyone would agree, the nursing home should have canceled the Mardi Gras party, which was held the day after Mardi Gras, on February 26. It was Ash Wednesday, and a man from a nearby parish had come by in the morning to mark some of the residents' foreheads. Also that morning, Life Care's infection-prevention nurse, in consultation with its medical director, had decided that it was probably time to declare a respiratory outbreak. Administrators sent a memo out to staff, asking them to scrub down the common spaces and close the dining rooms because of all the residents who were getting sick. But then the jazz band arrived.

The party went ahead. Nursing aides hung ribbons on the walls, and the chef made a king cake, with green and yellow and purple food dye. There were Cajun sausages and cups of Sprite. There were jester hats and plastic beads. The residents, as one partygoer recalled, sat wheelchair to wheelchair. Some sang along to the music. Others were a bit out of it; they nodded off or spilled their drinks. Aides wove through the crowd tidying people up. Some of them wore masks, but some didn't.

Twilla always went to the parties. Because of her dementia, she had forgotten how to like a lot of things, but she still liked music. When she heard it, she swayed in her wheelchair. Helen did not like parties, and she stayed in the room. She was never social at the nursing home: never one to join in or clap along. For some people, the Life Care parties could be a bit hammy. The costumes, the playlists, the syrupy, exaggerated way that everyone seemed to speak to the residents. Are you having fun? But on that day, there was something else. Helen told her daughter that a caregiver had come into her room and warned her not to attend. "Stay in your room, Helen," she said. "People are sick."

Helen had come to the Life Care Center after a fall. Six years earlier, she'd had an apartment in an assisted-living facility, where she managed just fine despite the arthritis that made her joints ache terribly. She was "so independent, it was ridiculous," her daughter said. She had been all her life: from back when she was a Minnesota farm girl and her mother died and she was left to fend for herself through the Depression and the war. But then one day, she fell in the bathroom. She spent six hours trying to crawl to the phone. "Oh boy," Carolyn told her. "I don't think you can stay here." Helen did not want to go to Life Care, but she did not cry. She hadn't even cried when her husbands died—in fact, her daughter had never seen her shed a tear.

Helen did ask how much the nursing home was going to cost. She had been surprised to learn that Medicare would cover only short-term rehab at a nursing home—a few weeks of therapy— but not the long-term care that she would need afterward. She would have to pay for everything herself: $11,400 a month, until she ran through her savings (nursing-home administrators call this "spending down") and qualified for government assistance. By Carolyn's calculations, her mother would be eligible for Medicaid when she was 103. When Helen heard the arithmetic, she balked. She had always worked and lived frugally, and the plan had been to leave her money to her children; she didn't want to blow it all on a nursing home. But Carolyn told her that there was no other way. "It's your fault for living so long," she joked.

Twilla's undoing also came with a fall—and given the way she was living, her daughter later thought, it might just have been a matter of time. She had stopped taking care of herself after her husband died. She ate McDonald's and Taco Bell and didn't exercise. She tripped while getting up from her recliner. Then things went south in the usual way. A string of infections. Incontinence. Disorienting moments, when she confused her daughter for her sister. Her kidneys started failing. Debbie, who had recently lost her job at a grocery store, moved into Twilla's spare bedroom.

She taught herself about nutrition so she could cook her mother healthful meals. She installed a safety bar in the bathroom. She tried to keep seeing her increasingly fragile mother as she had once been: a bookkeeper who taught herself to trade stocks as a hobby.

Before long, Twilla needed more, and then too much. Debbie wanted to hire a professional caregiver to spend just a few hours each day with her, but she couldn't afford one, and Medicare wouldn't pay for it (Medicare rarely covers at-home caregiving), so she started looking at nursing homes. Tony Chicotel, an attorney with the California Advocates for Nursing Home Reform, told me that low-income and middle-class Americans are far more likely to end up in nursing homes before they need to be there, while people with more means are able to stave off institutionalization by hiring private caregivers or renting apartments in less medicalized assisted-living complexes. "People end up in nursing homes because that's where the government funds go," he said.

Debbie first picked a nursing home near Twilla's house. The day she moved in, the two women cried and cried. "It's the doctor," Debbie told Twilla. "He says I can't take you home. They won't let me." That wasn't true, strictly speaking, but it might as well have been—and it was easier to say. Two years later, Debbie relocated her mother to the Life Care Center so she could be closer. By then, Twilla's mind was more mixed up. She thought she was a little girl being sent away to school.

The other Life Care residents had similar stories. They fell. They got dementia. They had strokes. They got urinary-tract infections that were overlooked for so long that they caused delirium —and then finally were diagnosed correctly and treated, but not before the whole family got freaked out and decided that *it was time* to start looking at facilities. Some were just very old and very weak and had fallen into what doctors sometimes call a "failure cascade": one node in a body system breaking down and, in turn, causing the breakdown of another, and on and on. One resident lived alone until her daughter found her in the bathtub, where she had been sitting all day long, cold and shaking and suddenly unable to stand. Some of the residents had been okay while living with their spouses but had unraveled in widowhood. The Life Care Center of Kirkland, a tidy town of 93,000 just east of Seattle, seemed good enough. Nice, even. On a federal government website, the nursing home was rated with five stars, which was the most stars that a nursing home could have.

Neither Helen nor Twilla had wanted a roommate, and, from the looks of it, nobody had given much thought to the pairing. Inside Room 10, Helen sat calmly in her wheelchair in the smart knit slacks and 100 percent cotton blouses that she always wore and that her daughter laundered for her every week. The wall above her bed was covered with photographs and a map of the world so that nurses could point to the countries where they were from and Helen could ask them what life was like there. When Carolyn visited on the weekends, Helen wanted to know what was happening in the world, in this country or that. She wanted to know how many people were homeless and what the government was going to do about it. She asked how Shiites were being treated in such and such a place compared with Sunnis. At first, her curiosity startled her daughter; Helen had never seemed so interested in the world before. But after moving into Life Care, she had stopped talking much about the past or the future. Now, there was only the present to know.

Across the room, the walls were mostly bare. On a good day, Twilla might be calm. She might nap in the special bed that was meant to prevent bedsores. She might wheel the facility's halls. But on a bad day, she might shriek at the top of her lungs. Some of the things she said made no sense at all. Other times, they were awful things, sometimes racist things, hurled straight at the nurses who cared for her. When Debbie heard Twilla yell that way, she thought about how humiliated her mother would have been by the sight of herself. Other times, Twilla yelled at Helen.

"I'm going to kill you," she said once.

"I don't think you are," said Helen.

Whenever Debbie visited, she was a bit annoyed to see the decoration over Helen's bed. She thought the photos made the space look cluttered. Still, she was impressed by her mother's roommate. Helen could bathe herself at the sink in between showers. She could put on makeup and comb her thick hair and use a cell phone. She could still ask for what she wanted: for her toenails to be cut a different way or for her dinner vegetables to be served more simply—for the chef to give her mashed potatoes without the gravy. Why was she in a nursing home? "It struck me as odd,"

Debbie said. "But you know, some people's kids just do that." Debbie thought that Life Care was a dismal place and that it smelled like urine. She thought the nursing aides were too rough with Twilla and that they were giving her bruises. Because of all that, by late 2019, Debbie had stopped visiting. She didn't even come on Christmas. Later, she would say that she was afraid of crying in front of her mother. If she cried, Twilla would want to know why she was sad, and what was Debbie supposed to say? Because you're here.

After the holidays, people started getting sick. But then again, people always got the flu around the holidays. The year before, Life Care had shut down to visitors for a few days because of a bad influenza outbreak. That was normal. This year was strange, though, because all the flu tests were coming back negative. In mid-February, a number of residents were diagnosed with pneumonia. They were given oxygen or antibiotics or nebulizer treatments, which deliver medication to a person's clogged-up lungs and make it easier to breathe but can also send virus-filled particles swirling around the room. Nurses would later conclude the nebulizers were a bad idea.

On February 12, Life Care's infection-prevention nurse and physician assistant met to discuss the growing number of respiratory infections. The nurse thought they were dealing with a weird seasonal flu, and the physician assistant agreed. By law, Life Care is required to report the existence of an infectious-disease outbreak to county public health officials within twenty-four hours, but neither filed a report. A week later, on February 19, Life Care staff held a routine "Quality Assurance and Performance Improvement" meeting. According to a later federal investigation, the infection-prevention nurse did not attend, as she was required to do, and the respiratory infections were not discussed. The nursing home's medical director was a no-show too; he had also missed the previous meeting in late January. By February 23, six more Life Care residents had fallen ill. In response, some staff started wearing masks, but others didn't. "Some people are being cautious," a nurse explained when a visitor asked what was going on.

When Helen told her daughter that she had sinus pain, Carolyn took a deep breath because every so often her mother got a sinus infection, and it was always the same: Helen would refuse to tell the nurses until two or three or even five days had passed and she was so sick that Carolyn would have to threaten to call them herself. "She likes to be a martyr," Carolyn said. On the phone, Carolyn told her mother that she had better let the nurses know before her weekend visit.

On February 27, a Life Care nurse called the Seattle–King County Public Health Department and left a voicemail saying that she had noticed "increased numbers" of respiratory illness at the facility. Later, public health authorities would say that "the message had little detail . . . and made no mention of Covid-19." But then again, authorities hadn't asked nursing homes to be on the lookout for Covid, even though a man in nearby Snohomish County had tested positive for the virus back in January after returning from a trip to Wuhan. It was also on February 27 when the Centers for Disease Control and Prevention expanded its strict testing criteria to allow for symptomatic people to be tested even if they had not recently traveled to China—and that doctors at nearby EvergreenHealth hospital, who had for weeks been receiving a steady stream of feverish Life Care patients, got permission to send out their first tests for analysis.

The next day, visitors to Life Care saw signs on the wall noting that there was a respiratory outbreak in the building. By midday, they were told that they could come inside only if they wore masks, even though some of the staff weren't wearing them. That evening, around nine, a paramedic unit arrived at the nursing home to collect another sick resident. According to KUOW, a local radio station, the team saw Life Care nurses moving room to room without wearing any personal protection equipment. "Hey, you guys are supposed to be in self-quarantine," one firefighter told them.

"No, we're not," a nurse answered.

What the hell? the firefighter thought. *Who is not telling you that you have two suspected coronavirus cases?*

Debbie got the call the next day, on February 29. "I would just need to let you know that we do have coronavirus in the facility," the administrator said. She said that the nursing home was locking down but that Twilla was okay. "She's just not eating much." Carolyn didn't get a call. She found out everything when she arrived to visit her mother and saw Life Care surrounded by news crews.

Other family members tried to call the nursing home but

couldn't get through, and so over the next several days, they drove to Life Care and gathered outside on the front lawn. Scott Sedlacek brought a whiteboard so he could write out messages and hold them up to his father's window "to let him know why we're not visiting him." Bonnie Holstad came with a hand-printed sign, asking staff to please take her husband's temperature. Katherine Kempf shouted through the window at a nurse who was tending to her father. "Why don't you cover his legs up?" Some relatives talked about storming the facility to "bust them out"—but then where would they go? The other nursing homes in the area didn't want Life Care residents.

Some people did get through to a nurse, and when they did, they demanded that their mother or father be sent to the hospital right away. A number of them were told that unless residents had three specific symptoms—high fever, cough, and difficulty breathing—they were not supposed to be sent out, and that if they were sent out, the hospital would just send them back. Amir Medawar said that when he spoke to a Life Care nurse, the nurse told him that his mother, Odette, wasn't sick enough to go to Evergreen-Health because her temperature had gone down after she took some Tylenol. But when Amir called EvergreenHealth's hotline, a nurse said that he should get Odette to the emergency room as soon as he could and that he shouldn't wait until morning. (An EvergreenHealth spokesperson declined to answer questions about the hospital's involvement with Life Care patients.) Amir called Life Care back and was again discouraged by a nurse from moving Odette. "We are done talking," he said. Amir called an ambulance. Odette was admitted to the hospital, where she tested positive for Covid-19 and remained in isolation, on oxygen, for three weeks.

Nancy Butner, the Life Care Centers of America's northwest divisional vice president, told me that nursing staff never refused to send a patient to the hospital. "There was never pushback. If the patient needed to go, the patient went." The problem, she said, was that some people were panicking and wanted to hospitalize their parents when they didn't need to be. There was something else too. "During that time, very early on, the concern was: We're going to fill up the hospital," Butner said. "Everybody wanted to go to the hospital, but it would create too much of a surge." There was, in fact, plenty of room in nearby hospitals, but because no-

body was triaging patients at the county level, paramedics continued to bring Life Care residents to the nearest hospital, which was
EvergreenHealth, which was slowly becoming overwhelmed.

This was a bad moment for a Washington nursing home to have
a respiratory outbreak. Already, PPE supplies were running low
across the state. That season's influenza had been especially viru-
lent, and regular supply chains from China had been cut off because of the virus in Wuhan. Some nursing homes had run out of
masks and gloves. But in other ways, the state was always going to
be hit hard. According to the National Health Security Preparedness Index, which was created by the CDC, Washington scores below average when it comes to the preparedness of its long-term
care facilities. In 2016, the state ran a pandemic-response drill and
wrote a ninety-page response plan. The plan made only several references to long-term care facilities—and then only in general lists
of health-care facilities. By contrast, the report had several sections
devoted to state veterinarians.

Life Care had its own emergency plan, as all nursing homes
are required to have. It was printed out and looped into a thick
plastic binder in the nursing home's back office. But the plan
had more to do with hurricanes and floods and earthquakes and
power outages and terrorist attacks than with pandemics. As it was,
Life Care's most immediate problem was staffing. People were out
sick and scared to come in, and then the replacements were getting sick too. Not even local staffing agencies, once eager to supply
the nursing home, could find people who were willing to go to
Life Care. The nursing home's parent company initially wanted
to bring in nurses from facilities in nearby states, but Washington
only allows nurses licensed elsewhere to practice within its borders
after a lengthy application process. A Life Care executive called
the state licensing board to see if the applications could be expedited but was told, "No."

On February 29, officials at company headquarters sent out a
message to all Life Care properties within Washington, asking for
volunteer nurses. Chelsey Earnest, a forty-seven-year-old nurse,
saw the message on her way to church, around 40 miles south of
Kirkland. By the time services were over, she had decided to go.
Chelsey's husband was a combat medic who sometimes left home
for weeks-long deployments, and now, she thought, in a way, it
would be her turn to go to war. Nursing homes hadn't always been

Chelsey's calling; fifteen years ago when she was looking for work, local nursing homes had simply been the only places that were hiring. But then she had come to love "my elderly people." Chelsey arrived in Kirkland on March 1 in the afternoon. She thought the place would be flooded with public officials and maybe even the National Guard. Instead, she found an understaffed facility, where ambulances were arriving every few hours to transport gasping residents to the hospital. She was assigned to the night shift.

Life Care staff asked the county's health department for Covid-19 tests so they could start assessing residents on-site. They were surprised when officials said that there were no tests to be had; later, they said that it wasn't their responsibility to provide them. Life Care also warned that the facility was dangerously short on nurses. On February 29, just after midnight, a county official emailed Life Care to ask for a list of staff "that you NEED" and to promise that she would "get staffing help for you." But then no help came.

Later, a health-department spokesperson told me that the department had no responsibility to help with staffing and that its role was limited to "surveillance and investigation of communicable diseases." "We usually don't send staff to shore up private entities," Dr. James Lewis, the department's Covid-19 acute health-care system support lead, said. I asked him if he saw a conflict between the role of private nursing homes and the responsibility of public health officials to stop a global pandemic. "Yeah," he said. "We're still working on that, and there are ongoing growing pains."

On the night shift, Chelsey tried to keep her head down. "I mean, we were just trying to get by," she said. "Find sick people. Do the vital signs. Turn them over if you can. Get them changed if you can. Then wait for help." But there weren't enough staff, and there weren't enough medical gowns. Chelsey made sure that all the doors to the rooms were open, so that she could at least glance inside while rushing down the hallway to see that nobody had fallen out of bed. She started getting blisters inside her new sneakers.

On March 1, a team from the CDC arrived in Kirkland: an eighteen-person group of epidemiologists, lab experts, and infection-control professionals. At a press conference announcing the deployment, Seattle–King County health officer Dr. Jeffrey Duchin said that the team would "assess each and every one of these people and provide the appropriate guidance around isolating, and all of these people will be tested." But soon, it was obvious to anyone inside Life Care that the CDC was not there to help but rather to study. Most of the CDC people never set foot inside the nursing home; they asked questions over the phone. The ones who did come inside "were there with clipboards, sort of watching and observing," Tim Killian told me. "This was more of an academic exercise." (In an email, a CDC spokesperson confirmed that its role was "not to provide treatment," but rather to offer "technical guidance.")

"I think we felt like we were . . . umm. I don't know what," Chelsey told me. "Ignored? No, that's not the right word. Like we needed things, and they weren't coming."

More helpful was an infection-prevention expert named Patricia Montgomery, who was sent to the nursing home from Washington's Department of Health, along with two nurses who worked the night shift for a couple of days that week. Montgomery had been shocked to learn that Life Care, of all places, had the virus. The facility was one of the best she had worked with, and she thought its infection-prevention nurse was as good as they come. But when she arrived at the nursing home, she was startled to see that not everyone was wearing a mask. Some nursing aides didn't even know how to use PPE properly. At one staff-training session, Montgomery and her colleagues asked staff to rub a lotion on their hands, asked everyone to wash up at the sink, and then turned on a black light to expose any spots that had been missed. On some hands, spots had been missed.

The patients were getting sicker and in peculiar ways. Chelsey had never seen anything like it. "I saw some strange phenomena." There were patients who seemed absolutely fine, who didn't even have a fever, but then would fall into acute respiratory distress within the hour. Once or twice, Chelsey said she took a patient's forehead temperature and found it to be normal but then touched his chest and found it hot to the touch. Then there were the red eyes. The first time Chelsey saw them, she asked her colleague if the resident had been crying, but she hadn't been. "That was the only symptom she had," Chelsey said, "and that patient died in the hospital about five days later."

The night shift was when the creeps started calling. When Chelsey answered the phone, people told her that they were priests and healers and that they knew the cure for the virus. They begged her not to hang up—to listen for just one minute, please—while they explained themselves. Their cures were always lunatic. One guy told Chelsey to mix baking soda and lime juice and rub it in her patients' eyeballs. Other callers wanted only to curse the Life Care staff for "bringing Covid to America" and for "killing residents." One evening, a Life Care administrator was followed home by a man who said he had learned about the facility from Reddit.

On March 2, Life Care's medical director, Dr. Dhirendra Kumar, was at the nursing home, assessing patients, when he started feeling sick. He left right away and went into quarantine at home, promising that he would stay available to consult with nurses over the phone. As medical director, Kumar was a paid employee of the nursing home, responsible for overseeing the general medical care of patients. But he was also the primary care physician for about 90 percent of the residents. This is a common setup at nursing homes, and there are no rules against it, but the system effectively places a doctor in charge of his own supervision.

Between March 3 and March 5, at the height of the Life Care outbreak, there were no doctors in the nursing home to evaluate and treat the dozens of residents who needed to be assessed—and officials at the county, state, and federal level knew it. Life Care's physician assistant later told federal inspectors that in Kumar's absence, she had sometimes made medical decisions on her own. Other nurses had just sent patients to the hospital; around forty of them went in a single week. "I didn't want them to pass away while I waited for them to get sicker," one nurse, who asked me not to use her name, said. "When in doubt, get 'em out."

It was Chelsey who noticed, in the early hours of March 3, that Twilla had a fever. She gave her some Tylenol. Another nurse called Debbie and left a voicemail: "We anticipate that she, too, has the coronavirus. We do not anticipate her fighting this." When Chelsey found Twilla dead, less than twenty-four hours later, she cried. Her own tears amazed her because she had worked in nursing homes for so long and had seen so much death and had become so used to it. "This Covid outbreak jarred me out of it," she said. "Around the fourth day, I started bawling, and I didn't stop for two weeks."

A few hours after Twilla's body was taken away, Helen woke up wet with sweat. "I'm just drenched," she told the nurse. She declined Tylenol and asked for a cold cloth to dab against her face. She called Carolyn, who spoke to a nurse who said that Helen's temperature was hovering around 100. "You better get her to Evergreen," Carolyn told her. But the nurse said that Helen wasn't sick enough to go. Her fever was still mild. "If my mom gets to the point where she has a temperature of 103, 104 and she's coughing and she can't breathe, she'll be dead in twenty minutes," Carolyn said.

It wasn't until March 4, five days after Life Care went on lockdown, that western Washington's Central District Disaster Medical Coordination Center—responsible for directing patient movement during emergency incidents—was activated to help the nursing home. The spur to action was a call by the region's paramedic chief, who said he needed *someone* to do *something* about all the 911 calls coming from the nursing home. "We have an internal disaster at Life Care," he said. "We have a multiple-casualty incident, and we need to do something fundamentally different."

Until that point, Dr. Stephen Mitchell, director of the Coordination Center, said he had not understood the situation at Life Care to be urgent. "It was very much a slow-moving disaster," he told me, "and with slow-moving disasters, it's hard to step out and to say, 'Oh my God, there is a disaster.'" But as soon as he looked, there it was. "Their backs were up against the wall. They were pleading for help." Later, it would seem to Mitchell that "nobody had done the work, if you will," to prepare the region's nursing homes for medical emergencies. "In general, the flow of patients is from hospitals to skilled nursing facilities. They are meant to be opportunities to offload patients who don't need acute care in hospitals. They aren't usually where the disaster originates." On March 4, Mitchell took part in a conference call with representatives from the county, the state, and the CDC. "We need to get physicians into the building," he said.

The next day, King County sent two physicians to Life Care to help triage residents. They arrived in the early morning, when it was still dark out. Dr. James Lewis, one of the doctors, walked the halls of the facility and wondered if the whole place should just be evacuated. It was wild, really. There were so many patients, and they were so sick, and there were so few nurses. In the end, he decided instead to identify the fifteen sickest patients and transfer them to eleven nearby hospitals.

On March 6, Life Care got its first cache of Covid-19 tests from the state Department of Health. But when administrators looked inside the box, they saw that there weren't enough tests for every resident and that there were none for staff. "The day before, a cruise ship got two hundred test kits airdropped to test people who were trapped on the boat," one staff member told me. "I had ninety residents and could only get forty tests? Somebody want to explain that to me? I have to play judge and jury here: Who is going to get tested and who is not?" By then, the Life Care death toll stood at ten.

At the end of Chelsey's shift, nurses decided to start testing in Room 1 and work their way up over the following days. When they got to Room 10, they saw Helen lying in her bed with a plastic nebulizer mask strapped around her face because she was having a hard time breathing. The nurses administered her test and put the kit in a cooler and then moved on down the hall.

The story goes that in the 1950s, a young man by the name of Forrest L. Preston—the son of a Seventh-Day Adventist pastor living in Walla Walla, Washington—started selling Electrolux vacuum cleaners to help make ends meet while studying at a local college. He was plucky and had a knack for sales. Although he intended to become a physician, he struggled with inorganic chemistry, so he continued as a salesman instead. Eventually, he joined his brother's printing business, which sold pamphlets and marketing materials to hospitals and nursing homes.

In 1970, Preston opened a nursing home of his own, in Cleveland, Tennessee, which he called Garden Terrace Convalescent Center. "These lights will never go off again until the second coming!" he reportedly proclaimed on the facility's opening night. Half a century later, Preston's company, Life Care Centers of America, is the largest privately held long-term care corporation in the United States, with more than 200 nursing homes and senior-living centers in 28 states and approximately 40,000 employees. It is headquartered in Cleveland, a modest city near the border with Georgia. Preston, who at eighty-eight remains the sole owner of the company, is a billionaire.

The modern American nursing home grew out of the nineteenth-century almshouse, a kind of public, charitable organization that was set up to help the "worthy poor" (originally, widows of good social standing who had fallen into destitution). The almshouse system expanded until the 1930s, when officials at the U.S. Social Security Board began to worry about the "increasing dependency" of "the aged"; they feared that old people would bankrupt the country with their expensive infirmities. They made efforts to shut the facilities down, and they proposed that the government start a small pension, what would become Social Security benefits.

In place of the almshouses came pay-to-stay "rest homes" and, later, more medically staffed nursing homes, all competing in a private marketplace for eldercare. By 2000, nursing homes were a $100 billion business, and the little mom-and-pop shops that had once dominated the industry were being fused together and swallowed up into larger entities. For a time, it seemed like nothing could stop the growth. It didn't matter when, in the early 2000s, five of the country's top-ten nursing-home chains entered into Chapter 11 bankruptcy proceedings after undertaking a string of heavily debt-financed mergers and acquisitions. The companies were restructured, and sometimes rebranded, and then continued on their way. Today, around 70 percent of nursing homes are for-profit, and more than half are affiliated with corporate chains.

The modern nursing home has adapted itself to the freakish architecture of Medicare (for people over sixty-five) and Medicaid (for those on low incomes or with disabilities) and the vast gaps inside and between them. Specifically, the facilities benefit from a patchwork insurance landscape that often pushes older Americans into institutional living. Take, for example, falls—like the ones that precipitated Helen's and Twilla's move into Life Care. Each year, about 30 million older Americans fall, resulting in 300,000 broken hips and 30,000 deaths. Nevertheless, many elderly people are not assessed to see if they are at risk of falling and could be helped to avoid it—in part because there is a shortage of geriatricians trained in the practice but also because, until recently, primary care doctors could not bill insurance for the assessments and so didn't do them. Medicare, however, is willing to pay for weeks of costly post-fall, post-surgery rehab at a nursing home, and Medicaid is there to take over the cost for the many patients who are never able to walk again and need to remain.

Still, in the early 2000s, a number of large nursing-home operators came forward to say that they were in financial distress and at risk of failure—and that the most decisive reason for this was low Medicaid reimbursement rates. While Medicare often pays nursing homes handsomely for providing skilled rehab and therapy (sometimes more than $1,000 per day), state Medicaid programs pay much less (on average, around $200 per resident per day) for "long-term" care. Nursing homes said that they were bleeding out money because of Medicaid patients.

Most facilities, however, found a way to tip the balance in their favor. Many reserve beds for more-lucrative rehab patients, though it is illegal for them to discriminate based on payment source. Some rush patients through therapy schedules: declaring them fit to leave as soon as they have maxed out their most highly reimbursed Medicare coverage days, and then filling the bed with someone new. In a number of states, reports of illegal nursing-home evictions—often of residents on Medicaid or about to go on Medicaid—have risen. The phenomenon is so common that there is now a catchphrase for the practice: "resident dumping." Residents are sometimes packed into vans and then abandoned in low-budget motels, or homeless shelters, or even onto street corners —or, in one reported instance in Maryland, into a storage facility.

In 2012, the U.S. Department of Justice filed a case against Life Care Centers of America, accusing the company of Medicare fraud. Two employees, in two different states, had come forward to say that awful things were happening at company nursing homes. According to court documents, Life Care therapists "canvassed the facility looking for residents they could provide therapy to in order to increase billing." Sometimes, this resulted in old, sick people receiving needless rehab sessions up to seven or eight times in a single day. According to the Justice Department complaint, one resident who could not walk was allegedly carried up and down the hallway so that the nursing home could bill Medicare for walking therapy. A ninety-two-year-old man who was dying of metastatic cancer was allegedly given forty-eight minutes of physical therapy, forty-seven minutes of occupational therapy, and thirty minutes of speech therapy two days before he died, despite the fact that "he was spitting out blood." At one Life Care facility in Florida, the entire rehab staff had signed a letter declaring that they had "been encouraged to maximize reimbursement even when clinically inappropriate." They also said that the command to boost rehab billing had come straight from Forrest Preston, who had allegedly intervened to thwart the work of his own internal compliance officers.

Although Life Care and Preston denied the charges, in 2016, the company agreed to pay $145 million to settle the case. At the time, the settlement was the largest ever between the U.S. government and a nursing home. But it hardly set Life Care apart. All five of the country's largest nursing-home chains have been accused of fraudulent practices by the federal government. (In addition to Life Care, two others settled "false claims" cases for tens of millions of dollars.)

All the while, nursing-home chains continued to get bigger, until just five companies owned more than 10 percent of the country's 1.7 million licensed nursing-home beds. Private equity also entered the sector, buying up four of the ten largest for-profit nursing homes. "There's essentially unlimited consumer demand as the baby boomers age," Ronald E. Silva, president of Fillmore Capital Partners, told the *New York Times* in 2007, after paying $1.8 billion to purchase a large nursing-home chain called Beverly Enterprises Inc. "I've never seen a surer bet." These new ownership groups changed things in ways that people who lived in them could feel. Earlier this year, a Wharton School–New York University–University of Chicago research team found "robust evidence" that private-equity buyouts lead to "declines in patient health and compliance with care standards." When nursing homes are bought by private-equity groups, the team concluded, frontline nursing staff are cut, and residents are more likely to be hospitalized.

But the most consequential change may have happened within the for-profit companies themselves. It all started, most undramatically, with a 2003 academic article in the *Journal of Health Law*. In "Protecting Nursing Home Companies: Limiting Liability Through Corporate Restructuring," its authors—two health-care lawyers—made note of two financial threats to nursing-home operators: lawsuits by nursing-home residents (for, say, negligence) and efforts by the government to recoup overpayments (for, say, false claims on Medicare billings). The solution, the authors suggested, was in restructuring. Specifically, nursing homes should split up into separate limited-liability corporations, one for real estate and one for operations. This new structure, they wrote, would keep assets safe from litigious family members and retributive bureaucrats. It would also attract money from real-estate investors who were keen on nursing homes but wary of the liability risks. By 2008, the top-ten companies had all split themselves into real-estate and operations LLCs.

Then many companies went further, creating networks of sub-companies called "related parties" that could trade and transact with one another. What had once been a nursing home became a corporate cluster, including separate entities for real estate, insurance, management, consulting, medical supplies, hospice, therapy, private ambulances, and pharmacy services. By 2017, three-quarters of nursing homes did business with related parties, according to a study by *Kaiser Health News*. There was nothing inherently wrong, and certainly nothing illegal, about these increasingly complex formulations. The owners said that they were only creating a vertical supply chain for eldercare. By 2015, nursing homes were spending $11 billion a year on contracts with related parties.

But the structure had an additional benefit that the authors of the article had not pointed out: it allowed companies to siphon profits out of their nursing homes through sometimes exorbitantly overpriced transactions with their sister companies. Instead of hiring salaried managers to oversee a facility, a nursing home could now contract with expensive related-party management corporations and consultancies. Instead of owning the land around a nursing home, a company could lease it from a related-party real-estate business, sometimes at a higher-than-market rate. In this way, a nursing home could appear, on its accounting sheets, to be operating on slim margins, or even at a loss, but only because that loss was offsetting gains within the same company. When federal prosecutors charged Life Care Centers of America with overbilling Medicare, they described the company's nursing homes as "severely undercapitalized." (A Life Care spokesperson described this claim as "not true.")

"No one begrudges a company for making profits," Dr. Michael Wasserman, president of the California Association of Long-Term Care Medicine, told me. "This is capitalism. This is America." The issue, he said, is that doctors and nurses are pressed to cut costs while related parties are getting rich. "If the real-estate entity is making significant profits and the operation is break-even, then there's a problem. I would compare today's nursing-home real-estate owners to slumlords."

The related-party structure has another obvious benefit: opacity. According to Ernest Tosh, a plaintiff's lawyer in Texas who advises law firms on nursing-home finances, many companies hide profits in related parties because owners know that it would look bad for them to get rich off nursing homes that provide substandard care. "It's a kind of money laundering," he said. The bookkeeping trick allows them "to go to the state legislature, to Senate sub-hearings, and say, 'I have all these nursing homes, and they barely break even. We need more Medicare money. More Medicaid. We need bigger reimbursements. You guys are killing us!' The thing is, it's not true. The balance sheet and the income statement from a nursing home are fictitious documents. They say whatever the owner wants them to say." Tosh believes that industry claims about widespread financial distress are bogus. "You think investment trusts whose only purpose is to make money would invest in an industry that was losing money?"

By law, nursing homes must disclose to government regulators both their relationship to related parties and the dollar value of transactions with them. They also have to indicate how much it costs the related parties to provide their services, theoretically allowing regulators to spot incidents of gross overpayment. The transactions can be and sometimes are audited by the federal government. But Tosh says that in practice, and in general, this system fails, since the central company is the one providing the related-party cost reports. "You have the same person signing the contracts on both sides."

Tangled financial frameworks make things hard for regulators to follow—with the result, according to the Government Accountability Office, that tracking compliance problems across large companies "can be ad hoc." Although private nursing homes receive billions in public funds, they are not required to publish public financial statements. A nursing-home resident who wants to understand her facility's financials will have to file a Freedom of Information request with the Centers for Medicare & Medicaid Services, an agency within the Department of Health and Human Services, which provides printed-out Excel spreadsheets.

The Life Care Center of Kirkland is housed in an old, single-story stucco building, lined with hedges, that has functioned as a nursing home for more than thirty years. For a while, the Life Care Centers of America leased the business from another nursing-home operator, which owned the real estate, but around fifteen years ago, it purchased the facility outright. With Tosh's help, I reviewed the nursing home's financial data and found that the facility has followed industry trends. In 2018, the nursing home carried out around $2.5 million in related-party transactions with corporations that were mostly owned or completely owned by Forrest Preston, including a management company, a health-insurance company, a workers' compensation and auto insurance company, a real-estate company, a third-party administration company, and an interior-design firm. "I've looked at thousands of financial statements. I've never seen an interior-design firm before," Tosh said. That year, the Life Care Center of Kirkland claimed a net income of around $80,000.

For days after Twilla died, Helen lay in her bed, in her now-empty room, and listened to the nurses running in the hallway. On the phone with her daughter, she wanted to know where on earth the government was. "Why aren't they helping these people?" Sometimes, Helen spoke for thirty or forty minutes about this or that happening in the nursing home—and only then, at the last minute, before the two women were about to say goodbye, would she reveal something important about herself, like how she still wasn't feeling well.

On March 7, federal backup finally arrived. It came in the form of a Department of Health and Human Services "strike team": twenty-eight military doctors, nurses, technicians, and aides. Five days had passed since Life Care made a formal, written request to the county, which passed it to the state, which passed it to Health and Human Services on March 3. Later, I asked a department spokesperson about criticism that the federal government had waited too long to act. She said the question was based on misunderstanding. "We can't just send federal people into a state."

Chelsey had been promoted and was now Life Care's acting director of nursing. She met the strike-team members on the night they arrived. "I could have used you guys about five days ago," she told them.

"Well, we're here now," one of the nurses said. "So let's just move forward."

Yeah, okay, Chelsey thought. *I have post-traumatic stress disorder. I don't know about you.*

By March 9, there were 129 confirmed cases of COVID-19 associated with the nursing home, including 81 residents and nearly 50 staff and visitors. Chelsey was grateful for the strike team. They brought Covid tests with them, for one: not enough for the staff, but finally enough to test every resident. Still, with everyone spread out over three shifts and four facility wings, there was often nothing that anyone could do for a sick patient except send him to the hospital and hope that someone there could save him.

Chelsey spoke with a doctor at the Disaster Medical Coordination Center and told him what she was learning: that if she waited until a patient had a super-high fever—103 or 104 degrees, say—before calling for an ambulance, it would be too late, because that person was about to crash. Chelsey and the doctor agreed to lower the threshold for hospitalization. Two low-grade fevers and a resident was out, unless he didn't want to go, like the one man who said that if he had to die of the virus, he would rather die where he was.

Sometimes, there wasn't even time to clean up after a resident was sent to the hospital. One nurse said it weirded her out to see old hairbrushes and oxygen tubing lying on abandoned beds and bedside tables. Looking back, she thought, "it was potentially hazardous. Someone should have gone in there and cleaned the rooms up." The nurse said that residents often asked her about their friends in other rooms: how this resident was, whether that resident was still alive. She couldn't tell them much because she worried that would be breaking patient privacy laws. "Is Twilla coming back?" one wanted to know.

During the night shifts, Chelsey thought about all the things she wanted to teach the world about Covid, the things that experts didn't seem to know yet, like how the virus appeared to spread between patients, even if they had no symptoms. But it seemed to Chelsey that there was nobody to tell and that nobody was asking her. It didn't help that many members of the Health and Human Services team had never worked in long-term care before or even treated older patients. The doctor in charge of the night shift was a pediatrician. He was super nice, but still, Chelsey said, it took hours to explain things to him, like how nursing homes work.

For a week, officials at the county health department had been telling Life Care staff to group sick residents together to protect the healthy ones, but the advice didn't seem to make sense because none of the Covid tests had come back yet. Chelsey and her colleagues tried to separate the patients anyway. They gathered all the sick-looking people together in the back unit to keep the rest of the wings "clean." But a few hours into the experiment, a woman in the clean wing started coughing and her eyes turned red. Then she got upset because she had dementia and was confused by her new room and missed her old roommate, and Chelsey had to assign a nurse to sit beside her through the night to settle her down again. Now, the clean wing was contaminated. It would be weeks before the CDC issued any guidance on how to "cohort" patients. Chelsey didn't see how cohorting was going to work anyway if staff couldn't get tested and healthy-looking residents could actually be sick. In the two months she spent treating residents in Kirkland, she said, she was never tested for Covid.

While the strike-team doctors assessed patients, some Life Care nurses spent hours just trying to get residents to stop crying—and to eat *something*. "We had one lady who was just fearful every day," a nurse told me. "She cried a lot. We found out she liked ice cream. Well, the ice cream at the nursing home is in small foam cups, and it's not really that good. We found out she liked banana splits, so we would put our money together and order her a banana split. For like a week, we gave her a banana split every day." Other nurses brought iPads into the residents' rooms and helped them call children or grandchildren. But sometimes that made things worse. One ninety-two-year-old woman with dementia confused the screen for real life and got frantic when she couldn't escape through it. "Come get me! I don't have my purse," she told her daughter-in-law. "I can't get back to Long Island. I can't find my room."

The nurses were also starting to lose it. In the evenings, Chelsey lay awake in her hotel room and ran through all the conversations she'd had that day. She thought about the families and their sadness and their anger. Once, she dreamed that she forgot to bring a patient water and that the patient died because of her mistake. In the mornings, Chelsey woke up to dozens of Facebook messages. In an interview with CNN, she had mentioned the strange red eyes that residents sometimes developed when they got Covid, and now people from all over the world were sending her close-up selfies of their faces. "I need to know," they would write. "Do my eyes look like I have it?"

From their wall-mounted TVs, Life Care residents could see the way that Covid was moving from nursing home to nursing home across the country: so fast that even President Trump would admit, later, that nursing homes were "a little bit of a weak spot." On the news shows, some people called Covid the "Boomer Remover." Some said that old people were "sitting ducks." Others seemed relieved that, at least, it seemed to kill *only* the elderly. The lieutenant governor of Texas went on Fox News to insist that grandparents would absolutely be willing to die of the virus in exchange for keeping the country open for their grandkids. If it was between life and quarantine-free liberty, he said, the Greatest Generation would choose liberty.

By then, Debbie was certain that her mother had died of Covid, and so she was surprised to see that the virus wasn't named on Twilla's death certificate. Debbie said she called the nursing home and asked to speak to Dr. Kumar, who had signed the paperwork, and was told that he had not been at the facility for days. She said she also asked Life Care for a copy of her mother's medical records because she wanted to see for herself if Twilla had been coughing and if she'd had those red eyes. But the nurse who answered the phone said that she couldn't find Twilla's file. Debbie asked the Kirkland medical examiner to test Twilla's body for Covid, and the test came back positive, and the death certificate was changed.

Life Care Vice President Nancy Butner told me that there must have been a misunderstanding; she said that Twilla's records had likely just been removed from the nursing station after she died. Still, the call made Debbie wonder if "maybe Life Care is covering up something." Within weeks, federal inspectors would determine that the nursing home's overworked physician assistant had stopped keeping track of everything: that well into March, she was still typing up notes from late February.

It was also around that time that Helen's test results came back negative. When Carolyn spoke to her that evening, Helen seemed unmoved. "I can't believe I never got that virus with all the virus spewing out of Twilla's mouth," she said. Maybe she'd had a sinus infection, after all.

The Health and Human Services team left just a week after arriving. It was replaced by a group from AMI Expeditionary Healthcare, a private emergency medical company that has a standing contract with the State Department and that previously worked on the U.S.'s Ebola response in West Africa. On the first day of the handover, the new team gave out PPE. The company, unlike the state of Washington, had a decent supply. Back in February, its managers had concluded Covid-19 was going to spread across the world and that they would need to source as much PPE as possible, and soon, so they called up the government of Sierra Leone and asked if they could buy up the supplies left over from Ebola. In order to conserve the PPE that they had, AMI doctors taught Life Care nurses some tricks that they had learned in Liberia, where basic medical resources were also scarce: like how to spray medical gowns with a decontaminating 0.5 percent chlorine solution so that they could be reused.

Dr. Ryan Azcueta had been living in Dubai, coordinating AMI efforts in the Middle East and Africa, when his boss called and told him that he was deploying to suburban Seattle. When he got to Kirkland in mid-March, Azcueta was amazed to find things as dire as they were; of Life Care's remaining residents, about half were infected, and there were only a few dozen staff members available to be spread out among shifts. "Our heads changed," he told me. "It wasn't an emergency response. It was clearly a humanitarian emergency mission." The doctor was also surprised that the federal strike team hadn't moved patients around to create Covid-only and non-Covid hallways, even though, by then, the residents had all been tested. "It was quite shocking," he told me. "It took a while to stabilize things." (Life Care's Nancy Butner later said that staff and the strike team had already begun to group sick residents together. "We moved them as we could.")

Federal inspectors had also arrived at Life Care: a handful of agents from the Centers for Medicare & Medicaid Services, who had been sent to evaluate Life Care's handling of the outbreak. Over several days, "surveyors" reviewed documents and pulled nurses aside for interviews. Life Care later said that its staff spent approximately 400 working hours answering questions. To Chelsey, the inspection was "the height of ridiculousness." Patients were still getting sick, and she was busy. "They're walking around with their fingers pointing. 'Oh, you didn't cross this t or dot this i.' 'I don't care about the t's and i's. Get out of my way.' "

On March 23, the agency announced the results of its inspection. Life Care, it declared, had failed to prevent and contain the virus. As a result, it had put its residents in "immediate jeopardy." Specifically, the facility "failed to identify and manage sick residents, failed to notify the state health department and the state about sickness among residents, and failed to have a backup plan for when their staff doctor became sick." The nursing home's staff had "contributed to acutely ill residents needing to be transferred out of the facility, and deaths of residents," some of whom had "died without sufficient medical evaluation."

The report described interviews with several Life Care nursing assistants. One had worked at the nursing home for more than fifteen years but didn't know how to properly use bleach disinfecting wipes and said he had never been trained to. It also described a laundry-room worker who was observed, on March 7, moving room to room without changing her mask or gown or gloves. When the agency announced it was fining Life Care $611,000, news of the penalty was reported by journalists around the world.

Within days of the fine being announced, Life Care residents and their children were getting emails and calls and Facebook messages from attorneys, offering to represent them in wrongful-death lawsuits against the nursing home. When Debbie agreed to sue, she said her lawyer promised that her case would go big, like O. J. Simpson big. In an ABC News story announcing the suit, Debbie was photographed half in shadow, holding an old framed picture of her mother. "It's not just Life Care," she told the reporter. "All nursing homes need to wake up and get their acts together."

Carolyn called Helen in the evening to talk about the lawsuit. She wondered if maybe Debbie just felt bad about abandoning her mother at a nursing home and then not coming to visit her. "Sometimes, instead of being remorseful, people get angry."

Over the phone that evening, and over many evenings since then, Helen and Carolyn talked about the people who lived and the people who didn't. "Have you seen anybody new come back from the hospital?" Carolyn would ask. "Did Peggy survive? Is Doug still there? What about Patty?" Every night, a variation of the same conversation. "Mom, think about this," Carolyn sometimes said. "This virus is traveling down the hallway going: 'Oh, Room 10. Let's kill Twilla tonight and leave *you*,' who is ninety-eight, and you're not even blinking your eyes about it?"

Less often, it was Helen's turn to be amazed. "Boy, can you believe I survived that thing?"

*

In December 2016, President-elect Donald Trump received a letter. It was nine pages long, and it was written by Mark Parkinson, a former Democratic governor of Kansas and now CEO of the American Health Care Association. The industry group, the largest representing for-profit nursing homes, devotes around $4 million a year to political lobbying—in addition to the hundreds of thousands that individual nursing-home companies like Life Care Centers of America spend in Washington, D.C., and state capitals. "Congratulations on your victory last month," the letter began. "Part of the public's message was asking for less Washington influence, less regulation, and more empowerment to the free market that has made our country the greatest in the world. We embrace that message." Parkinson went on to describe the purpose of nursing facilities, the populations they serve, the people they employ. And then came the warning: "The long-term care profession is on the brink of failure. That is not an overstatement."

In Parkinson's telling, the cause of the near-collapse was twofold. One was underpayment for skilled nursing services. This, he counseled, could be swiftly remedied: government need only increase Medicare and Medicaid reimbursement rates. The other cause was more intrinsic to the relationship between the state and its nursing homes. "We are being inundated with rules and regulations." Parkinson went on to explain how soon-to-be President Trump could eradicate soon-to-depart President Obama's regulations and replace them with a more "collaborative effort between the federal government and providers." A useful regulatory model, he suggested, was the Federal Aviation Administration, which takes "a highly collaborative approach with their industries" and yet still boasts "a remarkable safety record." The letter marks one more turn in a larger history that runs alongside the rise of the Big Nursing Home: the story of government deregulation of long-term care.

In 1987, the Nursing Home Reform Act—pushed through by a Democratic-run Congress and signed by Republican President Ronald Reagan—came into effect, and for the first time set federal quality standards for nursing homes. Until then, nursing homes had functioned with only nominal government oversight, and many had become execrable places. They were reviled as "warehouses" for the elderly: not-quite-medical institutions where, in about half of states, nursing aides didn't have to have any certification at all. The 1987 law set minimum requirements and penalties for failing to meet them. Its passage was the inspiration for a still-much-repeated canard: "The nursing-home industry," nursing-home owners like to say, "is the second-most-regulated industry in America, after nuclear!"

Now, more than thirty years after the Reform Law passed, Toby Edelman, one of the lawyers who helped craft it, considers the bill a failure largely because of the way it has been enforced. To illustrate this point, she often begins with a discussion of the five-star rating system that the federal government uses to rank nursing facilities. It is important for people to know, Edelman told me, that the five-star system is "a fiction." This is because some of the data that feeds into the five-star arithmetic—elements known as "quality measures"—are self-reported by facilities. Quality measures include statistics like the number of long-stay residents who get pressure ulcers while living in a nursing home (more than 90,000 a year, nationally), or the number of residents who become so agitated that they need to be put on anti-anxiety or hypnotic medication. Edelman says that because the data is rarely audited, it is often fabricated, leading to "five-star inflation."

"We've had decades of very lax regulatory oversight," said Molly Davies, a long-term care ombudsman in Los Angeles. She thinks the effect is especially evident when it comes to infection prevention and control. "These facilities, on a good day, pre-Covid, are not good at universal infection prevention. Now, all of a sudden, when we need them to be working like a Cadillac and they can't, we're surprised. We shouldn't be." According to a May 2020 report from the Government Accountability Office, 82 percent of nursing homes were cited for an infection-control deficiency (or several) between 2013 and 2017—and 48 percent were cited in multiple, consecutive years. Even the best-ranked facilities are struggling; according to an analysis by *Kaiser Health News*, around 40 percent of five-star nursing homes have recently received an infection-control citation. "Many errors are rudimentary," the article explained, "such as workers not washing their hands as they moved to the next patient."

Not even Covid has changed this. As late as March 30, inspectors from the Centers for Medicare & Medicaid Services found that more than a third of nursing homes were not following proper hand-washing protocols. This means that in the midst of a global pandemic, more than a third of nursing homes had staff who did not wash their hands properly while inspectors were watching them.

Davies thinks that Covid could not have played out any other way. According to research in the *American Journal of Infection Control,* "healthcare-associated infections"—infections acquired during a stay at a medical facility—result in almost 388,000 nursing-home deaths every year. "As a society, we have allowed these facilities to provide substandard care," Davies said. "Nobody has been outraged, and our nursing-home residents aren't well enough to speak out or to protest."

Despite all these deaths, it wasn't until November 2019 that nursing homes were even required to have an "infection preventionist" on staff, as a result of a 2016 Obama-era reform. Even then, the preventionist only had to be a part-time employee, with the precise meaning of "part-time" left undefined. Today, there is no federal guidance on what qualifications an infection preventionist must have to merit the title.

During an April 2019 inspection, inspectors found that the Life Care Center of Kirkland "failed to consistently implement an effective infection-control program." One resident was spotted in a wheelchair with her bare feet resting directly on the ground, even though one foot had a pressure ulcer so wide and deep that a layer of fat was visible on her right heel, which was emitting a "foul odor and yellowing discharge." In another instance, inspectors watched staff enter a patient's room without wearing PPE, even though the resident had a suspected respiratory infection. Life Care was not fined for either violation. This is generally true of infection-related deficiencies, which state inspectors almost always classify as low-level, "minimal-harm" offenses. Typically, Davies says, facilities are told to reeducate staff and to do better next time.

Industry defenders will emphasize that nursing homes are not hospitals. Their staff are not trained like hospital staff or paid like hospital staff, and their interactions with residents are inherently different. Nursing aides are in close physical contact with residents for hours each day: bathing them, feeding them, helping them in the bathroom. Their touch is medically necessary but also, often, loving. Staff hug residents and kiss them and make sure their hair looks nice when their kids come to visit. This isn't done with hospital-level sanitary protocol, the defenders concede, but can we expect it to be and would we want it to be?

"Nursing-home inspections tend to be very checkbox-y," said David Grabowski, a public health researcher at Harvard University. "I went back and had a look at the inspection report from Kirkland. They had everything in there from laundry to dining to patient care. It struck me that some of it wasn't all that useful in terms of really pinning down how well they did on infection control. It felt very imprecise." In fact, just a week before the first Life Care resident tested positive for Covid, inspectors visited the facility to assess a suspected tuberculosis case and found that "the infection-control facility policy was being followed."

Infection-prevention nurses are supervised by nursing-home medical directors, but the contours of this role are also fuzzy. Beyond being physicians, medical directors don't need to be certified or credentialed. There are no firm federal guidelines on how much time they must spend on the job and no obligation that they have any experience in geriatric medicine. Some work as high-paid contractors for numerous independent nursing homes, in addition to holding other clinical and hospital appointments, because there are no limits on how many patients they are allowed to oversee at any given point. When nursing homes are found to offer substandard care, medical directors are rarely cited or fined or punished.

Researchers at Harvard University and Vanderbilt University have found that three-quarters of American nursing homes were understaffed before Covid hit. By federal law, facilities must have a registered nurse inside the building for eight consecutive hours each day and a licensed practical nurse available at all times. Beyond that, they must provide "sufficient" staff—with the standards of sufficiency left largely to states or, in states that don't set minimum standards, to the companies themselves. In Washington, nursing-home residents are required to receive just 3.4 hours of staff care each day—well below the 4.1 hours that a federal report recommended back in 2001. According to payroll data, in the last quarter of 2019, the Life Care Center of Kirkland was offering 3.63 staff hours per resident per day.

In late February, when nurses in Kirkland started getting sick and staying home from work, their absences infuriated Carolyn because it reminded her of the summer before when, she said, Life

Care was so short on aides that Helen wasn't given a shower for three weeks. Carolyn said that when she showed up at Life Care's front office to complain, an administrator told her, apologetically, that she was scrambling for staff. "I don't care what you are," Carolyn said. "My mother smells. Give her a shower right now. Period." (Life Care's Nancy Butner said she was not aware of the incident and that the company routinely monitors staffing levels at its facilities.)

"Carolyn!" Helen had admonished, because she hadn't wanted to cause a fuss. "Don't be screaming all over the place."

In June and July, two peer-reviewed studies found links between low nursing-home staffing and the likelihood of a Covid outbreak. One, in the *Journal of the American Geriatrics Society,* found that every additional twenty minutes of registered nurse staffing, for each resident for each day, was associated with 22 percent fewer confirmed Covid patients. Several researchers have contested these findings. The papers, however, build on previous studies, which found that large for-profit nursing-home chains like Life Care Centers of America are more likely to be understaffed than smaller or nonprofit competitors.

For nursing-home aides themselves, the work is poorly paid: though rates vary by state, the national average is $13.38 an hour, or $22,200 annually, in most cases without benefits and little opportunity for advancement. Nearly 13 percent of nursing aides live below the poverty line, and almost 36 percent rely on some form of public assistance. "We are competing with McDonald's and Burger King for the individuals who are coming in and working for us as certified nursing aides," said Robin Dale of the Washington Health Care Association. As a result, the nursing-home workforce is fluid; minimally paid and minimally trained aides come and go, leaving residents to be cared for by a rotating army of strangers.

Even today, many nursing aides are not paid if they contract Covid on the job and go out sick. This includes workers at Life Care Centers of America, who must use accrued sick and personal days to cover their quarantines. If these run out, they might apply for workers' compensation or unemployment. "We couldn't sustain paying everyone's salary," Life Care Centers' Nancy Butner said. This continued to be company policy despite the fact that the Life Care Center of Kirkland has received nearly $919,571 in federal pandemic relief.

In July 2017, seven months after the American Health Care Association first wrote to Donald Trump, the Centers for Medicare & Medicaid Services made sweeping changes to the way that nursing homes are fined for harming and endangering their residents, in a manner that saved the industry nearly $50 million in penalties in just eighteen months. Then, in 2019, the agency proposed to go further. Its administrator, Seema Verma, promised to remove requirements on the nursing-home industry that are "unnecessary, obsolete, or excessively burdensome," and in doing so to save facilities more than $600 million a year. In a 32-page document published that July, the government referred to nursing-home regulation as "burdensome" or a "burden" more than 100 times. It also recommended, just months before Covid appeared, that existing infection-prevention measures be relaxed—that the role of the "part-time" infection preventionist be reduced.

Helen now spends her days alone in Room 10, watching CNN. She sits in her wheelchair or lies on her bed, propped up on a pillow that has two pillowcases, just as she likes it. The bed itself is an old one—it adjusts only a little in each direction—but it suits Helen's back and calms her sciatica. Sometimes, a nurse leans against the doorframe and watches the TV with her, making sure not to cross the blue line that has been taped across the floor, in front of the closet, marking the spot beyond which staff can't go without a mask and gown and plastic gloves. Once, a nurse who had really loved Twilla came into the room and stared at her empty bed for a while, looking sad.

By the end of May, Life Care finally had enough supplies to test all of its residents and staff at once, and everyone tested negative. There were thirty-eight residents left, and some found it hard to acclimate to the new pandemic way of living: cloistered and apart. Chuck Sedlacek, who lost thirty pounds during the outbreak, had awful pain in his back and knees and hands; after spending so many weeks in bed, unable to get up, he now struggled to sit in his chair—even just for a little while, if only to eat. "I want to come out," June Liu told her daughter every day, because she had dementia and forgot what she was told the day before. That month, more than eight weeks after Twilla died, the federal government finally ordered nursing homes to inform the CDC of their Covid fatalities, so that the government could at least know how many people inside facilities were dying. A month later, in June, the House of Representatives announced an investigation into the country's five largest private nursing-home companies, including Life Care, and requested information from each on its preparedness for the pandemic.

By then, Debbie still hadn't gone back to Life Care to pick up Twilla's things. A teddy bear. An old ring. If she were honest, she was a bit scared to. It wasn't that she thought her mother's old clothes and trinkets were contaminated with the virus because, after all, they'd been sitting in boxes for months and months. But . . . maybe? Really, Debbie didn't want to be anywhere near a nursing home. She'd read all the ghastly stories. There was that nursing home in New Jersey where police officers, responding to an anonymous tip, found seventeen bodies stored in a room. (Andover Subacute and Rehabilitation Center II was the largest nursing home in the state with 543 beds. It is owned by affiliated entities of a Chicago-based firm called Altitude Investments, which leases it to an operations company called Alliance Healthcare, whose owner New York health inspectors have previously denied a license on the grounds of "character and competence.") There was the nursing home in Virginia where more than fifty people died. (The Canterbury Rehabilitation & Healthcare Center was purchased last year by Marquis Health Services, a subsidiary of a private-equity investment group called Tryko Partners.) All those stories made Debbie think about how she might end up in a nursing home one day, especially if she didn't get her knee fixed and get her weight under control. If she did end up in a facility, she thought, she would die badly. It was sort of inevitable.

"A lot of this gets into the larger cultural narrative about nursing homes," said Tim Killian, the Life Care spokesperson. He believes that Life Care staff did everything right and everything they could—and that, despite this, the facility has become a kind of metaphor and so a target for all the animus that Americans feel about aging and dying and nursing homes. But a nursing home isn't a metaphor. In June, representatives of Life Care appeared before the Department of Health and Human Services to appeal its fine. Chelsey testified and couldn't stop crying. After nearly three months in Kirkland, she went back to her original Life Care nursing home, which then had its own Covid outbreak. "I'm really sick of it, to be honest," she said.

When and if Debbie's lawsuit moves forward will depend, in part, on whether Washington follows at least twenty other states in granting nursing homes immunity from most lawsuits during the pandemic. Those immunity provisions were passed after weeks of campaigning by health-care lobby groups, including the American Health Care Association, which are also advocating for a federal immunity statute. Senate Majority Leader Mitch McConnell, for one, has vowed not "to let health-care heroes emerge from this crisis facing a tidal wave of medical malpractice lawsuits so that trial lawyers can line their pockets." (Life Care declined to comment on Debbie's lawsuit.)

As states decide how to proceed, the American Health Care Association has let its members know that it will not be chastened. In a June letter titled "WE WON'T BACK DOWN," the organization's CEO, Mark Parkinson, wrote: "Rather than recognizing that long term care providers were helpless to identify pre-symptomatic carriers who were spreading the virus, we have been blamed." Parkinson announced "an historic media campaign to fight back," including $15 million for social media and cable TV ads in Washington, D.C. In July, the association asked for an additional $100 billion in federal aid. "There's no question that some money is needed, but it is critical that there is accountability to that money," says Dr. Wasserman of the California Association of Long-Term Care Medicine. But, already, the federal government has promised Covid relief with few questions and "no strings attached."

As the virus continues to spread, some advocates have looked for solutions to the larger problem of the American nursing home. This is, maybe, an attempt at the did-it-have-to-be-this-way? searching that often follows tragedies—only this time, in real time, because nursing-home residents are still dying. Some reformers have faith in a design fix. They think things will get better if nursing homes are made smaller, or cleaner, or homier, or more compartmentalized, with private rooms instead of double and triple and quadruple ones. On the other end, nursing-home abolitionists are making the case that long-term care facilities have failed in their most basic duties and so should be shut down. Some advocate a slow deinstitutionalization, through increased funding to home-based and community-based care. Others want nursing homes emptied now.

Some are less optimistic about the promise of a fix. They see, in all the tens of thousands of nursing-home deaths, signs of a deeper cultural abdication: something that transcends any mistakes made by Life Care, or the nursing-home industry, or its regulators. According to a survey by the AARP, the vast majority of Americans over fifty want to age at home. Other surveys and studies, some more scientific than others, have found that many people would rather die than live in a facility. Some nursing homes are nice, and others are not nice, and some nursing aides are kind, and some are not, but either way, many nursing-home residents feel, as the geriatrician and writer Dr. Louise Aronson writes, that they are "in prison for the 'crime' of growing old and frail." And still, we allow nursing homes to be built. And still, we put people there and imagine that we could never end up there, and then we ourselves end up there.

Of course, this isn't just about nursing homes. Covid and our response to it have revealed something rotten in modern medicine. Look anywhere and there is proof of ageism. Hearing loss is one of the most common symptoms of senescence, and still Medicare and private insurance won't pay for hearing aids, even though they cost, on average, $2,300 an ear—and even if they would help keep a person living independently and living vividly and, in this moment, would allow them to speak with quarantined family members over Zoom. Those same patients already struggle to find specialist physicians, since few new doctors choose geriatrics. By some estimates, the United States will need 33,000 geriatricians by 2025. Today, there are just 7,000, with only half of them practicing full-time.

Or look at medical research. The National Institutes of Health has required that clinical trials include women and people of diverse ethnic backgrounds since the 1980s—because drugs and their remedies may affect different demographics differently—but it didn't issue a similar directive for older people until recently. Even now, dozens of Covid-19 drug and vaccine studies are excluding participants over eighty, or seventy-five, or even in some cases over sixty-five, raising the possibility that a vaccine developed to stop a disease that disproportionately affects older people may not be proven safe for them to use.

As the country inches forward (and then back and then forward again) toward reopening, the interests of older Americans continue to be held apart, cordoned off, quarantined. Our solution to Covid, however temporary and desperate, has been to sequester away the old people and try our best to carry on. This part of the story is bigger than nursing homes—bigger, even, than medicine —and maybe most clearly encapsulated in that refrain from the earliest days of the outbreak: it *only* affects old people. Decades from now, will we be haunted by that "only"?

Helen says she isn't scared. But then again, she won't leave her room. Visitors are still barred from visiting Life Care, but residents are now allowed to walk and wheel the hallways, provided they wear masks. One woman does laps around the internal courtyard with her walker. Another, who has dementia, asks nurses where her suitcase is, and how she will get to the airport, and when her mother is coming to collect her. Helen, though, stays inside Room 10.

"Mom, I have not laid eyes on you since February," Carolyn tells her. "I think it's time you come out of your room, come down the hall, come to the big foyer where it's all glass. I can see you through the window and wave to you. I think it's time you do that."

"I'll let you know when I'm ready," Helen tells her.

JULIA CRAVEN

It's Not Too Late to Save Black Lives

FROM *Slate*

MARIA HAS NOT kissed or hugged her children since March 28. It was around then that she started to develop shortness of breath, chest pressure, and a scratchy throat. Maybe, she thought, it was strep, so she ventured to a local urgent care to get tested.

It came back negative.

Maria went home, where her chest pressure persisted into the night. It was too much, and she went to the emergency room at the hospital in metro Detroit where she works in nursing. There, she was told to assume that she had contracted SARS-CoV-2, the virus that causes Covid-19, and to go home. She burst into tears. "I told [the doctor] I have a young baby at home," she said. "I need to know what my status is, if I should be quarantining. I need to know what I need to do. So they swabbed me, just to be nice, I guess." (Maria asked that her name be changed because of fear of employer retaliation.)

The swab grazed the inside of Maria's nose. She told the clinician administering the test that she'd heard this procedure was painful. This didn't hurt. She questioned if she was being tested properly. But the clinician reassured Maria that everything was fine.

Four days later, her results came back negative, which was enough for her employer to insist she come back to work. Within two weeks, Maria's shortness of breath had disappeared, but she started to experience diarrhea, nausea, vomiting, and chills in its place. Soon, she lost her ability to smell. But, still, no fever. So her employer wasn't hearing her out when she asked for time off.

Six days after the loss of smell presented itself, a friend of Maria's encouraged her to visit her local health department for a test. It was a completely different experience. The test used to identify the presence of the coronavirus is unpleasant at best. A six-inch swab is inserted deep into the cavity between the nose and mouth where it is rotated several times for about fifteen seconds. This time, Maria said, it was distressing.

On April 18, her test came back positive.

Maria's situation—facing resistance and uncertainty about access to care and testing, even while working an essential and risky job—is a study in how the Black community is being pummeled by the coronavirus. Data from across the country show that Black Americans are more likely to contract Covid-19 and are disproportionately dying, often younger, from complications of the virus. In conversations with *Slate*, multiple Black women described their frustrations about them, or someone they loved, not being able to receive a test for Covid-19 even though they were frightfully sick. One woman lost her stepmother. Another was almost discouraged by a paramedic from going to the hospital even though she was gasping for breath.

The medical catastrophe of the pandemic has met the slow-motion disaster of everyday health disparities. Inadequate testing has been a marquee issue throughout the crisis, and that failure is exacerbated in Black neighborhoods where a lack of testing sites is inflamed by residential segregation. Black folks are overrepresented among populations that cannot practice proper physical distancing, that live in densely populated areas or multigenerational homes, work an "essential" job, suffer from food apartheid, or have an illness that worsens outcomes should they contract the coronavirus. They are also more likely to be poor, which can further cause disparate health outcomes.

But this doesn't have to be fatal. There are initiatives that could be taken up on a local, state, and federal level to mitigate the impact and increase the possibility that people who are infected survive. Several experts mentioned four planks that could, and should, be addressed: reducing exposure, adequate testing and contract tracing, getting reliable information out to Black communities and combating the flow of medical advice through the grapevine, and forming a coordinated response to address preexisting inequities. The first two are the most amenable to policy and could be accelerated initiatives.

"We live shorter, sicker lives, and we die prematurely," said Shawnita Sealy-Jefferson, a social epidemiologist at Ohio State University. "That is tied to the history of this group in this country."

Black folks have been in this country for just over 400 years and more than 85 percent of that history came before the civil rights era, setting the patterns for the poor health and inadequate medical care experienced by Black Americans. Black patients are often underprescribed pain medication or not offered an opportunity to join experimental drug trials that could aid in managing their illness. Or, like Maria, they're refused certain medical procedures or more likely to be candidates for destructive ones.

Individual Black patients, generally, may not be able to concretely prove that their concerns were dismissed because of their race, but they often have a hunch that they weren't treated the way they should have been following an experience with a physician.

If something were done about this, Sealy-Jefferson said, the impact of the coronavirus could potentially be mitigated.

"If we are serious about decreasing the disparities by race and ethnicity and class and multiple systems of oppression, if we are committed to doing something about the ways in which these systems have been set up to disproportionately impact communities, we have to do something about racism. That is the fundamental cause of the disparities in health that we've seen."

From the beginning, Sophia Caldwell's experience at Mizell Memorial Hospital in Opp, Alabama, was unsettling. She took the seven-minute trip to the emergency room from her home in Elba early on the morning of April 23. She had chills and a high fever—another worried addition to the wheezing and shortness of breath that began the day before, a stark progression from what had been a little cough on April 21.

Caldwell was initially treated outside in the parking lot, which didn't sit right with her, before being wheeled into a narrow hospital hallway for chest X-rays. The doctors and the technician barely acknowledged her, she said, much less explained what they thought was going on. They simply looked at her scans, said they'd seen all they needed to see, and sent her back to her car.

"Neither one said anything was in my lungs," said Caldwell. "He sent me back to my car and then he told me that I had bronchitis. He said, 'Step back outside and then we'll discuss some things.'"

Mizell has not responded to requests for comment.

Caldwell, a certified nursing assistant of seventeen years who has worked in nursing homes, was insistent that whatever she was feeling wasn't bronchitis. "I told my husband that I wasn't crazy. I said 'This is not bronchitis. I've got all the classic symptoms of Covid,'" she said. "I know what bronchitis does."

The clinicians didn't listen. Caldwell wasn't tested for the coronavirus, despite her symptoms and tests for flu and strep coming back negative. She was prescribed Norco to treat the cough and told to get some Delsym to suppress it. The doctor also wrote her a prescription for a Z-Pak. She took a double dose from the Z-Pak while sitting in the hospital parking lot, and then she took another double when she got her prescription filled later that day.

But her symptoms worsened. On Sunday morning, she woke up shaky. "My legs didn't want to carry me," said Caldwell. She went to an urgent care clinic in Troy, forty-five minutes away, bypassing Mizell since she didn't want to risk being dismissed a second time. On the ride over, Caldwell was terrified. Not once was she able to catch her breath.

"I honestly think if I had gone to bed on the twenty-sixth," she said, "I would have been dead on the twenty-seventh."

Her experience in Troy was completely different. She was allowed into the urgent care clinic, and the staff didn't act as if they were afraid to touch her. Caldwell learned that her airways were indeed restricted, and doctors asked if she wanted to be tested for the coronavirus. She begged them to do so.

Her test came back positive two days later on April 28.

"When I got there, urgent care treated me like somebody," said Caldwell. "You treated like just something horrible down at Opp. I can understand their fear, but they gotta understand that we're the sick ones. We need to be made to feel comfortable too."

During the 1700s and 1800s, enslavers did anything they could to put off allowing a physician to assess the health of an enslaved African. Often, they accused sick people of "malingering," or feigning an illness in order to get out of work. But in the event that a doctor was paged, they operated as enthusiastic agents of the systemic mistreatment of the enslaved. In *Medical Apartheid,* Harriet Washington recounts an instance where a physician began his evaluation of an enslaved man with "remembering simulation was a characteristic of his race." Doctors, Washington explained, would share notes detailing ploys used to get enslaved Africans back into the fields as quickly as possible. Violent medical tactics or blatant physical abuse were cited as most effective.

Malingering became a maxim of medical racism, and the clinical experimentation to which enslaved Africans were subjected aided in building the institution of Western medicine. The legacy of that abuse is a primary cause of iatrophobia, the fear and disdain many Black folks still have toward medical institutions, according to Washington. Poor health outcomes were treated as a deficiency inherent to Black people instead of the result of economic disparity, poor housing, abhorrent work conditions, and extreme levels of stress. Clinical beliefs formulated during enslavement are why physicians still believe that Black patients are more difficult to deal with, feel less pain, have thicker skin, or exaggerate symptoms— beliefs that provided a convenient excuse for the continuation of enslavement. Political institutions have allowed for the severe socioeconomic gaps between Black Americans and their white peers to persist so Black populations remain at higher risk of being in poor health and increased risk of mortality.

During the coronavirus pandemic, these factors collided and resulted in a painfully high infection rate: 27 percent of cases based on incomplete data from the Centers for Disease Control and Prevention, within a demographic that makes up just over 13 percent of the U.S. population.

Sealy-Jefferson, the social epidemiologist, stressed the importance of telling the truth about this history in order to put forth any viable solutions. "It's not about behavior, it's not about some genetic susceptibility to poor health and mortality," she said. "A lot of published research, even in 2020, is based in eugenics, this false narrative that genetics predispose people to poor health."

A number of stories that focus on the disproportionate number of Black folks contracting Covid-19 have leaned into the notion that there's something wrong with Black people—whether it be admonishing them about food choices or not properly distancing. U.S. Surgeon General Jerome Adams chided Black Americans, and others of color, singling them out in a call to "avoid alcohol, tobacco, and drugs" for their "abuela," "granddaddy," "Big Mama," and "Pop-Pop."

Not as much energy has been dedicated to addressing white Americans or the root cause of any health disparities.

"If we don't tell the truth about how racism is structuring our society and how racism is privileging some people and disadvantages other people, we aren't going to make any headway into mitigating disparity," said Sealy-Jefferson. "Health providers who took an oath to do no harm, should take that seriously and understand all of the ways the implicit biases are playing out in their care."

Around the time she got sick, Maria's hospital started to receive its first Covid-positive patients. It was early in the upward swing of Michigan's infections and the hospital was still tinkering with the procedures. Maria said she and her colleagues were not wearing proper personal protective equipment. They were given surgical masks instead of the N95s the CDC recommends when caring for patients infected with the coronavirus, which were being reused for six days then cleaned under a UV light. Maria and others started questioning whether staff should be masking in the hallways but were discouraged by administrators.

"We were just kind of flying by the seat of our pants," she said.

Caldwell thinks she contracted the coronavirus while providing in-home care for an elderly woman whose son was a truck driver. She didn't think to take any extra precautions since she and another person who cared for the elderly patient were already practicing physical distancing in order to protect the woman receiving care. Soon, the woman developed a nasty cough and high fever, and she eventually passed away. It was not confirmed if she had contracted Covid-19, but Caldwell said her co-worker also tested positive around this time.

Providing essential workers—anyone from health care professionals to janitors or in-home care providers—with adequate personal protective equipment as a condition of being at work would be a great start toward tempering the effects of the pandemic, said Nancy Krieger, a social epidemiologist from Harvard. Workers, particularly those outside of the health services field, should also be trained on how to properly use and dispose of it. This will help decrease exposure in the workplace.

Further interruption of an outbreak lies in identifying people who are infected and preventing them from spreading it. This is more complicated, but it is amenable to policy. Isolation wards can be set up for those who live in crowded households. For example, officials in New York, Chicago, Seattle, and California have leased hotel rooms for Covid-19 patients who cannot physically isolate themselves at home.

"The thing that can absolutely be mitigated is the risk of exposure," said Krieger. "It's going to get trickier once people are exposed. There may be differentials in the mortality because of preexisting health injustices and that's not going to be as simple. That's why the primary prevention is key."

A lack of testing, however, could pose a problem to the efforts to contain exposure. But right now officials in states that have not been overwhelmed by the coronavirus have an opportunity to acquire enough tests and prioritize who gets tested based on where their disparities are. "They can look at who's died from Covid, who's had severe cases of Covid, and you can map that out based on geography," explained Dr. Lisa Fitzpatrick, a medical epidemiologist and infectious-disease physician from George Washington University. "Where do these people live, where have they been, where did they work?"

People who live in congregate settings like nursing homes, incarcerated populations, people who live in public housing, and others who are unable to physically distance should be at the top of the list when it comes to providing robust access to testing. Once positive cases are determined, officials can begin contact tracing and prompt anyone who's been in contact with someone infected to get tested.

"If you send somebody home who's tested positive, you've basically ensured that at least some of the family members are also going to get sick," said Maureen Miller, an infectious-disease epidemiologist at Columbia University, reiterating the need for more tests and feasible isolation initiatives.

It's been nearly eight weeks since Maria's children have seen her bare, unmasked face. At times, her oldest child will run in the opposite direction if she's seen her mother from an angle where she doesn't appear to be wearing it. For the first seven days, she quarantined in her basement, away from her husband and children.

But she hasn't been able to get retested, and she isn't willing to risk the health of her family. "I was just fearful for every contact that I had with my own children," she said. "I've heard stories of people that have been positive for forty-five to fifty days. And you don't know if they're still contagious or just shedding viral particles, but still I can't take the chance."

Caldwell has been left with a nagging cough. She's no longer short of breath and her legs have quit shaking. The chills and fever are gone, but she still feels off. "I can tell that I'm better," she said. "But I can also tell that I'm not well—if that makes any sense."

She was able to get retested in early May, but her results came back positive. Her third test was originally scheduled for Monday, but she pushed it back until Friday. She wanted to be absolutely certain the virus was out of her system, and since the test for Covid-19 is so painful, she doesn't want to come back in for another. Caldwell was angry because she has family members who could have been infected. An urgent care doctor in Troy told her to act as if she was infected—even though her initial results hadn't come back in. She has isolated herself from her family, keeps her mask on, and sprays anything she touches with disinfectant. Doctors at Mizell never told her to assume she was infected in order to protect her family, just in case.

"And that bugs me because I could have hurt somebody that I care about," she said, her voice cracking. "I just think about how I could have hurt somebody that means a lot to me because they were too stupid to even offer the freaking test. I've been an emotional mess through this whole thing."

"I just thank God I feel better," she continued. "I'm just ready for it to be out of my system and I know God is going to heal me because it could have been so much worse. My children could have been with me and I'm just—I just thank God I'm still here."

BROOKE JARVIS

The Scramble to Pluck 24 Billion
Cherries in Eight Weeks

FROM *The New York Times Magazine*

CONSIDER THIS CHERRY, actually, this one here, hanging off the tree at the very end of a long, deep green row. Look at how its red and gold skin shines in the bright sun. It's a famous hybrid variety, a Rainier, which means it has sweet yellow flesh and that you'll have to pay a premium price to eat it. If you do, it will be delicious, the very taste of summer. But first it will have to get to you.

So far, this cherry has been mostly lucky. No disease has come for its tree, though there's a bad one, little-cherry disease, stalking nearby orchards. No frost kept its springtime blossoms from giving way to fruit. No excessive rain has fallen in the short time since it ripened.

That could have been a disaster, because water likes to pool in the little divot by the stem. There it seeps into the flesh, making the cherry swell. Too much, and the cherry will burst through its own skin, causing splits; whole harvests can be lost this way. So dangerous is poorly timed water that cherry growers rely on fans, wind machines, and even low-flying helicopters to dry ripe fruit before it is lost. Yet wind presents its own peril: it can knock cherries against one another or into branches, bruising them so that they're rejected on the packing line, where fruit is sorted for size and quality with high-tech optical scanners. Rainiers, because of their color, are particularly prone to showing their past with telltale "wind marks," tiny incursions of brownness on that golden skin. This cherry has just a few.

But it's not to market yet. The window in which a sweet cherry can be picked for sale is excruciatingly narrow. Cherries don't continue to ripen once they're off the tree, the way a peach does, and once picked they don't store for very long, even when refrigerated. If they're too ripe, they won't make it to the packing house, the truck or the airplane, the grocery-store display, your summery dessert. The sugar content must be Goldilocksian—neither too high nor too low. Wait even a couple of days too many, and it may be too late.

Paige Hake, the second generation of her family to farm this orchard, considered the cherry. Then she considered its neighbors, with their own wind marks, in the lambent heat of a June afternoon. She looked down the long green row of trees, lined with its strip of white plastic fabric, meant to reflect sunlight onto the undersides of the cherries, helping them color evenly. She consulted with her father, Orlin Knutson, who has been growing fruit on this stretch of dry sagebrush steppe near Mattawa, Washington, for forty-one years, the last thirty-one of them organically. There was a refrigerated truck waiting by the gate, with a growing stack of full bins next to it. There was rain in the forecast, as well as more heat, and sugar levels in the cherries were rising as they spoke. They wanted to get these cherries harvested today; they were far enough along that it was probably now or never, a whole year of investment and work leading to this one afternoon. But it was getting late, and there were a lot of other cherries that needed to be picked, and today the crew of people available to pick them was smaller than they would have liked. She turned to me and pointed to the wind-marked cherry, still unsure whether it would be worth the cost of trying to get it to market. "Would you buy that at Whole Foods?" she asked.

The yellow cherry was one of a great many across the orchards of Washington State that were just beginning to ripen. Karen Lewis, who works with growers as a tree-fruit specialist for the agricultural extension service of Washington State University, has tried to calculate exactly how many individual cherries need to be picked during a whirlwind season that Jon DeVaney, the president of the Washington State Tree Fruit Association, calls "eight weeks of craziness." Multiplying all the millions of boxes by the number of cherries they can hold, Lewis determined that as many as 24 billion individual cherries must be plucked, separately, from their trees and placed carefully into bags and buckets and bins, each and every one of them by human hands.

Lewis thinks that people who aren't used to thinking much about the source of their food, or who assume that the food system is as mechanized and smoothly calibrated as a factory, spitting out produce like so many sticks of gum, ought to spend some time contemplating that figure and what it means. "I'm here to tell you that people do not think we harvest everything by hand," she says. But hands, belonging to highly skilled workers, are needed for every last cherry. During the harvest, many thousands of people are out picking by dawn, nearly every day, their fingers flying as they watch out for rattlesnakes under dark trees. (Compounding the labor crunch, this is also the time when workers in the region must hand-thin more than 100 million apple trees, so that the remaining fruit can grow larger.) Later in the season, many of the same hands will pick and place each peach and plum and apricot, every single apple—five and a half billion pounds, just of apples, just in Washington, just last year. "I think those numbers are staggering," Lewis said.

The cherry industry has done everything it can to squeeze every possible bit of extra time into the season. Growers plant at a range of different elevations: every 100 feet above sea level, one orchard manager says, buys you an extra day until maturity. And they choose different varietals that ripen at slightly different speeds—most red cherries are marketed to the public simply as "dark sweets" but are actually a genetically distinct array, whose different sizes and tastes and unique horticultural personalities are intimately known by growers and pickers. If everything bloomed and matured all at once, Lewis said, there's no way there would be enough bees, enough trucks, enough bins, to make the scale of the current cherry harvest possible. Most of all, there wouldn't be enough people. There already aren't.

For years, the tree-fruit industry in Washington—like the salad industry in California, the blueberry industry in New Jersey, the tomato industry in Florida, and countless other sources of the things that we eat—has been struggling to find the workers it needs to keep producing food. Across the country, the number of farmworkers is dwindling. Current workers, who are often immigrants without legal permission to work in the industries that are reliant on them, are getting older; those who are able to are leaving an industry that's poorly paid and physically damaging and often exploitative; and crackdowns at the border mean that there are fewer new arrivals to take their place. To cope, some growers have turned to a ballooning visa-based "guest-worker" program, which comes with its own significant problems, while many others have simply buckled under debt and rising costs, going under or selling their orchards to ever-bigger companies. "Everyone's squeezed pretty much to the limit," Knutson said, surveying the dark leaves, the shining fruit, the clear blue sky. "It's kind of an ugly time."

Such was the state of things before the coronavirus pandemic arrived, bringing with it a host of new troubles. When I called Lewis early in this year's cherry harvest, she had just sent out a newsletter that, along with the latest updates on cherry disease and apple varieties, included information on suicide prevention. Piled on top of everything else, she said, "this is enough to take people to their knees."

In March, when the United States began to lock down to slow the spread of the new virus, some workers noticed a change in how the government talked about them. As leaders planned for closures, it became clear that many of the lowest-paid and least-respected jobs in America were, in fact, the most important: the ones that could not be paused or interrupted or bypassed if society was to keep functioning. You could not, as Knutson put it, simply close the door to a farm for a month and then reopen it. People who had regularly been called illegal suddenly found themselves rebranded as essential.

Harvest seasons were under way or rapidly approaching across the country; without enough workers, the nation's food would not be produced. Immigration and Customs Enforcement announced that it would "temporarily adjust its enforcement posture," narrowing its focus to people involved in criminal activity rather than arresting anyone who was undocumented. In California, where labor-intensive fruit-and-vegetable crops account for about 85 percent of the state's crop sales, farmers handed out letters that workers who feared attracting the attention of law enforcement by going to work during lockdowns could carry with them: not papers by the usual definition, but a paper to show that they were, informally, and just for now, legitimate by virtue of being indispensable.

In Sunnyside, a city of about 17,000 people in the heart of the fruit belt that follows the Yakima River across south-central Washington, this change in perspective felt belated and insufficient. Israel S., a father of four who has worked seasonal jobs in fruit and hops for the last decade, and who asked that his last name not be used because he is undocumented, had a trove of videos ready to pull up on his phone. One was a news report, in Spanish, about crops spoiling in Alabama after a harsh anti-immigrant law went into effect in 2011 and undocumented workers fled the state. (A similar law in Georgia that year may have cost as much as $391 million in unpicked crops, according to one University of Georgia study, even after the state tried to fill the gap with prisoners.) Another video showed endless rows of ice-covered Pink Lady apples, frozen in place on their branches before they could be picked, while mournful piano music played. (It reminded Israel of the previous fall, when there had been so few apple pickers that he sometimes stayed out working until dark, the apples hard and frigid in his hands.) Other farmworkers he knew shared the videos on Facebook, and Israel understood why. They knew their own importance, even if much of the country did not.

It was nearing dinner time, and Israel was in the kitchen of a rented house with his wife, Guadalupe, who was cooking tortillas on a hot plate, and their eighteen-year-old daughter, Nayeli, who was stretching tired arms. Israel and Nayeli woke up at 2:30 that morning so that they could drive an hour and twenty minutes to a cherry orchard near the Oregon border, arriving well before the sun crested the hills to begin ten hours of picking. They had done the same every day for ten days, and would do the same the next day, and again and again, for weeks, until Israel's eyes started to droop as he drove. Each morning, Guadalupe would be up even earlier. When schools and day care centers were open, she would join her husband in the fields, but with them closed, the cost of a babysitter for the younger children would negate most of her day's wages. Still, she got up every day to make fresh tortillas to pack for her husband and daughter's lunches. Nayeli, she teased, didn't like the store-bought kind.

The family was worried about getting sick—other than work, they went out only to pick up food, and a large bottle of hand sanitizer took pride of place in the middle of the plastic-covered kitchen table—but they were also ineligible for stimulus checks or unemployment benefits. There had been no question that those who were able to do so would keep working. There was no question for many people living and working in the valley, with its orchards and vineyards and fruit-packing houses and dairies and meatpacking plants. The virus first spread in more populous and affluent Seattle, on the other side of the mountains, but a lockdown there brought cases down quickly. Here, in Yakima County, the curve of the virus never really flattened; outbreaks spread in meatpacking plants, which the Trump administration prohibited from closing, and the warehouses where workers pack fruit for shipment and sale. (Workers at seven packing houses went on strike to demand more safety precautions and hazard pay, and at least one of the strikers, David Cruz, died of the virus.) In June, cases were rising faster than anywhere else in the state: though the county was home to just over 3 percent of Washington's population, it would by the next month have 20 percent of its cases. "If you stay home," Israel said, "there is no money for rent."

Israel and Guadalupe are both from Michoacán, one of the poorest states in Mexico. They used to live in New Mexico, then California, then Oregon. Israel worked in restaurants and construction until jobs disappeared in the 2008 recession and friends told him there was opportunity in the fields of Washington. By driving long distances, he can find work most of the year: trimming, thinning, trellising, harvesting. But he wouldn't call it opportunity, exactly. Because of the pressure to work quickly, both he and Guadalupe had been injured falling from ladders and now lived with chronic pain: Israel in his shoulder, Guadalupe in her back. "Supposedly it was better here," he said, "but it's not. There's more work for less money."

Israel was still wearing his work clothes—he was careful not to wash them with his children's clothes, because of the pesticides, which burned his throat—and there was cherry juice smashed into his pants. He'd brought home a large bucket of the fruit that he and Nayeli picked that day, a dark red variety called Coral, which was a nice kind to pick because the cherries are large and fill buckets a bit faster. For each bucket they filled, climbing up and down a ladder they carried from tree to tree, they earned about $3.75. Cherry work is sometimes compared to a casino: if the trees are full and there are few split cherries to pick around, the money can be unusually high for fieldwork, leading people to travel from

California and elsewhere just for cherry season. But, Israel said, if trees "don't have much, you're just walking around for hours." Today had been a good day. Not counting the three hours of driving or the cost of gas, he and Nayeli estimated that they each averaged $20 an hour. "You have to go very quickly to earn that," Israel said.

Earlier, wanting to show how hard it was to stretch that money, Guadalupe gave me a tour of the house, the best they could find within their budget. "First, here," she said, pointing down at rough cement as I stepped through the front door, "there is no floor." Israel had pulled up the carpet, so full of rat excrement that the children were getting sick. He'd made other repairs too, but there was much more that needed fixing: a leaking roof, with mold visible in the ceilings and windows; no working electricity in the back of the house; a shower, kitchen sink, and toilet that all drained directly into a puddle in the yard. It cost a thousand dollars a month, plus utilities, and the landlady, who wanted to move into the house herself, was threatening to have the sheriff evict them. Guadalupe concluded the tour in the unlit back hallway. "We are essential, but we are in the shadows," she said. "No one sees us."

She dished out bowls of rice and stew for dinner. The family joked about the familiar indignities of different crops they had worked—the way hops can make your clothes smell like marijuana; how onion cutting requires a dangerous knife "as sharp as a mother-in-law's tongue" and can leave a smell in fabric that lasts a full year; the time a friend of Israel's was so exhausted during the cherry harvest that he fell asleep while picking and knocked his head on a tree branch. Nayeli made fun of the ancient boombox she and her dad sometimes took to the orchards, saying it was nearly the size of an air-conditioner yet still struggled to compete with the music of other workers. "In the orchard, you don't play English music," she said. "The language is Spanish."

Nayeli, who gave a speech at her virtual graduation from high school the week before, had her first job picking cherries, after school and on weekends, when she was nine. But she had no plans to keep following her parents' path. She had won a scholarship from the University of Washington and was saving her income from the cherry season to cover the cost of moving away.

Another steppe, another orchard. This one, between the Snake and Columbia Rivers, held 210 acres of cherries, 220 of apples, and 40 of peaches and nectarines. It was the first day of the year's second harvest. Paul Carter, who manages the orchard for the large fruit company Stemilt, surveyed the morning shadows, the sun on the trees. "We've invested, we've pruned, we've irrigated, and we've not got one dime back," he said. "Everything's out there," unpicked and vulnerable.

Carter remembered when he was young and his parents and grandparents used to come to Washington from their home in Arkansas to pick fruit, like lots of people did back then. These days, he added, "I'm the only white boy around here." He was also one of the few people in the orchard who lived locally. Today there were nine pickers who lived in nearby towns, though all came by way of other places. (One woman descended a ladder to explain that she lived in Pasco, but wasn't very comfortable being interviewed in Spanish—she was originally from Guatemala, and her native language was Mam.) Forty-five others had come from Mexico just for the season, to live and work at the orchards on temporary agricultural visas.

About ten years ago, the agricultural labor crisis in Washington became so acute that Lewis compared it to a pressure cooker: "We were about to burst." Growers began turning to a little-used program that grants temporary work permits, known as H-2A visas, to foreign farmworkers. A present-day version of the midcentury Bracero program, for years it was used to bring Mexican fieldworkers to California, South American shepherds to the West, and Caribbean sugar-cane workers to Florida. Washington State is now the second-largest user of the system, which has also expanded rapidly across the country: last year, the State Department issued 202,025 H-2A visas, compared with 31,892 in 2005. In order to employ workers on H-2A visas, farms are required to advertise for workers locally and come up short; they must also provide housing, transport, and a minimum wage to guest workers and local ones alike, so as not to depress wages. These costs mean that growers view H-2A as a last resort, but one that they are turning to more and more. "It's horribly expensive," said Carter, as masked workers picked another block of Rainiers, nestling them carefully into small, protective bins. But the alternative, he continued, is "pure disaster. You're just on the prayer system" at a time when "a farmer's whole destiny is in his ability to get a whole bunch of people."

In March, as the virus continued to spread around the world, the State Department announced that it would suspend visa processing, effectively shutting off the flow of seasonal agricultural workers coming from Mexico. Industry groups and Sonny Perdue, the secretary of agriculture, pushed back with such urgency that the decision was quickly reversed; the State Department announced that the H-2A program would be continued as "a national security priority." It's meaningful, said Daniel Costa, a lawyer and immigration expert at the Economic Policy Institute, that the H-2A system is one of the only parts of the immigration system that hasn't shut down because of the pandemic: "The government has moved heaven and earth to make sure companies can keep employing these workers." Even though the visas were reopened, said Kristin Kershaw Snapp, the director of corporate affairs for Domex Superfresh Growers, another major Washington fruit company, the threat that they might not be was enough to realize how precariously the system was propped up. "That was a very scary twelve hours," she said.

In light of the pandemic, Washington's governor, Jay Inslee, convened committees to draft new safety requirements for agricultural work sites. Social distancing was a particularly fraught issue for H-2A workers, who usually travel in employer-provided vans and live close together in shared housing or hotels. (Other cherry workers are often squeezed into tents, which the cherry industry, in an effort to forestall even worse conditions, has been given unique permission to use for housing.) Unexpected fragility was everywhere: if the state, for example, banned the use of bunk beds, as Oregon had, some farms could lose half of their emergency workforce. In the end, Washington allowed bunk beds, though only if workers were organized into "cohorts" of fifteen, kept separate from other groups, and slept head-to-toe to create more distance between their faces. The farmworkers' union Familias Unidas por La Justicia filed a lawsuit, arguing that the rules were a product of industry pressures and not supported by best practices in public health. But the court, which didn't find that the state had acted in an arbitrary or capricious manner, upheld the decision. In May, Lauren Jenks, an assistant secretary for the state health department, told the *Los Angeles Times* that the cohort system wouldn't remove all risk. "This is just sort of a terrible math right now," she said.

In the Stemilt orchard, a young man from Oaxaca, working on

his first H-2A visa, told me the work is "marvelous," the pay far more than he could make at home. When the crew stopped for "lunch"—it was 9 a.m., a sign of how early they started—a fifty-seven-year-old from Michoacán, working his twelfth annual contract, said that he was glad for wages that had helped pay for surgery for his son and that he had grown used to being able to see his family only part of the year. When he first started coming to the United States, he often worked alongside local workers, but now he mostly sees other guest workers: "We were few, and with time it's been growing and growing and growing." (A month later, in July, a group of H-2A workers who worked for Stemilt in 2017 sued the company for using what they said was an illegal production quota system, threatening workers with termination if they failed to pick enough apples in a day. A Stemilt spokesman declined to comment, citing unresolved legal proceedings.)

Dan Fazio, a co-founder of the Washington Farm Labor Association, which helps growers bring in H-2A workers, sees the program as a good start toward a legal, well-regulated workforce, though he adds that the costs and standards can be burdensome, especially for smaller-scale growers; he would like to see costs for housing passed on to workers, instead. When this year's extra, last-minute requirements were added, he said, "a lot of people threw up their hands and said, 'I can't do it.'"

Some worker advocates, by contrast, call H-2A a modern system of indentured servitude. Workers often enter the program heavily in debt, because of travel costs (employers are required to reimburse these, but a recent survey of workers found they often don't) and because of the exorbitant fees they pay to recruiters in their home country. Upon arrival, they are dependent on their employers for their right to be in the United States, which, coupled with the debt, makes it difficult for them to stand up to unsafe working conditions, wage theft, or retaliation. A few years ago when H-2A workers at a blueberry farm in northern Washington called for a work stoppage to protest conditions after a twenty-eight-year-old worker fell ill on the job and then died, dozens of them were fired and then deported.

"They're seen as loyal," Costa said, "but it's because they have no other options." This spring, as companies rolled out temporary "hero" pay for frontline workers, and the Trump administration offered billions in aid to farmers and ranchers, it suggested the opposite—a pay cut—for H-2A workers. (So far, there has been no cut.) Florida's governor, Ron DeSantis, blamed his state's skyrocketing cases on "overwhelmingly Hispanic" farmworkers, not stopping to question why they might be unable to stay home or what would happen to the food system if they did.

As workers kept Stemilt's orchards going in April and May, the company used proactive testing and quarantines to contain an outbreak of asymptomatic cases at its H-2A housing; there were no reported deaths. But when harvest season arrived, cases spread at other orchards. In late July, after the virus broke out among workers at Gebbers Farms, one of the state's largest cherry and apple growers and an employer of more than 3,000 H-2A workers, the state began an investigation into conditions at its labor housing, where bunk beds were still in use despite the farm keeping workers in cohorts of forty-two instead of fifteen, as mandated by the state. (A company spokeswoman said that Gebbers applied for a variance to use a larger cohort system and didn't receive a response. After the outbreak began, it reduced its cohort size to fourteen people.) As of early August, at least 200 workers showed symptoms or tested positive, and there were two confirmed deaths: Juan Carlos Santiago Rincon, a Mexican in his thirties; and Earl Edwards, a sixty-three-year-old Jamaican. Reached by phone, Edwards's wife, Marcia Smith-Edwards, said that he was placed in the isolation unit on the farm and received no medical care during his illness. (In a statement, the Gebbers Farm spokeswoman said that safety officers checked on sick employees and that thermometers and transport to a hospital were available.) When Smith-Edwards last spoke to her husband, on July 31, he told her, "I am being treated in America like hogs and pigs." Later that evening, she received a call that he had died.

Many local farmworkers have mixed feelings about the system that employs the people they call *los contratados*. Some told me they resent that H-2A workers have their housing and transport paid for while they struggle to pay their own bills, that they hear about cases where employers, illegally, pay local workers less than workers on visas. "What about those of us who are already here?" asked Josefina Luciano, a farmworker and advocate who started working in dairies after years in fruit, asparagus, and onions. "Why don't you value my work?" Still, when she talked about the *contratados* —staying away from town in isolated housing, afraid of reprisals for speaking up—her eyes filled with tears. "They're disposable," she said. "Same as us."

One June morning early in cherry season, the sun rose on hundreds of cars and trucks, many of them parked along the edge of a long orchard off a dusty road. In the rows of trees, many people were unmasked, working quickly, and clusters formed where they poured cherries from their buckets into bins. Erik Nicholson, a national vice president of the United Farm Workers, stayed on the road, which was public property, and called out to the pickers in Spanish as they carried their ladders from row to row. "How many buckets for you so far?" he asked. "Are they giving you masks?"

I met Nicholson, who lives in eastern Washington and has worked with farmworkers for three decades, in the parking lot of a corner market at four that morning. (Farmworkers were specifically excluded from the National Labor Relations Act, and now fewer than 2 percent of them are unionized, but the UFW sees its role as broad advocacy of their rights.) Nicholson had led negotiations with growers' organizations to help craft the state's new safety regulations, and now he was spending his days on the road, checking on orchards to see if the regulations were being followed. That meant tapping into the informal network—text messages, word of mouth—that local workers use to find the best places for picking. A man who stopped to get coffee had shown Nicholson a text message with the address of this morning's orchard. The drive was at least forty-five minutes from his home in Prosser, but word was the picking would be good.

Nicholson checked the portable toilets and hand-washing stations, required this year to be placed every 110 yards instead of the usual 440 (for workers, a long walk is a costly use of time), and found an unusable sink, stuffed with wet paper towels, and bathrooms without any toilet paper. He pointed the mess out to a supervisor, who said a cleaning service would arrive soon; about twenty minutes later, it did. In a few hours, after the state's Department of Labor and Industries opened for the day, Nicholson would file a safety complaint against the orchard owner, Finley Cherries, one of more than thirty reports that he and two colleagues would file about infractions they saw at various farms. (More than six weeks later, he received an official reply to this one: "Could not be substantiated.") At the public talks he sometimes gives, he asks people to hold a strawberry and then to think about the hands that last handled it: Did that person have access to sick leave, to clean bathrooms, to soap and water, to health care? Often, the people he's talking to react by putting the strawberry down.

Many of the orchard workers who talked to Nicholson—cautiously, while supervisors in idling white trucks weren't pointedly watching—were more interested in discussing their pay than the pandemic. A man who looked to be in his sixties complained that the per-bucket rate was lower than he expected. "It's robbery," he said, "but what are we going to do?" At 7:20 a.m., after two hours of work, a group of young men climbed into a pickup truck and drove away. The pay was too low, they said, the picking standards too exacting. Tomorrow, they would return to California, hoping for something better.

Growers and pickers talk differently about the price of farm work. Labor represents an ever-higher percentage of growers' spending—these days, Lewis said, cherry growers spend as much as 75 percent of their annual production costs in the few weeks of their harvest—and their revenue per bin is declining. Fazio shared a common joke in apple circles: farmers lose money on every box, but don't worry, they're going to make up for it in volume. But pickers, too, are losing. They haven't seen their per-bucket pay go up for at least a decade, even as their costs of living have risen substantially, and despite shortages, farmworkers remain one of the lowest-paid groups in America. Philip Martin, a professor of agricultural economics at the University of California, Davis, says that when piece rates do not rise with minimum wages, slower workers are squeezed out, leaving the rest to work fast and with higher risks. With H-2A workers, Martin added, "we're getting the NFL of pickers."

Without the power to set prices, growers have struggled to compete with cheaper production costs elsewhere in the world, and more and more of them are leaving the industry. What's left, increasingly, are companies with the size and the money—or, as is becoming more common, with the necessary cadre of outside investors—to adopt expensive new technologies meant to reduce their reliance on people. This trend toward consolidation and automation is long-standing, says DeVaney, of the Washington State Tree Fruit Association, but most pronounced in bad years; he expects the financial stress of the pandemic to speed it up. Nicholson says he wonders where that will leave rural communities that have already been hollowed out.

After being asked to leave the orchard, we bought breakfast tacos and ate them in a park, where Nicholson gestured at the trailers across the road, the quiet, dingy downtown down the street. When people are replaced with technology, he said, "the money is not going to stay here. It's going to go to Mountain View or Palo Alto." For fruit such as cherries, Martin expects some combination of three possible futures: more mechanization and ever fewer people; a major expansion of guest-worker programs; and the replacement of American produce with imports. This happened not long ago to the Washington asparagus industry, which largely collapsed in the face of cheaper Peruvian asparagus. Around the world, the portion of food that is imported is rising fast.

It wasn't just the United States that panicked when the pandemic threatened migrant labor. In Germany, the government encouraged students and unemployed people to step in after borders closed but ended up allowing experienced workers to be airlifted from Romania and Bulgaria instead. In Britain, where Brexit's impact on migration left 16 million apples to rot last year, the government's patriotic, pandemic-era "Pick for Britain" campaign mostly failed, also leading to airlifts.

Martin quoted a dictum from the economist Varden Fuller that he thinks of often: that modern agriculture depends, for its seasonal labor force, on poverty at home and misery abroad. He has calculated that the cost of raising the wages of farmworkers in the United States by 40 percent, bringing workers above the poverty line, would—even if it were footed solely by consumers—increase a typical household's spending on fresh fruit and vegetables by just $21 a year.

In the early days of the pandemic, there was outrage at images of all the food wasted—rivers of dumped milk, fields of plowed-under potatoes—when restaurants and schools closed, airplanes and cargo ships failed to depart, and the workings of the food system were thrown out of whack. DeVaney wondered if shortages and back orders would make people realize how much work and how much vulnerability is hidden when they click an "Order" button and groceries show up at their door: Would there be a real effort to make things more resilient? Or would people go back to being "complacent and oblivious"? Kershaw Snapp, of Domex Superfresh Growers, was used to thinking of farmers as eminently adaptable, simply because they dealt so frequently with disaster. (She remembered a time a warehouse burned down overnight, and an alternative plan for storing freshly picked fruit was in place by 8 a.m.) After the virus, though, things suddenly looked different. "Everything is more fragile than I thought it was," she said.

In Sunnyside, Israel and Guadalupe found a new small house, where the plumbing and electricity worked and the deposit was within reach. The workdays stayed long. They had little hope that even the small changes since the pandemic—things like more access to soap in the orchard—would last after it was over. But they could still imagine that things might be different.

One day while talking with the family, I told them that I was also interviewing other people who interact, in different ways, with the cherries they pick: growers, consumers, vendors. What should I ask them? Israel wanted to know if there could be a path to legalization; Nayeli if wages could be raised. But it was Guadalupe who answered the quickest. "If we are important," she said.

Outside, the heat of summer beat down on dry hills and lush orchards alike. Every day, hands filled buckets, buckets filled bins, and, one by one, across the state, the cherries were plucked. By the thousands, then the millions, and finally the billions—an exuberant bounty corralled into neat, sellable stacks by dusty roadsides.

Eventually, the cherries began to fade, but the peaches and apricots and nectarines grew heavy and ripe and the first of the apples reddened. The virus continued its spread, heat advisories kept children inside, and still the fruit was relentless. Every day workers carried their ladders from row to row. Every day the fruit of their labor was stacked into refrigerated trucks and sped away.

Connections

SUSAN ORLEAN

Rabbit Fever

FROM *The New Yorker*

MOST RABBITS HAVE, in their skill set, the ability to pretend that they're healthy even when they're quite sick. It's sort of the inverse of playing possum, but done for the same purpose, namely, to deflect attention from predators, who would consider a sick rabbit easy pickings. As a result of this playacting, rabbits often die suddenly—or what appears to be suddenly—when, in fact, they've been sick for a while.

This past February, a pet rabbit being boarded overnight at Manhattan's Center for Avian and Exotic Medicine, the busiest rabbit veterinary practice in New York City, died. The fact that the rabbit had seemed fine and then expired without warning was chalked up to the rabbit habit of feigning good health. Later that evening, another rabbit at the clinic died. The coincidence of the additional death was strange, especially because the first rabbit that died was elderly, and the second was young. A third rabbit that died the same night was middle-aged; even though she was known to have had an abdominal mass that compromised her well-being, there had been no reason to think she was about to perish. Two deaths might have been a fluke; three seemed ominous.

The clinic's staff wanted to get the remaining fifteen or twenty rabbits out of the building immediately, but many of the owners were away and unable to retrieve them on short notice. This group happened to include Dr. Alix Wilson, the clinic's medical director, who was on vacation and whose own rabbits, Captain Larry and Dolly, were boarding there. In the meantime, the staff threw out all the clinic's rabbit food and bedding, in case something in them had poisoned the three rabbits. Within several weeks, eight more rabbits that had been at the clinic in February succumbed. Captain Larry was thriving, but Dolly, a medium-sized Lop that Wilson had just adopted to keep Larry company, died.

One of the lagoviruses of the family Caliciviridae causes a highly contagious illness called rabbit hemorrhagic disease. RHD is vexingly hard to diagnose. An infected rabbit might experience vague lethargy, or a high fever and difficulty breathing, or it might exhibit no symptoms at all. Regardless of the symptoms, though, the mortality rate for RHD can reach a gloomy 100 percent. There is no treatment for it. The virus's ability to survive and spread is uncanny. It can persist on dry cloth with no host for more than 100 days; it can withstand freezing and thawing; it can thrive in a dead rabbit for months, and on rabbit pelts, and in the wool made from Angora-rabbit fur, and in the rare rabbit that gets infected but survives. It can travel on birds' claws and flies' feet and coyotes' fur. Its spread has been so merciless and so devastating that some pet owners have begun referring to it as "rabbit Ebola."

According to the U.S. Department of Agriculture, RHD is a "foreign animal disease": one that is "an important transmissible livestock or poultry disease believed to be absent from the United States and its territories that has the potential to create a significant health or economic impact." All foreign animal diseases are "reportable." This means that any incidence needs to be logged with a state animal-health official. In most places, that's the state veterinarian, who, like a governor, oversees local policy. (Many animal issues are decided at the state level.) It must also be reported to the USDA and to the World Organisation for Animal Health, which is headquartered in Paris and tracks viruses globally.

Alix Wilson was familiar with RHD, and she wondered in passing whether it might have been responsible for the deaths at her clinic. "But then I thought, *No, impossible,*" she said recently. "Rabbit hemorrhagic disease isn't in the city." Her staff sent tissue from one of the dead rabbits to a lab at Cornell University, which subsequently transferred it to the federal foreign-animal-disease lab, on Plum Island, in New York. When the diagnosis came back as a variant of the virus, called RHDV2, Wilson was astonished. The clinic immediately stopped taking in any rabbits, and began a deep cleaning, which included replacing ceiling tiles and discarding thousands of dollars' worth of equipment that couldn't be sterilized. Todd Johnson, the USDA's emergency coordinator for New York and New Jersey, helped oversee the cleanup, and a veterinary epidemiologist and an intern from the department contacted 155 owners of rabbits that had been in the clinic during the previous few months, in an effort to identify Rabbit Zero. The bewildering thing was that, as it turned out, rabbits had already been dying of RHDV2 in Washington State, and soon were dying in other states, including Arizona, Texas, New Mexico, and Nevada.

The rabbit-hemorrhagic-disease virus was originally identified in 1984, in Jiangsu Province, China. First, it killed Angora rabbits being raised commercially for wool, and then it burned through pet rabbits and rabbits farmed for meat, all of which are members of the same species, *Oryctolagus cuniculus*, or what is commonly known as the European, or domestic, rabbit. During the initial outbreak in China, some 140 million rabbits were killed by the virus. The disease soon ravaged rabbit populations elsewhere in Asia, and then in Europe, the United Kingdom, Australia, and the Middle East.

There were only a handful of cases of the original variant of RHD in this country, and they were quickly contained. Still, the majority of rabbit products—meat, fur, skins, and live rabbits—imported to the United States came from countries where the disease had been widespread. The USDA still classified it as a foreign animal disease, but a department report, written in 2002, warned that RHD "has emerged as a growing concern for the rabbit industry following outbreaks in 2000 and 2001."

In the universe of human-animal relations, rabbits occupy a liminal space. They are the only creatures we regularly keep as pets in our homes that we also, just as regularly, eat or wear. Fitting into both the companion-animal category and the livestock category means that rabbits are not entirely claimed by either. A number of animal statutes—particularly, felony-cruelty provisions—are specific to dogs and cats, but not to rabbits. Laws protecting livestock, such as the Humane Methods of Slaughter Act, don't apply to rabbits either, even rabbits being raised for meat, because the USDA does not officially recognize them as livestock. There is probably no other animal that is viewed as diversely, and valued as differently, by its various partisans. Simply being a rabbit person doesn't mean that you look at rabbits the same way as another self-identified rabbit person. Any of the almost 20,000 members of the American Rabbit Breeders Association are just as likely to be raising a prized Jersey Wooly that sleeps in their bed and is primped for rabbit shows as they are to have hundreds of caged rabbits that will end up as stew.

A few years ago, a lawyer named Natalie Reeves, who volunteers at a rabbit shelter and has lectured on rabbit law at the New York City Bar Association, was having trouble untangling the hair of her pet long-haired rabbit, Mopsy McGillicuddy. She found an Internet group for long-haired-rabbit owners, and posted about Mopsy's hair troubles, expecting tips on conditioners and brushes. On the site, she noticed that a common response to similar problems was to kill the rabbit and start fresh with another.

Rabbits are everywhere among us. They live on every continent except Antarctica, in a wide range of environments. They have been domesticated for hundreds, if not thousands, of years. There have even been noteworthy rabbit booms. During the Victorian era, the most sought-after, bidded-up bunnies were a domestic breed, the Belgian Hare, which was developed to look like a wild rabbit. Belgians had sleek, black-ticked chestnut fur, comically long ears, and long bodies; they were the spitting image, in fact, of Peter Rabbit, who debuted in Beatrix Potter's *The Tale of Peter Rabbit* in 1902, in the thick of Belgian Hare mania.

A shipment of 6,000 Belgian Hares that arrived in the United States from Europe in 1900 attracted interest from tycoons including the Rockefellers, the DuPonts, and J. P. Morgan, who considered them an equity investment. (A male sold for $5,000 —the equivalent of more than $150,000 today.) According to the Livestock Conservancy, nearly every large American city soon had a Belgian Hare club. Fans in Los Angeles alone had 60,000 Belgians. Rabbits being rabbits, the number of Belgian Hares grew exponentially, and the market for them eventually buckled. People moved on to other breeds, and superfluous Belgians began to disappear. By the 1940s, there was a worry that they might even become extinct.

Rabbits have an unusual history with viruses. The first virus ever deliberately used to eradicate a wild-animal population was myxoma virus, which causes myxomatosis, a disease fatal to domestic rabbits. It was deployed in 1950, in Australia, where a dozen or so domestic rabbits released on a hunting estate in 1859 had outperformed all mathematical modeling and become many hundreds of millions—the fastest-known spread of any mammal on earth. The rabbits wreaked ecological disaster as they ate their way across the country. Shooting them made only a temporary dent in their numbers. Myxoma virus was introduced in the hope of controlling the population; it soon killed an estimated 500 million rabbits. (In parts of Australia, it is still illegal to have a pet rabbit.)

Two years later, a French doctor, annoyed by rabbits stealing from his garden, caught two of them and injected them with myxoma. They bolted, surviving long enough to carry the virus to other rabbits. The disease eventually bloomed across Europe and the United Kingdom, killing almost every rabbit in its path. Eventually, a myxomatosis vaccine was developed, and the disease was more or less brought under control.

Myxomatosis did travel to the United States, but for some reason it never got much of a foothold here. Had it established itself, it would have been disastrous, because at that time, in the middle of the last century, rabbits were a significant food source. Most people associate rabbit meat with European diets, but it was once a staple in this country. There were many huge commercial rabbit farms in the United States, and rabbit meat was readily available in supermarkets. The biggest processor was (and still is) Pel-Freez, founded in 1911 by a man who was given a pet rabbit that was pregnant, and, according to Pel-Freez's corporate history, turned "the dilemma into an opportunity." Rabbit meat, which could be cooked like chicken, was appealingly high in protein and low in fat. Back then, it was also much cheaper than beef. And raising rabbits is easy. A rabbit can give birth every thirty days; a young rabbit reaches what is known as "fryer age" in just sixty days. But, even during their heyday as a market product, rabbits weren't treated like other livestock. Because the USDA doesn't classify them as such, it has never required rabbit meat to be inspected or graded.

After the Second World War, the demand for rabbit meat began to decline. The number of cattle being raised domestically nearly doubled, and beef, which had previously been something of a luxury, became affordable. The cattle industry, which brings in some $70 billion a year, began promoting beef as the patriotic mainstay of the American dinner table. In the same years, chicken also became more widely available. Soon, it became the white meat of choice, and rabbit was marginalized as an occasional dish.

Eric Stewart, the executive director of the American Rabbit Breeders Association, also lays the blame for the decline of rabbit meat on Bugs Bunny. Created in 1940, Bugs, a sassy, man-size gray-and-white cartoon rabbit, had a leading role in the Warner Bros.' *Merrie Melodies* and *Looney Tunes; The Bugs Bunny Show* began in 1960 and ran on network television for forty years. Stewart believes that the generations that had grown up watching Bugs could not stomach eating rabbit. Then, in 1981, pet-rabbit ownership got a huge boost, with the publication of *Your French Lop: The King of the Fancy, the Clown of Rabbits, the Ideal Pet.* The author, a rabbit owner named Sandy Crook, argued that pet rabbits, which were then typically relegated to a hutch in the backyard, made perfect house pets, just like cats and dogs. Rabbits could be kept inside, because they could be trained to use a litter box. Many people consider the book the foundational text of the house-rabbit movement. Another best-selling manifesto, Marinell Harriman's *House Rabbit Handbook: How to Live with an Urban Rabbit,* was published four years later.

The number of rabbits kept as house pets has grown ever since. We don't know exactly how many rabbits there are in the United States, but they are the third most popular pet in the country, behind dogs and cats, and the most popular small pet, beating out hamsters, guinea pigs, and mice. The USDA estimates that there are more than 6.7 million pet rabbits, but the total number of domestic rabbits would depend on whether you're counting only pet rabbits or including rabbits raised for slaughter. Further complicating matters is the category of rabbits raised as, say, a 4-H project, which, once the project is done, might segue from pet to meat.

Rabbit-related activities are also on the bounce. There are about 5,000 ARBA-sanctioned rabbit shows each year; the largest one features more than 25,000 rabbits. There are rabbit fashion shows, which are especially popular in Japan. One show in Yokohama has featured rabbits dressed like Sherlock Holmes, Amelia Earhart, and Santa Claus. In New York City, pet owners organize rabbit playdates in Central Park. Across the country, there are rabbit speed dates, which are opportunities for rabbits to meet and see if they like one another, if an owner is thinking of getting a second, or third, or fourth. The typical American owner has more than one rabbit, so speed dating is important, because rabbits, despite their prodigious ability to multiply, don't always get along.

Within five years of the emergence of the original form of rabbit hemorrhagic disease in China, a vaccine protecting against RHD had been developed. A number of manufacturers produce vaccines against this strain, including Filavie, in France; HIPRA, in Spain; and Merck, which is headquartered in New Jersey but made the vaccine only for the European market. The vaccine was never offered in the United States. After all, there were only a few RHDV1 cases here, including one in Pennsylvania, which was theorized to have come from an Oktoberfest party, where imported rabbit meat was served. If the meat was infected, the virus could have spread to vegetables prepared in the same kitchen; the vegetable scraps were then fed to rabbits.

Most countries affected by the virus began offering the vaccine, and within a few years the spread appeared to have been tamped down. But, in 2010, rabbits in France began dying of what turned out to be a mutated version of the virus. The vaccine for the original virus was ineffective against the new strain; this was RHDV2. Soon, it was rampant throughout Europe and Australia. In England, the spread was so vigorous that parents were advised not to let their children bury their dead rabbits in the garden, because, "while comforting to children," the practice "may help circulate rabbit virus." The mortality rate of the new variant appeared to be slightly lower than that of the original, and at first this seemed like good news. But, in fact, RHDV2 was even more efficient at spreading, in the sense that more infected rabbits were surviving, and, because they might not show symptoms, they weren't being isolated, and passed along the disease. Vaccines guarding against RHDV2 were developed. (In some cases, they were produced in combination with the vaccine that prevents RHDV1.) By 2016, they were available across Europe, and vaccinating rabbits became common.

The new variant, like the original, at first seemed to stay away from the United States, except for a few isolated cases. But, in July 2019, a pet Norwegian Dwarf male on Orcas Island, near Seattle, died with a bloody nose. A veterinarian who saw the rabbit in a clinic was aware of RHD, and knew that bloody noses are a symptom, so she called the Washington State Department of Agriculture to report the death. Susan Kerr, an education and out-reach specialist there, was alarmed, because she knew there was an RHDV2 outbreak in British Columbia, so the clinic sent the rabbit's body to a lab for a necropsy.

While waiting for those results, Kerr's co-workers got calls from a number of people on San Juan Island, about a dozen miles southwest of Orcas Island. San Juan is famous for its rabbits. In the 1930s, a commercial breeder there went out of business, and released 3,000 rabbits into the wild. They multiplied and became a tourist attraction, and rabbit hunting on the island was so cel-ebrated that, in the sixties, *Sports Illustrated* ran a story about it, titled "Hippity Hop and Away We Go." By 1971, San Juan Island, which covers just 55 square miles, was home to an estimated 1 mil-lion feral domestic rabbits.

Kerr's department also received calls from a few nearby islands reporting rabbit deaths. Soon, the lab confirmed that the Orcas Island rabbit had died of RHDV2. Kerr then got news that most of the 145 rabbits at a facility on the Olympic Peninsula, across Puget Sound from the islands, had died in a three-week period. Their symptoms sounded like RHDV2; the virus was traveling. As a precaution, rabbits and rabbit products were banned from the ferry system that services Puget Sound.

A colleague of Kerr's posted about the outbreak on an online animal-health newsletter, and was inundated with requests from owners who knew that a vaccine existed, asking how they could get their rabbits inoculated. But the vaccine for RHDV2, like the vac-cine for the original virus, was available only overseas. No compa-nies had a license to distribute it in the United States. The USDA opposed importing it, except for limited special circumstances. One problem is that attempts to produce the vaccine on cell lines in a laboratory have failed. Merck produces a vaccine in cells, but it's a live, genetically modified vaccine, which is not permitted in this country. The other companies that currently manufacture the RHD vaccine produce it by infecting live rabbits with RHD. When those rabbits die, vaccines are made from their livers. According to a spokesperson for Filavie, one rabbit yields several thousand vaccine doses.

The USDA also maintained that vaccinating some rabbits would make it difficult to distinguish between sick rabbits and those with antibodies produced by the vaccine, although, since most sick rabbits died, the distinction would actually be very clear. In the rabbit community, the department's mulishness was infuriating. Some people said that it reflected a bad attitude toward rabbits, seeing them as disposable goods, easily replaced. Others thought that the department simply didn't want to manage all the paperwork required to bring a vaccine from overseas, or just didn't want to acknowledge that the virus was present.

The USDA finally agreed to consider requests for the emergency importation of limited amounts of the vaccine, but only if veterinarians applied first through their state veterinarians. Leaving the question of the vaccine to the states, though, meant that there could be fifty different decisions on it—a patchwork of guidelines for a disease that would travel with no regard to borders. A number of veterinarians said that they were interested in applying to import the vaccine, but, once they discovered the headaches involved, most of them gave up. Alicia McLaughlin, one of the medical directors of the Center for Bird and Exotic Animal Medicine, in Bothell, Washington, was the first veterinarian in the country to obtain the vaccine, which she ordered from Filavie. She, too, had heard about the RHDV2 outbreak in British Columbia, so she started researching how to get the vaccine, and compiled a list of several hundred clients who were requesting it. "I knew the virus would get here," she said recently. "Once it was in British Columbia, it was just a matter of time."

She applied to the Washington State veterinarian; was shuttled for a month between the state agriculture department and the USDA; had to manage language and time-zone barriers; then had to hire a customs broker to shepherd the vaccines across borders. Finally, more than four months after she applied, she received 500 doses of Filavac VHD K C+V, which protects against both RHDV1 and RHDV2. By the time McLaughlin was finally able to administer the vaccine to her clients, in April, concerns over Covid-19 had meant that she could offer only curbside service, and had to struggle to find personal protective equipment, which she needed, because she was interacting with patients and handling their animals.

Around this same time, the Center for Avian and Exotic Medicine, in Manhattan, had been thoroughly sanitized after its RHDV2 outbreak. To be absolutely sure that it was uncontaminated, Alix Wilson brought two rabbits to live at the clinic as sentinels. Because the virus is so contagious, the rabbits would almost certainly come down with RHDV2 if it was still in the facility. "No one wants to bring animals in to die," she said. "But it's one sure way in veterinary medicine to prove that a cleanup has worked." The rabbits survived. Wilson then applied to import the vaccine, and received a letter from the USDA saying that "without evidence of widespread infection" the risk was low, especially since, as the department maintained, household rabbits have no contact with others.

Some of the clinic's clients were furious that there had been a few days' delay between the first rabbit deaths and when the clinic put the word out about the diagnosis. According to Wilson, the clinic couldn't have done it any faster, since it had to wait to hear the results of the necropsy. The fact is that even the mention of RHD panics rabbit owners. Thousands have joined a Facebook group to exchange knowledge, vent, and worry. Useful information is interlaced with dread. A chief concern is whether it's safe to let a veterinarian know if you think that your rabbit might have RHD, since the veterinarian is obliged to report it to the state veterinarian. The fear—which, according to the New Mexico state veterinarian, Ralph Zimmerman, is mostly justified—is that, if your rabbit does have the virus and you have other rabbits, you will be required to "depopulate"; that is, you will have to euthanize them. There is also persistent chatter that the vaccine actually caused the disease, as part of a global plot to rid the world of rabbits. Recently, a member of the Facebook group proposed that rabbit owners sue Australia, perhaps conflating the past use of myxoma virus there with the outbreak of RHD. "No," another member replied, "we cannot sue Australia."

As the clinic in New York was reopening, thirty dead rabbits were found near Fort Bliss, Texas. An unusual number of dead rabbits were also found in Arizona, New Mexico, and Colorado. Along with pet rabbits and a small rabbit-meat industry, the Southwest has large populations of wild black-tailed jackrabbits and cottontails. Although they resemble domestic rabbits, these are different species entirely; they can't interbreed with domestic rabbits, nor are they susceptible to all of the same diseases. Wild rabbits seemed immune to RHDV1. But RHDV2 made a cross-species leap and, in March, jackrabbits and cottontails throughout the Southwest began dying in droves. "I've gotten reports that it's in the thousands," Zimmerman said recently. "I'm sure next I'll be hearing that it's in the tens of thousands." He has scraped together money from New Mexico's state budget to import 500 doses of the vaccine, which he will distribute to veterinarians around the state. He assumes that those doses will go to "high-dollar breeding animals."

But nothing can help the wild rabbits. Some vaccines, such as the one for rabies, can be distributed to wild animals by putting them out as food, but the vaccine for the rabbit-hemorrhagic-disease virus has to be given by injection and repeated every year. The concern is not only that many wild rabbits could be lost; what happens to them reverberates in other animals, including foxes and bobcats and wolves and hawks, since rabbits are their chief protein source. "Once they run out of rabbits," Zimmerman said, "cats and poodles will become their preferred food." Or, if there aren't enough wayward cats and poodles to go around, they'll starve.

In the past three months, RHDV2 has shown up in seven western states. Now that it has jumped to wild rabbits, most veterinarians I've spoken to believe that it's here to stay, and that the USDA should change its designation from a foreign animal disease to one that's endemic. There have been a few lucky breaks. For instance, the big rabbit shows scheduled for the spring, which would have brought together tens of thousands of rabbits—a recipe for contagion—were canceled because of Covid-19. Nevertheless, RHDV2 is advancing unabated.

The National Assembly of State Animal Health Officials, which represents all state veterinarians, recently formed a working group to evaluate the vaccine and make recommendations to the public. The group hasn't released its report yet, but recently I got an email from Annette Jones, the California state veterinarian, who is the vice president of the assembly, saying, "Yes, we should have [the vaccine] available . . . We are hoping a manufacturer steps up and applies to USDA for a license." No American manufacturer has applied so far, but at least one has contacted the American Rabbit Breeders Association to feel out interest. "It's hard to get these companies interested in rabbit medicine," a veterinarian told me. "All the money is in cats and dogs." The fact that people tend to own rabbits in multiples makes the economics of selling the vaccine complicated. Pet owners with one or two rabbits might not shy away from an annual vaccination that could cost $30 or so, but many rabbit owners have ten or twenty or two hundred rabbits, so the cost becomes prohibitive.

Cost and availability aren't the only challenges. "I'm worried that there is the controversy that a bunny died to make the vaccine," Alicia McLaughlin told me, sighing. "This vaccine is the only option we have at the moment, so it's frustrating." The House Rabbit Society, a rabbit-rescue and education organization, posted a letter on its website noting that rabbits die in the manufacturing process, and urging members to consider whether "the RHDV2 vaccines are right for you and your family." Natalie Reeves had been eager to vaccinate her current rabbit, Radar, as soon as possible. She has never skimped when it comes to her rabbits—she spent thousands of dollars on veterinary care for Mopsy McGillicuddy, her rabbit with the tangled fur, which suffered from lymphoma. "I just learned information tonight that is very disturbing," she wrote me in an email, after reading the House Rabbit Society letter. "I don't know whether I could in good conscience vaccinate my rabbit, knowing that I would be contributing to the death of others."

As RHDV2 is poised to become endemic in the United States, the vaccine, which is the one thing that might stop it, is now caught up in the contradictions of rabbits. When I first spoke to Reeves, she had mentioned that rabbits are the most discriminated against of all domestic animals: ridiculed for their lustiness; viewed as expendable; lumped in with oddball animals like chinchillas and prairie dogs; always subject to the question of whether they're pets or meat. She rued that the vaccine was just one more example of the peculiarity of their place in the world. Would dog owners be expected to use a medicine that was produced by killing a dog? Once again, rabbits seem to be betwixt and between. As Ralph Zimmerman put it, "Rabbits are just a real conundrum."

An Atlas of the Cosmos

FROM *Longreads*

WHEN I WAS eight, I noticed an atlas on the bookshelf in my room. I had just started amassing large art books from family museum trips but this was the first abnormally sized book in my possession—it was so oddly shaped its pages spilled over the edge of the shelf. One day I used all my strength to wiggle it down off the bookcase. I sprawled on my bedroom floor and began sifting through the long pages. It must have been from the '50s or '60s. It smelled old but it was clearly a book that had been cared for over the years. Its pages were a mix of pastels so dizzying and complex; in how pinks separated from light green and the skinniest blue rivers cut across the pages. Once I was old enough to read, my grandpa started ceremoniously gifting me books from his shelves.

One by one, every time I saw him, a piece of his library became mine. He had traveled all over the world and knew how much it could change a person. And whenever I'd visit him, I'd browse the books on the lower shelves and run my fingers along the spines like a car's wheels over speed bumps, each cover sort of yellowed from years of his cigarette smoke and constant reading. Once this book and I were formally introduced, I began having regular dates with the atlas. Each day I would lay on my stomach and then sit cross-legged hunched over the pages, running my fingers down the rivers in Africa—the Nile, Limpopo, I'd take a trip to France or Chile. I would attempt to pronounce Czechoslovakia and many other long words that threw me into a joyous tizzy. Every mountain range, every body of water, every large city I would look at longingly wondering one day when I got older, how many of these mysterious places I would see with my own eyes. My wanderlust grew as I grew. There was so much to be explored, there was so much space that existed around my little home in Los Angeles. There was so much I didn't know.

For as long as we have existed, humans have been trying to understand themselves in the context of their physical location. Granted we associate value and identity with where we come from, where we live, and even many times where we know we are going. One of the oldest maps of the world on record in human history comes from the seventh or sixth century in Mesopotamia. The Imago Mundi is a simple clay tablet carved with cuneiform writing. Its eight sections describe a region around the Euphrates River in Babylon. But it is much more than a physical representation of where the Babylonians lived. Surrounding the Euphrates is a circle meant to symbolize the ocean or "bitter river." The sections outside the circular ocean are called "nagu" or distant regions, some of which are also mentioned in the Babylonian *Epic of Gilgamesh*. This relationship between the parts of the map shows us that the ancient Babylonians were trying to place themselves and their location into the greater unknown regions beyond their understanding. By this time human civilizations had spent many years examining the stars and planets and marking down the movements of these objects in the sky. But the Imago Mundi is our first record of a map of our direct surroundings. "It's a really unique text because we have a lot of descriptions of the world, but we don't have a lot of drawings of it," says Assyrian scholar Dr. Moudhy Al-Rashid. "We have drawings of buildings, building plans, that sort of thing but not an attempt to explain the world visually," she says. "There are drawings of stars, there are star maps but not maps of the world."

Two scissor lifts, with men strapped to the metal banister, flank the side of the telescope. There's an oil leak near where the 5,000 robots are supposed to move around and they can't stare at galaxies if there's an oil leak. My visit to the Mayall four-meter telescope at Kitt Peak National Observatory in Arizona came on a very regular working day in September 2019. The drive up the mountain started clearly, with no other cars in sight. There is a long narrow highway that drives straight toward the mountain, with two lanes that cut through the middle of the Tohono O'odham National Reservation. At first I could make out the greenery of shrubs and trees and rock layers folded together like neapolitan ice cream with milky chocolate ripples, rosy-tan, and white. These features were formed during the Triassic period, some 200 million years ago.

Each passing minute up the 6,883-foot climb became more and more opaque—I was driving into a cloud. When I parked and got out the wind knocked me onto my car, and when I looked up to stare at the dome, I couldn't see the top. At eighteen stories high, the dome housing the Mayall four-meter disappeared into the sky above.

The Kitt Peak team were nearly done installing the DESI instrument that would search the universe for dark energy—the elusive force responsible for expanding our universe outward at speeds of tens of thousands of miles per second—70 kilometers per second per megaparsec, to be precise. DESI, the Dark Energy Spectroscopic Instrument, was being attached to the telescope to officially hunt galaxies in the early days of 2020. On October 22, 2019, they opened the double doors of the dome and the telescope collected its first official light as the DESI mission officially began. Their goal is an ambitious one—to make the most detailed 3-D map of the universe. They will do this by looking back in time as far as 11 billion years ago when the universe was very young, galaxies were just beginning to form, and the contents of the universe were nestled much closer together.

For thousands of years people have climbed mountains and crossed rivers to create maps to understand their place in the context of things. In a way this 3-D map of the universe is the last map humans can make. Sure it won't be the last map of the universe, but we've traced the boundaries of land, marked rivers and oceans, countries and species. We've mapped Mars, the Moon, the solar system, even our own galaxy. Which means there is only one thing left to understand in this symbolic way and that is the entirety of the cosmos.

DESI is a conglomerate of 500,000 parts moving in a synchronized ballet. Collected in tubes that run 40 feet from the top of the telescope to the bottom are 5,000 fiber-optic cables. These threads of pure glass are as thin as a strand of hair and act as conduits for light. At the top of the telescope all 5,000 strands fan outward where each single cable will be assigned to an individual galaxy. Every twenty minutes the team will point the telescope at a new patch of sky while each one of the 5,000 cables locks onto a different galaxy. It will take only a few seconds for each robot to do-si-do and swivel to a new object. On an average night the team expects to collect the light from 150,000 different galaxies and will rarely look at the same galaxy twice. While this might sound daunting, the tricky element with DESI isn't mapping nearly 40 million galaxies over five years, but the 5,000 miniature robots that move each individual hairlike strand inside the telescope. "This is a very complicated instrument," says Michael Levi, DESI director. "It has a half a million moving parts." Not unlike the way complex internal clock mechanisms work, DESI's 5,000 robots are so small that if one thing goes wrong each time they shift, it risks the entire operation, and a setback in data collection.

DESI's 5,000 eyes will spend five years looking back in time at ancient light to better understand the story of the universe. By collecting this data they will be able to decode the light's journey across the void. While we don't have any functioning DeLoreans just yet, we do have telescopes and telescopes are real time machines. It is easy to forget that the images we see of space are never from the present—that light has taken billions or millions of years to reach us. When we study deep space, we are studying the past, objects as they once were, not as they are now. We do this by studying photons. A photon is one of the lightest particles in the universe which happens to be responsible for what we know of as light and they play a vital role in how DESI will help us understand dark energy and the expansion of the universe.

Because this light takes so long to reach us, each photon has a story to tell of where it comes from and where it's been. These photons have spent billions and billions of years traversing the cosmos to get to Earth but when they enter the mirror of the Mayall telescope their journey will not be quite finished. As the light enters each fiberglass cable, it will travel down the length of the telescope through each individual thread of glass another 40 feet and across the white tile floor into a room that houses ten identical spectrographs. The instruments will break the light apart sort of like a mail sorting machine, only with the spectrum of light from each galaxy. Depending on the story of each individual collection of photons, it will appear in the instrument as either redshifted or blueshifted. As light travels, the colors within the spectrum appear at different wavelengths—if an object is moving toward us its light is crunched and appears toward the blue part of the spectrum, whereas if an object is moving away, the light is stretched out and appears red. After traveling many billions of years, the journey of the light from all 40 million galaxies will end in a clean room inside of a dome on a mountaintop in Tucson, Arizona.

In 1929 astronomer Edwin Hubble was studying the light spectra of galaxies and announced that his observations showed that many galaxies were redshifting—they were in fact moving away from us. But what he'd actually discovered was the expansion of the universe. Those galaxies weren't just speeding away on their own, the very fabric of space-time itself was ballooning outward. He didn't believe this was evidence of expansion; it would take another seventy years before scientists realized that not only was the universe expanding—the expansion was speeding up.

Nearly a decade before Hubble took to the telescope, Albert Einstein proposed a theory called the cosmological constant in tandem with his theory of general relativity. The idea being that the universe was a static place and the density remained constant. When Einstein saw Hubble's news about the redshifting galaxies he threw this theory away, except Einstein was sort of right, go figure. The universe is not a static place—we know it's expanding rapidly, but the density in the universe *still* remains constant. Think of it like this, imagine you're in your living room with a table and TV and some books and a cup of coffee. Now imagine if that room began to expand like a balloon and got bigger and bigger. The objects in your living room would not increase in density —they are what they are. This is the same with our universe, as it balloons out the density remains the same, hence, your cup of coffee is the cosmological constant.

This was a tricky thing for astronomers to accept for the longest time because there is *a lot* of matter in the universe. And because of gravity we know that matter clumps together, so shouldn't the universe be contracting? Newer estimates by astronomers say that there could be up to two trillion galaxies in the universe which are made up of two types of matter. The matter made of "normal" things—like you and me and your cat and desk and iPhone—represents only 5 percent of the matter in the universe. Dark matter, which we cannot see, is about 25 percent. That is a lot of mass, and a lot of mass that is gravitationally attracted to each other; however, despite this unfathomable amount of material and density, it is no match for the dominating force in the universe which accounts for 70 percent of everything in existence—dark energy. Ninety-five percent of the universe is made of things we cannot see and have no real understanding of. It's fair to say at this point in history that we and our 5 percent are not "normal" matter—we are the real anomalies in this universe.

The name "dark energy" is born out of ignorance—we call it dark because scientists simply cannot see it and they don't really have any idea what it is. Well, they have some ideas. "The simplest understanding is that it's a thing called the cosmological constant," says Dr. Risa Wechsler, an astrophysicist and professor at Stanford University. "This would mean essentially it's a property of space itself, which is a constant over all space and time." This is conveniently also the only way Einstein's density theory works. Score one for the current working model. But it's a bit of a Catch-22—the more space there is, the more dark energy there is and the more the universe expands. Therefore the more the universe expands the more space there is and thus more dark energy. "We are at a very interesting stage right now in cosmology," says Wechsler. "We have had what is essentially a standard cosmological model for the past twenty years. That model is basically still working but there are starting to be some signs that it's breaking. And I think right now we're in the stage where we don't know whether things are going to go away when the data gets better or whether when the data gets better, we're going to see a real sign that the model doesn't work anymore."

The survey data from DESI may show that the current understanding of the universe is wrong—this would not be the first time this happened. At least we know that dark energy is not a particle like dark matter. Some scientists think it could be another dimension leaking into our universe. But more than likely, it might be space itself. This would mean that space wasn't actually empty, we just can't see what it really is. But so what does "space itself" mean? We have no idea.

Given how new to the planet humans are on the grand scale, we've figured out quite a lot about the cosmos. We know the Earth is about 4 billion years old, we know that 13.8 billion years ago there was nothing and then there was everything. We know absolutely nothing about the first one-ten-millionth of a trillionth of a trillionth of a trillionth of a second but after that we have the time line worked out to minutes. Very shortly after the universe came into existence it inflated like a balloon and very quickly spread the matter around. But this inflation was brief and the universe continued to expand but not accelerate. So everything is fine, stars are being formed and eventually galaxies begin to coalesce and then galaxies huddle together and make galaxy clusters, and the objects in the version of the universe we know are born. So here is where things get weird.

All of this matter was gravitationally pulled toward each other, just as you'd expect matter to behave. As a result of all of this matter clumping together, the expansion of the universe *slowed* down. But then suddenly around 7 billion years ago, the expansion started to speed up and it's only gotten faster since. Sometime between 11 billion years and 7 billion years ago, dark energy turned on and began dominating the universe. For a sense of scale at how fast this expansion is going—our universe came into existence 13.8 billion years ago. Without dark energy and expansion, the diameter of the universe would then be 13.8 billion light-years wide, but we *do* have dark energy and because of this expansion the observable universe is now 91.32 billion light-years across.

Many scientists around the world, both on the DESI team and elsewhere, are desperately trying to understand why this shift suddenly happened. Why did dark energy suddenly turn on? All we know right now is that dark energy is winning.

Today's task is to rebalance the telescope. Because the DESI cables are so heavy, they've thrown the balance of the scope out of whack. I stand with David Sprayberry, my host and on-site manager of the project, while his team straps themselves into harnesses on a floor elevator eighteen stories up. Men my dad's age slip their legs into something fit for a person working on the outside of the Empire State Building. Some of these technicians have been working on the telescope since it was built over forty years ago and have watched it turn over to different scientific endeavors, DESI being the newest. They banter back and forth about their weekend plans and who is starting the elevator and are they all strapped in? Don't wanna die today! Up they go.

To balance the telescope they have to carry trays of solid lead weights the size of envelopes up into its belly. Two by two they will screw on each weight and then test it, then add more weight to certain spots and test it again. They are just weeks away from collecting what astronomers call "first light." The lucky first object of DESI's gaze will be a spiral galaxy called Triangulum, 2.7 million light-years away. This particular galaxy has been studied so much over the years that its spectra is very well known, making it a sort of galactic calibrator.

There's so much laughter despite the mundane part of this work. So here we are, scissor lifts and lead weights on a quest to understand the most mysterious force in the universe. They should all have shirts on that say TEAM HEAT DEATH but I am not in charge.

Before they start to fix the balance issue, we climb down a thin metal ladder and into the center of the telescope directly below the mirror. At four meters, twice the size of the Hubble Space telescope mirror, the Mayall might be a bit older but it's a serious telescope, which is good because it has a serious job to do.

The inscriptions on the Imago Mundi map describe some basic features—where the sun rises, where the mountains are. But then there is a line that refers to the four quadrants on the map as "The four quadrants of the entire universe." A fact they could not have known, but it still remains true that the boundary of the universe is limited by what we can see, even today.

Finally, the last paragraph of writing on the Imago Mundi is a poignant ending, "In all eight 'regions' of the four shores (kibrati) of the ea[rth . . .], their interior no-one knows." Ancient Babylonians were limited in their knowledge of what existed beyond the mountains to the East—while they knew the underworld was in one direction and a group of people that were enemies in another. The Babylonians were limited in their scope of understanding, and while our knowledge has grown exponentially, we are in many ways still in the same position—looking up and out and wondering what lies beyond ourselves.

For DESI this might seem a bit absurd, making a map of a universe that is constantly changing, forever expanding. There is something striking and a bit ironic about creating a map of a boundary that is constantly moving away from us, like attempting to map grains of sand on a beach while the tide continues to roll in. The "edge" of our universe will continue to expand until everything is gone. But this map of the universe is almost more about understanding our past than our future. In this case, in order to know where we are going, we must first know where we have been.

DESI's project director Michael Levy likens it to a space MRI. "It's a little like medicine when they switched from X-rays, which were inherently two-dimensional, to MRIs where you could take slices of the body." DESI will serve a similar purpose but for space, "we will now have time slices or distance slices of the universe."

By collecting data and slicing the light into periods of time, it will allow scientists to reconstruct the history of the universe. Eventually once the survey is completed we will be able to examine deep astronomical time the same way geologists have used fossils and minerals to tell the story of the Earth. Instruments like DESI allow us to breach the laws of physics. In this case by looking back in time, we can deduce what will happen in the future, even as far forward as the end of the universe.

We use our location as a way to think about our identity. In the case of the cosmos the timescale is well beyond our very short lifetimes or even beyond our comprehension. Some of the answers to these questions won't be solved while we are still here but will be left to the incoming generations and the truth is there are questions that will simply be passed on and never answered. The quest might seem a bit nonsensical. Why does it matter when or how the universe began? Why does it matter when or how it ends? It matters for the same reason your locations throughout your life carry context for who you are. We exist on a time line together—we pop into existence and then one day we stop. It matters for the same reason one of the first questions you learn to ask in another language is, "Where are you from?" To know where you are at any given time is a frame of reference in which to measure your life in some way and in many ways those locations, those slices of time, hold a great deal of meaning.

Before leaving we stop inside the main office where people pick up keys for their dorms. Many folks are sleeping during the day as astronomers have a tendency to do. A corkboard at least thirty years old is hanging from a wall covered in very old sun-faded cartoon clippings from newspapers that all have something to do with astronomy. At the very bottom is a black-and-white *Ziggy* cartoon with torn, frayed edges. It shows a person staring into a telescope, and it says, "I read that the universe is racing away from the Earth at over 15,000 miles per second! I wonder if it knows something we don't."

I left the telescope and DESI team around 4 p.m. that day and by the time I left the dome, the clouds had burned off and I could see all of Kitt Peak. Each ridge had clusters of telescopes all with their own quests for different answers, all of which lay overhead. I could see the roads snaking through the valley below—I could finally see where I was.

One day, googols and googols of years after you and I have died, the universe will end. Just like us, it is currently in the process of its death. It is expanding outward at unfathomable speeds, so much so that eventually all matter in the universe will begin to separate, growing further and further apart. As a result of this expansion the universe and all the matter in it will cool off until everything is the same temperature. This is one of the most popular theories for the end of the universe called the Heat Death—literally the death of heat. Over time, stars will die, galaxies and their solar systems, globular clusters and everything we've ever known will get consumed by black holes—the last things to exist in this universe. Eventually the matter inside of those black holes will evaporate until there is nothing left. (If you could move forward in time when the only thing left were black holes, where average temperatures hover just a fraction of a degree over absolute zero, you would be the hottest thing in existence. The radiation emitted from your body would glow hotter than anything else.) This goes far beyond Sagan's pale blue dot—everything we've ever known will be gone, every human ever born and died, every person we've ever loved, every work of art, every book, every planet, every galaxy, every star, every atom that was ever created will cease to be.

Meanwhile, you and I are going about our days on an average rocky planet in just one of trillions of solar systems. Our planet orbits around an average star that moves around the third arm of the Milky Way galaxy, local group Virgo supercluster in an ancient universe that is moving ever outward. Where are we? The answer is always changing.

KATY KELLEHER

Periwinkle, the Color of Poison, Modernism, and Dusk

FROM *The Paris Review Daily*

ON A STRETCH of rural road not far from my house, there is a small wood where, once a year, for just a few short and cold days, the ground turns a magnificent shade of purple. In a reversal of fortunes, the stand of gracious Maine trees becomes secondary to the ground cover below. When the periwinkles are blooming, it's hard to have eyes for anything else. The delicate mist is an impossibly soft color, like clouds descending into twilight, like the snowfall in an Impressionist masterpiece. It's a color that almost doesn't belong here—it's a plant that certainly doesn't.

Periwinkle goes by many names. You might know her by one of her more fabulous monikers, like sorcerer's violet or fairy's paintbrush. In Italy, she is called *fiore di morte* (flower of death), because it was common to lay wreaths of the evergreen on the graves of dead children. The flower is sometimes associated with marriage (and may have been the "something blue" in the traditional wedding rhyme), sometimes associated with sex work (because of its supposed aphrodisiac properties) and also with executions. I grew up calling her vinca, a pretty little two-syllable name, taken from her proper Latin binomial, *Vinca minor.* My mother cultivated periwinkle in our forested Massachusetts backyard, encouraging the hardy green vines to trail over the boulders and under the ferns. I would have been delighted to know even a fraction of vinca lore back then, but I knew nothing except she was poison. I could eat the royal-purple dog violets, but I was not to pick the vinca. Vinca was poison and poison meant death.

This, it turns out, is false. It's one of the many easy assumptions of childhood. I thought all plants that grew in my yard were meant to be there, and I thought all poisonous things were bad. Vinca —or periwinkle or creeping myrtle or dogbane, as she's also called —is invasive to North America. It chokes out other plants, stealing too many nutrients for native ground cover to grow. Many New England gardeners do not plant it for this reason. Yet I grow it, partially because I know what it can do, what it has done.

Vinca contains alkaloids, which can be terribly bad for you if ingested in the form of a flowering vine. If you're a dog and you munch several vinca vines, it could kill you. But let's say you have cancer. Let's say its lymphoma and you're my husband and I can't imagine the world without you, can't imagine what would happen if the small, hard tumors nestled around your collarbone took your life. For three hours every two weeks, you go and sit in a room with other patients, other sick people who have lost their hair and their eyebrows. Together, you get alkaloids injected into your veins. You live because there is a medicine made from Madagascar periwinkle (a close relative of *Vinca minor*) that can kill cancer cells and cure your blood disease. You live because something poisonous can also be healing, an invasive species can also be curative—for a land-scape and its people.

Vinca is a complex little plant, and periwinkle, named for its blossoms, is an equally complex color. A subset of violet, which is a subset of purple, periwinkle denotes a precise shade that appears somewhat brighter than lavender, bluer than lilac, clearer than mauve, and dimmer than amethyst. But it's hard to say with precision, because the purples are strange ones, polarizing, and violets are even more so. Few hues are more beguiling and more reviled than this grouping, the last stop on the rainbow and the tacked-on *v* at the end of that schoolchild's mnemonic, Roy G. Biv. According to the scholar David Scott Kastan, shades of violet exist within their own special category. *Violet* is, like *glaucous,* a color-word that denotes a certain quality of light. "Violet seems to differ from purple in whatever language—not so much as a different shade of color than as something more luminous: perhaps a purple lit from within," Kastan writes in *On Color,* his 2018 book on the subject. "Violet is the shimmering, fugitive color of the sky at sunset, purple the assertive substantial color of imperial robes."

This latter kind of purple—reddish, bold, saturated—has been bedecking the backs of the rich since its discovery by the Phoenicians, who were milking snails for their secretions long before the Common Era began. Known as Tyrian purple (supposedly for Tyre, in present-day Lebanon), Phoenician red, or imperial purple, it even has a heroic myth about its "discovery." According to Roman scholar Julius Pollux, Hercules's dog was the first creature to discover the pretty color hidden under those predatory shell-dwelling creatures. (Peter Paul Rubens painted his vision of this event in *Hercules's Dog Discovers Purple Dye.*) Hercules had been on his way to court a nymph named Tyro, and when he got to her abode, she took one look at the stained dog and asked for a gown the same color as his mouth. Thus, Hercules was granted the glory of "inventing" Tyrian purple. The nymph, meanwhile, went on to get raped by Poseidon. ("And by the beach-run, Tyro, / Twisted arms of the sea-god, / Lithe sinews of water, gripping her, cross-hold," wrote Ezra Pound in the *Cantos.*)

Tyrian purple was a difficult color to manufacture. Thousands of snails were required to create a single ounce of dye. In first-century Rome, a pound of Tyrian purple cost "about half a Roman soldier's annual salary, or the equivalent of the cost of a diamond engagement ring today," according to a 2019 exhibition from the Kelsey Museum of Archaeology at the University of Michigan. While it was possible to mix other dyes and pigments to create shades of purple, Tyrian remained the most significant color until the invention of "mauveine." This, too, was an accidental invention, though we have more documentation about the creation of mauve than we do of Tyrian purple's. In 1856, teenage chemist William Perkin was attempting to create quinine for a university assignment when he discovered his black, tarry mess had a purple tint. He patented the formula and soon it became the first chemical dye to be mass-produced. Samples of the early mauveine dye show it to be a bright reddish purple, vivid and intense. It is a bit less brown than Tyrian purple, but it clearly exists in the same color family. It's a purple, a true one.

According to the historian Sarah Lowengard, author of *The Creation of Color in Eighteenth-Century Europe,* "modern American English" tends to consider *purple* and *violet* synonymous, "as simply red plus blue." But that wasn't always true: "In eighteenth-century conventions, purple has more red (r + r + b) and violet more blue (r + b + b); one can have light and dark violet as well as light and dark purple."

Violet was deeply significant to the Impressionist painters of Europe—and deeply offensive to their critics. Kastan pinpoints Monet's *Impression, Soleil Levant* (1872) as the inciting incident in the critics' war against this artistic use of the shade. "Wallpaper in its embryonic state is more finished than that seascape," wrote Louis Leroy after his eyes were subjected to the wishy-washy scene.

It's a painting of the ocean, but it's a painting about color. It's about misty gray blues and light violets. The same could be said for Monet's *Water Lilies* series or Pissarro's winter landscapes or Renoir's crowd paintings or even J.M.W. Turner's turbulent marine paintings. The more adventurous art galleries in 1870s Paris were filled with blurry landscapes, portraits, and still lifes, tied together by their techniques (thick layers of wet paint applied to wet paint) and their strange, almost surreal colors. To twenty-first-century eyes, these images look ordinary, but critics were unimpressed. Human skin, lamented the members of the artistic establishment, had turned green and purple and orange. Naturalism had been abandoned in favor of these periwinkle monstrosities.

Periwinkle's first known appearance in English as a color-word was in the 1920s, but it has been in the painter's toolbox for far longer, nestled under the violet umbrella. *Periwinkle* is a Modernist word for a Modernist color. It's a word that has several meanings— in addition to being a flowering plant, a periwinkle is also a type of snail, though not, confusingly, one that secretes purple liquid. It's a nature word for a color most often found in nature. A dreamy word for a color that exists at the edges of the night.

While the Impressionists are perhaps the most beloved of the nineteenth-century artist-innovators—their vague flowers make for good merch—there were other movements bubbling alongside. One of these less-remembered movements was symbolism, an artistic practice that predated (but perhaps predicted) the surrealist boom of the twentieth century, combining elements of sublime Romanticism and Rococo drama with Modernist abstraction techniques to create works that were intense, often quite ornate, stylized, and, above all, dreamy. In contrast to the Impressionists, who painted from nature and labored to show exactly how we experience colors in the wild (hence all those violet sunsets), the symbolists thought you had to inject a little unreality in art in order to get the viewer closer to experiencing a universal truth. They wanted to show what love felt like or what madness meant, so they painted worlds that were stuffed full of references to stories and (naturally) symbols. They revered Greek mythology and were heavily influenced by pagan religions in general—for the symbolists, spirituality was far more important to art than naturalism. While the Impressionists (a movement based largely in Paris) and the symbolists (a movement that flourished in Central and Northern Europe) had very different goals, both groups relied heavily on certain colors—chief among them the secondary hues, the marigolds, the emeralds, the tangerines, and, of course, the violets.

My interest in symbolism arose alongside a newfound interest in sunsets. Both of these obsessions were quarantine-born. I once laughed at sunset paintings and sunset pictures—so obvious and ordinary. But lately, I've found myself waiting for the sun to go down, timing my walks so that I can be outside then, when the bats begin to swoop around the oaks and the mosquitoes hum around my face. It's not the golden hour (which occurs about an hour before the sun touches the horizon), it's the periwinkle window. It lasts only a few minutes in the summertime; dusk descends fast in the north. But for fifteen minutes, the sky is painted with various shades of violet, indigo, and mauve. At dawn and dusk, my tiny little dead-end road becomes another place, quieter than during the daylight hours, but visually much louder.

And after the sun had set, while trying to lull my baby to sleep, I immersed myself in the works of Nicholas Roerich, Edvard Munch, and Jan Toorop on my phone. Unlike Impressionist pieces with their heavy ridges of paint and texture, symbolist pieces seem made for a screen. They're often flat, with broad swaths of contrasting colors (think of Klimt's quilt-like surfaces or Gauguin's two-dimensional flowers). Some of these paintings are a bit cartoony, kind of childlike, something you might see in a children's book alongside a nursery rhyme. Sometimes, these paintings are heart-achingly lovely. Mostly they're a bit mad. Naked women dance in periwinkle twilight, demons garden in golden fields, one-eyed monsters rise from a backdrop of flowers, and lovers kiss in a flat, jeweled world.

I slowly came to love these images from the same reserve of feeling that I held for dusk. Scrolling through painting after painting felt a bit like picking flowers. Even the sinister pictures, the poison blossoms, were still so pretty. I spoke their language and understood their references. I could see where they came from, what they were trying to do.

Recently, after spending months thinking about this color and this flower, I emailed Kastan to ask whether he still loves violet and whether he had any thoughts on periwinkle. We'd met once at a color-related event, and struck up a friendship based on my color stories, his color book. He replied, writing from a house in Rhode Island, "Periwinkle seems the color of grace, and not least because of the flower's modest ordinariness." So much is changing, he wrote, "but there is always color—it is the promise of joy."

SABRINA IMBLER

The Unsung Heroine of Lichenology

FROM *JSTOR Daily*

IN THE MIDDLE of the twentieth century, a lichenologist named Elke Mackenzie labored to prepare a monograph on *Stereocaulon*, a genus of lichen. The monograph was to be exhaustive: a tome on how the lichen evolved and took its pale, knurled form, complete with detailed illustrations and photographs. Compared to some of its muted, flatter brethren, *Stereocaulon lichens* could be considered flamboyant. The lichen effloresce in clusters of knobby stalks called pseudopodetia, branching elongations of its body. One species, *S. arenarium*, resembles, among other things: a fractal cauliflower; a tangle of silly string; a brain coral, dusted in snow.

Due to a string of unfortunate incidents—windswept specimens, a fall down an elevator shaft, and financial woes—the monograph was never completed. But Mackenzie's meticulous, exhaustive attention to the genus remains the single greatest contribution to science's modern understanding of *Stereocaulon*—and other lichens that dwell in the cold, harsh regions of the Earth.

Mackenzie, who died in 1990, transitioned in the final years of her career. Despite this, almost all of Mackenzie's legacy exists under her former name. I chose not to share it here because I do not know Mackenzie's relationship to her former name, and she has no way of consenting to its inclusion in this story. When Mackenzie transitioned, she shared her news with little fanfare: a small announcement in the October 1976 issue of the *International Lichenological Newsletter*, where she noted she "should now be addressed as Dr. Elke Mackenzie." But four years earlier, in one of the last studies she published—"*Stereocaulon arenarium* . . . a hitherto overlooked boreal-arctic lichen"—in the 1972 issue of *Occasional Papers of the Farlow Herbarium of Cryptogamic Botany*, Mackenzie named herself in the acknowledgments of the paper:

> Grateful acknowledgement is also due to Dr. Eric Hultén (Stockholm) and Dr. Vsevolod P. Savicz (Leningrad) for advice on the situation of localities in Kamtchatka, and to Miss Elke Mackenzie for technical and bibliographic assistance in the preparation of this paper.

If other scientists were not willing to acknowledge her contributions by name, she would do it herself.

Elke Mackenzie was born in London in 1911, a time when the field of lichenology still held enormous mystery. Scientists had believed for centuries that the pastel encrustations were plants, occupying some interstitial space between algae and fungi on the tree of life. In 1867, the Swiss botanist Simon Schwendener proposed that lichens exist in duality as the physical incarnation of a partnership between fungi and algae, as the botanist Rosmarie Honegger writes in a 2000 article in *The Bryologist*. Schwendener's hypothesis was roundly rebuffed by the world's leading lichenologists until it was proven right at the end of the nineteenth century.

According to an obituary of Mackenzie by the lichenologist George A. Llano, published in a 1991 issue of *The Bryologist,* her fascination with lichens sparked in 1935, when she landed a job as an assistant at the Natural History Museum in London, then called the British Museum (Natural History). Mackenzie apprenticed under Annie Lorrain Smith, one of Britain's most revered lichenologists, who authored a textbook in 1921, *Lichens,* that remained the standard for much of the twentieth century. Though Smith worked at the museum for four decades, the museum did not hire women for official staff positions at the time, so Smith was paid via a separate fund. Under Smith's mentorship, Mackenzie began studying the museum's specimens of lichens collected from Antarctica. Little was known about these far-flung species, and Mackenzie took up the mantle.

A year later, Mackenzie married Maila Elvira Laabejo, who gave birth to their first child in 1940 amid a hailstorm of bombs dropped by German planes in a months-long campaign. In the final years of the war, Mackenzie stole away to Antarctica as a part of "Operation Tabarin," a top-secret expedition meant to strengthen

Britain's territorial claims on the southernmost continent, under the guise of protecting Allied ships from German U-boats.

There, Mackenzie wore many hats, including botanist and musher of dogs, one of whom was named Dainty. In Mackenzie's typewritten diary entries, she chronicled a rugged, hostile life briefly interrupted by moments of joy. There were the quintessentially dour encounters with frostbite and animal carcasses. But there were also afternoons of igloo-building, evenings of fried seal steaks, and nights watching the sunset fade into bright starlight. During one lunchtime pit stop, Mackenzie wandered away from camp to collect the lichen *Alectoria minuscula*, which tufted a metamorphic rock in "neat black rosettes," she wrote.

There were lichens in abundance: encrustments by penguin nests, sea cliffs, erratic boulders, shores studded by fossilized ammonites. Mackenzie recorded these encounters with resolute awe. She wrote in one January entry:

> At one place, where small rivulets of snowmelt water irrigate stony slope, there is a wonderful profusion of moss, mostly Polytrichum, of a most vivid green colour and forming cushions up to 2 feet across—the most luxuriant moss growth I have ever seen.

To ensure her specimens would survive the harried pace of sledging trips, Mackenzie nestled the lichens she collected in pemmican boxes filled with soft snow and dried them out on heated stones back at camp. There, Mackenzie's passion was palpable. A medical officer remembered that she encouraged even the most uninterested of her expedition companions to take an interest in lichens. By the end of the journey, Mackenzie had amassed 1,030 specimens, 865 of which were lichens.

When Tabarin concluded, Mackenzie took jobs in Argentina and Canada before becoming the director of the Farlow Herbarium at Harvard, where she would work until retirement. But the one constant in her life remained *Stereocaulon*. Any lichenologist will tell you that *Stereocaulon* is a handsome but difficult genus. Some of the 130 or more species of *Stereocaulon* grow in open boreal forests, on siliceous bedrock and boulders, on the sandy edges of roads. The species are often hard to distinguish from one another, Stephen Clayden, a curator emeritus of botany and mycology at the New Brunswick Museum in Canada, told me.

"Take all the hardwood trees, such as maples, elms, ashes, and birches, and shrink them down until they were an inch in size, and all you had to differentiate them was distinctions in the shape of the leaves," he said. "You can imagine the challenges." This minute subtlety renders *Stereocaulon* more vexing to classify than many other lichens, organisms that famously elude easy classification.

Mackenzie collected and desiccated *Stereocaulon* specimens for thirty years in preparation of the monograph she hoped would be the magnum opus of her career. She had built a reputation as a meticulous lichenologist and taxonomist, and the leading international authority on *Stereocaulon*, Clayden said. The genus's variable nature makes Mackenzie's research all the more impressive, as she classified species based largely on these tiny distinctions and how she suspected they developed and evolved. Researchers now rely on genetic analysis to elucidate differences between species.

But Mackenzie's years of hard work and dedication were foiled by what might be called bad luck. In one instance, after several weeks of collecting lichens in mountains near Tucumán, Argentina, a steely gust of wind untied the string holding bundles and bundles of her collections, which tumbled thousands of feet, littering the valley with white herbarium paper. In the 1950s and 1960s, burdened by a number of debts, Mackenzie was forced to sell a number of her collections and live in the Farlow.

Around this time, Mackenzie's mental health was rapidly deteriorating, and she was hospitalized for three weeks. After separating from her wife, Mackenzie found a specialist in New York City, who diagnosed her with dysphonia syndrome—a voice-box disorder. When Mackenzie told Llano, the lichenologist who wrote her obituary, about her plans to transition, Llano asked if she might have second thoughts. Mackenzie told him there were times when she had contemplated suicide, Llano writes. She also assured him it would not affect her research, indicating her clear desire to continue her scholarship.

In her final years at Harvard, Mackenzie joined a theater troupe in Cambridge, Massachusetts, directed by Laurence Senelick, a now-retired professor of drama at Tufts University. In 1969, for her final performance in Ken Campbell's comedy *Anything You Say Will Be Twisted,* Mackenzie asked to be credited as Elke. Soon after, she asked to attend one of Senelick's Oktoberfest parties in women's clothing, a "coming-out in an environment she assumed would be welcoming," Senelick told me in an email. "'Of course,' I said."

Mackenzie's obituaries note that she retired from the Farlow in 1972, a seemingly premature decision, as she was just sixty years old. They also claim that she lost interest in botany. While there is no record of the official reason behind her retirement, Senelick remembers hearing that Mackenzie's department was "appalled" to learn of her transition and offered her early retirement. "I admired her courage in deciding to live as she wanted, at an advanced age, jeopardizing a secure position," Senelick said.

After her early retirement, Mackenzie continued to work on her *Stereocaulon* manuscript in the library on nights, weekends, and holidays, according to Donald F. Pfister, the curator of the Farlow, who met Mackenzie when he was a graduate student. "This was really a life work," Pfister wrote in an email, adding that her research filled a critical gap in lichenography. "These were exceptional accomplishments."

Mackenzie published infrequently throughout her career, producing just four major works between 1939 and 1968. This pace did not reflect any kind of sluggishness, but rather a painstaking work ethic. Mackenzie intended for all of her work to be exhaustive, comparing specimens from as many locations as possible to avoid taxonomic redundancy. Her scrupulousness sparked an ongoing rivalry with Carroll W. Dodge, a lichenologist infamous for describing heaps of new, reportedly endemic Antarctic species based on lone fragments of the lichens. In Mackenzie's eyes, Dodge's taxonomy was reckless. (Clayden corroborates this account, adding that he often revised Dodge's identifications of specimens at the New Brunswick Museum.)

In 1977, Mackenzie published her final conspectus on *Stereocaulon*, which Clayden sees as "a kind of hypothesis on *Stereocaulon*'s major natural patterns long before anyone could have anticipated that molecular biology could provide insights." Many of Mackenzie's ideas and taxonomic groupings stand up to preliminary molecular research, Clayden says. "I have tremendous admiration for the work she did," he says. "I wish I'd met her."

Details concerning the final years of Mackenzie's life are sparse. In 1976, she retired to an A-frame bungalow in Costa Rica. When she returned to Cambridge in 1980, she struck up a new hobby.

"A chance visit to Salem's Peabody Maritime Museum started her interest in making reproductions of whaler's sea chests, including fancy knotwork for the handles," Llano writes. In 1983, Mackenzie was diagnosed with the neurological disease amyotrophic lateral sclerosis, or ALS, and passed away seven years later.

Today, Mackenzie's scientific papers, her presence in the archives, and a collection of her diaries at Tabarin published in 2018 are all attributed to her former name. That name remains the namesake of five species of lichens and one genus of green algae. She is misgendered in her scientific obituaries, one of which refers to Elke as the scientist's "alter ego." This may have been understandable if Mackenzie hid her identity from the lichenological community at large. But it is clear she did not. In her conspectus, the culminating work of her career, she cited herself in a footnote following her former name.

More than 1,190 of Mackenzie's specimens remain in the Farlow collections, a vast assemblage still consulted by professional and amateur researchers around the world. Her Antarctic specimens in particular have taken on significant newfound value as climate change warms the Earth's poles, according to Michaela Schmull, the Harvard University Herbaria director of collections. Her final specimen, or at least the final specimen recorded in the herbarium's database, is the brown alga *Pylaiella littoralis,* which she collected in 1973 from mudflats in Maine. Though the typewritten label header is listed under her former name, it cites another collector: E. Mackenzie.

Geologic time does not remember all kinds of life equally. The earth often preserves only hard creatures—or the hard parts of creatures—for future generations to learn from, leaving us with heaps of ammonites and dinosaur bones but precious little record of the softer life forms. "Unhappily the geological record can tell us little about how lichens evolved; lichens do not fossilize readily," Mackenzie wrote in an article published in *Scientific American* in 1959.

It is always hard to descry a person's experience from the scientific record, from vast notes on collections: obscure descriptions of even more obscure fungi. This task becomes near-impossible when the best-recorded parts of a person's life may not accurately reflect their identity. But Elke Mackenzie's fleeting moments of self-citation—writing herself into her own authorship, even in a footnote—illuminate the hopes of someone who, against ease and tradition, did not wish to separate her identity from her research. They offer a glimpse into what might have been the most peaceful parts of Mackenzie's career: a retired scientist, typing out the culmination of her scholarship from a self-built bungalow in tropical Costa Rica. A sixty-one-year-old woman wandering a muddy beach in search of patches of green, continuing her life's work.

JENNIFER SENIOR

Happiness Won't Save You

FROM *The New York Times*

MORE THAN FORTY years ago, three psychologists published a study with the eccentric, mildly seductive title, "Lottery Winners and Accident Victims: Is Happiness Relative?" Even if you don't think you know what it says, there's a decent chance you do. It has seeped into TED talks, life-hack segments on morning shows, even the occasional whiff of movie dialogue. The paper is the peanut-butter-and-jelly sandwich of happiness studies, a staple in any curriculum that looks at the psychology of human flourishing.

The study is straightforward. As the title suggests, the authors surveyed lottery winners and accident victims, plus a control group, hoping to compare their levels of happiness. But what the authors found violated common intuition. The victims, while less happy than the controls, still rated themselves above average in happiness, even though their accidents had recently rendered them all either paraplegic or quadriplegic. And the lottery winners were no happier than the controls, at least in any statistically meaningful sense. If anything, the warp and weft of their everyday lives was a little more threadbare. Talking to friends, hearing jokes, having breakfast—all of these simple pleasures now left them less satisfied than before.

There were flaws in the study—its design, alas, was as crude as an ax—but you can see why it became famous. It had an irresistible takeaway: *Money! It doesn't buy you happiness!* Perhaps even more fundamentally, it had a sexy, almost absurd, premise. What kind of mind would think to pair lottery winners and accident victims in a research paper? Who in academic psychology had such a cockeyed imagination? It was social science by way of Samuel Beckett.

The answer to that question is a fellow by the name of Philip Brickman, a thirty-four-year-old rising star at Northwestern University. He was warm, irrepressible, spellbinding to talk to; his mind was a chirping hatchery of ideas. Unlike so many of his peers, his preoccupations had little to do with cognitive processes. Rather, they had to do with matters of the heart: how we cope with adversity; how we care for others; how we form commitments, subdue inner conflicts, wrench meaning and happiness from this brief life.

"He wanted the world to be a more humane place," his closest friend, Jeffery Paige, told me.

So for Brickman to come up with a study like this one made perfect sense. It was idiosyncratic, humanistic and, above all, relevant: Does money fulfill us? Does irremediable damage to the body cause irremediable damage to the spirit? Can we simply adapt to anything?

What, ultimately, do we need to carry us through?

Not long after publishing that study, Brickman left Northwestern for the University of Michigan, where he'd become the director of the oldest and most storied arm of the Institute for Social Research. It was a prestige gig, an honor often reserved for academics at the pinnacle of their careers. Paige, a professor emeritus of sociology at the University of Michigan, told me he thought Brickman was destined for the National Academy of Sciences one day.

We'll never know. On May 13, 1982, at the age of thirty-eight, Philip Brickman made his way onto the roof of Tower Plaza, the tallest building in Ann Arbor, and jumped. It was a twenty-six-story fall. The man who'd done one of psychology's foundational studies about happiness couldn't make his own pain go away.

According to those who knew him, Brickman was not a man who struggled with ongoing, intractable suicidal impulses. Depression and feelings of deep inadequacy, yes. But *suicide?* Not that they knew of, not until the final weeks of his life.

"To imagine what could have driven him to do that—I almost had to imagine a different person," Vita Carulli Rabinowitz, one of his former graduate students, told me. "So it made me wonder: Was there an underlying disorder that we just didn't see?"

Most suicides are cruel mysteries, the suffering of the deceased "private and inexpressible," as Kay Redfield Jamison put it in her 1999 masterpiece, *Night Falls Fast: Understanding Suicide*. But Brickman wrote over fifty book chapters and academic papers in his short life, plus a book, published posthumously. So if Brickman was suffering from an underlying disorder, as Rabinowitz suggests, there was also an awful lot that was hiding in plain view. It is tempting, in hindsight, to wonder if his scholarship wasn't a trail of bread crumbs—one long, unconscious attempt to unknot the riddle of his vexed self.

It seems safe to say that much of it was. But it also seems safe to say that his scholarship wasn't enough.

Was there something he missed? If so, would it have made a difference if his insights had been complete? There will always be a gulf—bridgeable for most, but unbridgeable for a tragic few—between understanding what ails us and having the means or desire to bring those difficult feelings to heel.

It may even be worth asking whether understanding is quite beside the point—more of a requirement for the living than for the dead.

What are we to learn from this man?

As a professor, Brickman was an affecting combination of exuberant and awkward, exacting and underconfident. He was an awful lecturer. But he was an intoxicating conversationalist, the type who'd burst into your office whenever a new thought occurred to him, eager to discuss it, even more eager to collaborate. The first time he met Camille Wortman, now an emeritus professor of psychology at Stony Brook University, his opening conversational gambit was to ask whether she thought serial killers deserved compassion. He had just picked her up from the airport for a job interview.

"For me," she told me, "it was intellectual love at first sight." During their talks, she'd often have to stifle the urge to run back to her office and jot down notes.

Yet Brickman was bedeviled by insecurities, both physical and intellectual. He was homely, his face perforated with acne scars, his lip crowned with an extravagant walrus mustache. He was affectionate but quick to take offense, supportive but high-maintenance—tender in every sense.

"If he got a negative review on a publication he submitted," said Wortman, "he would go insane. Nobody likes to get a bad review, but it had a profoundly negative effect on Phil. He would rant for *days.*"

"You really didn't want to have a meeting with Phil on your calendar," said Rabinowitz, the recently retired executive vice chancellor and provost of CUNY. He was needy. It was odd. He was cherished, even if he taxed your patience. And so obviously brilliant. Yet he didn't seem to have half the admiration for himself that others had for him. "I don't think he felt appreciated enough," she told me. "I don't think he achieved the recognition that he thought he deserved."

Whatever he did achieve, he never considered it good enough. He wore his perfectionism like a hair shirt, and he expected it of others. He'd give people grief if they stapled a paper in the wrong place.

The irony is that, better than almost anyone, Brickman understood that the pursuit of stature, material bounty—and ultimately happiness itself—was a fool's errand. Early in his career, he grasped that the more we achieve, the more we require to sustain our new levels of satisfaction. Our gratification from the new is fleeting; we adapt in spite of ourselves. "Fulfillment's desolate attic," as the poet Philip Larkin once put it. You may as well chase your afternoon shadow. Happiness always looms ahead.

In 1971, Brickman and the psychologist Donald T. Campbell went so far as to coin a term for the pointless quest for more, more, more: "The hedonic treadmill." The term stuck. "There may be no way to permanently increase the total of one's pleasure," they concluded, "except by getting off the hedonic treadmill entirely."

Which is all very well. But what on earth do you live for, if not happiness?

Your commitments, according to Brickman. They were the true road to salvation, he decided, the solution to an otherwise absurd existence. He recognized that they didn't always give pleasure; they may even "oppose and conflict with freedom or happiness," as he wrote in his book *Commitment, Conflict, and Caring,* published five years after his death. But in many ways, that was the point: the more we sacrifice for something, the more value we assign to it.

"Happiness," he wrote, "involves the enthusiastic and unambivalent acceptance of activities or relationships that are not the best that might possibly be obtained."

"What Phil led me to realize is that a lot of psychology operates on this rational notion that, *Well, I do the thing that works best for me,*" said Dan Coates, a former student of Brickman's and the second author on the lottery study. "But I think ultimately Phil would have said happiness is not what maintains us. What really maintains us is *un*happiness."

Which is a liberating—even electrifying—idea, particularly if you find happiness elusive, as Brickman did. "I think he was feeling that there was something a little bit wrong with him," Wortman told me, "because he had achieved so much personal and professional success, yet it wasn't as satisfying as he had hoped."

There was only one problem with Brickman's theory. Commitments, too, can be fragile and transient. Maybe less fragile and transient than the dopamine high of getting a paper published or falling in love. But fragile and transient nonetheless. Relationships end; jobs don't work out. It's a very painful discovery to make. The bonds we often think of as ropes are really gossamer threads.

Jeffery Paige used to say that he envied Brickman's family life. And on the surface, there was certainly a lot to envy: three adorable girls, a lovely wife, an idyllic farm outside Ann Arbor. But that portrait of domestic serenity was hard won. Brickman didn't exactly come from a family where commitment came naturally. His father was a tomcat, forever destabilizing the household with his extramarital affairs, his sister, Julie, told me, and Brickman was an anxious, insecure little boy.

"He would try so hard," said Julie. "He bought a million joke books so he could learn how to tell jokes to be funny, to amuse people."

But the month he turned seventeen, Brickman went to Harvard and finally found his own kind, and when he got to graduate school, he met his wife, BB. He was smitten, utterly.

When things went south is hard to say. But they did, as was made clear in a series of letters uncovered by Benjamin Robert Wegner, a therapist who recently completed a dissertation about Brickman. One, from Phil to BB, about her disdain for social psychology: "If you are really hostile and contemptuous toward the field as you sometimes seem, you must either want me to change my line of work or wish you had another life partner for whose work you could have more respect."

Another, from Brickman to a friend, about how much condescension his wife claimed to feel from *him:* "Part of my job, to put it in terms that are far too blunt for an insider, was to keep BB from feeling inferior." He took self-punishing steps to do so, he added, including cutting off friends from whom she detected scorn, even freezing out his sister and minimizing contact with his parents.

BB is no longer alive to address this interpretation of events. She died from Parkinson's disease in 2017. But her younger brother, Rick Schaeffer, told me he's pretty certain she would have viewed the decline of her marriage differently.

Had Brickman been a faithful husband? Not entirely. BB told her brother that he'd strayed once while at a conference.

But the woman was a stranger; the lapse, while painful, proved surmountable. Far more challenging, Schaeffer told me, was that Brickman had become impossibly needy toward the end of the marriage, effectively a fourth child. The move to the Michigan farm and the strains of his new, high-profile job had undone him —dissolving his sense of humor, coiling him with anxiety, rendering him more demanding, more intense. He spent most of his time in his study, neglecting the few household duties he had.

"BB was there to solve his problems," Schaeffer told me. "And to do his laundry."

For the first time in his life, Brickman seemed to be experiencing an unfamiliar sensation: failure. Just before he died, he applied for a large research grant; he didn't get it. His new post required lots of administrative and organizational prowess; he didn't have it. His colleagues were forever giving him an earful. "He got consistent feedback that they were disappointed in him," Wortman told me.

Sometime over the summer of 1981, said Schaeffer, BB asked Brickman to move out. You can imagine his isolation—and devastation.

"His two biggest commitments were his family and his work," Coates told me. "And both of those seemed to be crumbling in front of him."

Brickman spends many pages in *Commitment, Conflict, and Caring* contemplating the pain of a commitment-less life, especially in his final chapter entitled "Commitment and Mental Health." The chapter was a mess, in chips and shards, when Coates first started to assemble it after Brickman's death. What struck him most was how much *despair* he found in Brickman's scribblings. Both men agreed that losing your commitments was an existential problem, robbing individuals of direction and value. "But he would always write about how *painful* that was," Coates told me. "I guess I never experienced that level of pain. He emphasized it."

In the final text, the pain won. The chapter includes quite a bit about suicide.

Brickman moved into a grim, generic one-bedroom apartment in Ann Arbor. Its one notable feature was the little notes he'd taped up everywhere—"aphorisms or psychological sayings or things to encourage himself and make himself happy or feel good," said Julie, his sister. "They were really sad to me."

The revered professor—whose associative imagination was legend among students, who plumped like a sponge at the very mention of a new idea—was now trying to buck himself up with the hokum of fortune cookies.

Toward the end of April, maybe three weeks before Brickman died, Jeffery Paige got a call from him, saying he was in trouble and needed to talk. They met for dinner at a local restaurant. Brickman was as despondent as Paige had ever seen him. "Everything went down this avenue to: *Everybody else is happy and I'm not,*" Paige told me. The spring weather was making a mockery of his misery.

Paige was so alarmed that he drove Brickman to the emergency room. But when a woman on staff finally interviewed him, she let him go, declaring he posed no danger to himself. ("She treated me like *I* was the one who needed a therapist," Paige recalled.) Paige, far less convinced, extracted a promise from Brickman that he'd see a psychiatrist the next day, and Paige found him one, through a mutual friend.

I asked Paige what he thought happened in that examination room. "I think he talked her out of it," he said. "For a man who was an expert in helping, he really resisted getting help."

But that's not exactly true. Brickman *was* seeking help toward the end of his life. At the point that Paige found him a psychiatrist, Brickman was already seeing a therapist, one of the best in Ann Arbor. And shortly after he had dinner with Paige, he walked into Camille Wortman's office and told her point-blank that he was thinking of killing himself.

To this day, she regrets her response, not because of anything she did—she hugged him, let him know how much he was loved —but because of what she *didn't* do. "When someone tells you they're thinking of committing suicide, you're supposed to ask if they have a plan," she explained. "And if they have a plan, you're supposed to ask if they have the means to carry it out."

At the time, she had no idea. As strange as it sounds, it was exceptionally rare in 1982 for psychology professors to receive training in suicide prevention. More than twenty years later, in 2003, a survey still showed that only 50 percent of psychology interns reported receiving training in managing the suicidal—and those were clinicians-in-training, not social psychologists.

The fact was, Brickman was unraveling in front of almost no one *but* psychologists. And none knew what to do.

Yet many of them had thought deeply about the complicated challenges associated with helping people, and so had Brickman himself. The month before he died, he published a paper called "Models of Helping and Coping."

It is painful to read today. The paper makes an earnest attempt to break down all of our helping and help-seeking behaviors into four categories: those who think they're responsible for both their problems and their problems' solutions; those who think they're responsible for neither; those who think they're responsible for the solution to a problem but not the problem itself; and those who think they're responsible for the problem but not the solution.

It's this last category that turns your blood to ice. It basically describes people who think they've made a mess of everything but are inherently powerless to fix it, and therefore must permanently surrender their fate to a higher power. Alcoholics Anonymous, as the paper says, would be the healthiest scenario. But Jim Jones, as the paper also notes, would be the worst. He led more than 900 people to their suicides at his cult compound in 1978.

"I've always thought of that model as the most hellacious," said Vita Carulli Rabinowitz, the second author on the study. "And I've wondered in retrospect if it's the one Phil applied to himself. I don't think he felt he was in control." But he did feel responsible for his problems. He blamed himself for the dissolution of his marriage, and no one could persuade him to see it any other way.

Brickman didn't surrender himself to a cult, obviously. He first spent ten days at Mercywood, a psychiatric facility in Ann Arbor. But he signed himself out on May 13. It's unclear who picked him up—maybe his parents, who were in town, knowing their son was in distress—or what he did next. Though his sister, Julie, told me that he had a couple's therapy appointment that day.

But whatever else happened, he eventually got to the roof of Tower Plaza. It is 267 feet tall, 22 feet higher than the deck of the Golden Gate Bridge.

Julie Brickman said that her father saw his son's body in the street.

Rabinowitz was sitting on the floor of her study, working, when a fellow graduate student phoned with the news. She was still flush with pride over the positive reception "Models of Helping and Coping" had gotten. It was based on her dissertation. She and Brickman were preparing to do a follow-up.

She never revisited the material again. "He couldn't cope," she told me, "and we couldn't help."

A note, for a moment, on suicide by jumping: almost everyone I spoke to made a point of lingering on this gruesome part of Brickman's story. It's such an operatic gesture, a metaphor made literal, an actual flight from the devil's choir of your own suffering. But it's also such a ghastly method—so graphic, so violent, so *public*.

"It was clearly F-you to everybody," said Roxane Cohen Silver, who worked with Brickman both as an undergraduate and a graduate student at Northwestern, now a professor of psychological science at the University of California, Irvine. "He chose to die in a place that everybody would see on a daily basis." Tower Plaza is catty-corner to the Institute for Social Research, probably visible from half its windows. "It was a way to say to people, *You let me down, and here I'm going to make you pay.*"

Julie Brickman thought it was a hostile message to her brother's wife. Jeffery Paige thought it was a hostile message to his shrink, who he's pretty sure had an office in Tower Plaza.

I, too, got sucked into thinking about the meaning of jumping to your own conclusion, in every possible sense. In his work, Brickman was captivated by something called "opponent-process theory," which noted that much of human experience was marked by a positive feeling followed by a negative one or vice versa—a drug addict experiencing a high followed by a low, a crime victim experiencing terror followed by relief. Almost nothing would seem to capture this paradoxical phenomenon like jumping from a height of 267 feet. It's 4.07 seconds of screaming sensation followed by pitch-black extinction.

But then Julie told me a crucial detail: this was Philip's third attempt at suicide. Shortly before, he'd tried to get enough drugs for an overdose, but failed; he'd also tried to jump from his parents' car into traffic.

I asked Coates, who in our discussions often took a wry, existentialist view of life, what *he* made of Brickman's jump from Tower Plaza. Everyone, including myself, was assigning so much psychological power and symbolism to it. But was there, really?

He only thought about it for a brief moment. No. "I don't think he was doing it for the drama," he said. "He was doing it for the efficacy."

Many people get divorced. Many people find their work disappointing. Very few of them attempt suicide.

Predicting who will die by suicide with any precision has long eluded mental health professionals. There are no refined algorithms for it. Matthew K. Nock, a professor of psychology at Harvard and the country's foremost scholar on suicide, did a famous study in 2010 showing that clinicians in a large city psychiatric ER were no better than a coin toss at predicting who, after leaving their care, would attempt suicide.

We do know which factors seem to make suicide more likely. Brickman checked almost every box. Male. Living alone. Recently separated. Two previous suicide attempts. And the period during which one is at greatest risk for dying by suicide is just after leaving a psychiatric institution.

But the most common feature among people who die by suicide is symptoms of mental illness. Nock says they're present in 95 percent of all suicides, at least in the days or weeks before the event. And in Brickman's case, they were present for far longer than that.

Depression chased Brickman his whole life. Julie noticed it in him even as a little boy, which made him difficult to envy, in spite of his academic accomplishments. In Brickman's archives, Wegner found a letter Brickman wrote to his father in 1965, and it contained this stunning paragraph:

> I see no reason for not saying that I am as happy now as I ever will be. And still I can wake up in the morning and be depressed. Very depressed. For reasons that sometimes make sense, and sometimes don't. About the war in Vietnam. About my work, or about BB. Or even about being depressed itself. BB feels she knows this and can learn to live with it. She is quite capable of getting depressed too. I don't know whether I can learn to live with it, but I do know that I'll never be without it.

It's hard not to read that last line in the context of Brickman's scholarly obsessions. He was preoccupied by how readily we adapt to life events—how we tend to return to some kind of baseline after fortune strikes, good or ill. A "hedonic set point," as psychologists like to call it. And if that's the case, it seems possible that Brickman could have wondered, in his darkest moments, if he was condemned to a lifetime of sadness.

He could win the lottery.

He could lose the use of his arms and legs.

But he'd always return to the same place.

I don't know whether I can learn to live with it, but I do know that I can never be without it.

Was that what the lottery study was really about?

Yet Brickman was not always suicidally low. He worked, loved, formed friendships and a family. He had the energy to publish prodigiously and mentor generously and chatter joyously, for days, about ideas that tickled him.

What pitched him into a gloom so thick he couldn't see his way out?

We can try to speculate about what took Brickman to the ledge of Tower Plaza. But we can't ever know for sure. The living are reduced to police sketch artists, working off scraps, wisps, dusky recollections. They'll never really know what the beast looked like to the dead.

In his book *The Savage God,* the critic A. Alvarez writes about the insufficiency of explanations for suicide. At best, he wrote,

> they soothe the tidy-minded and encourage the sociologists in their endless search for convincing categories and theories. They are like a trivial border incident which triggers off a major war. The real motives which impel a man to take his own life are elsewhere; they belong to

the internal world, devious, contradictory, labyrinthine, and mostly out of sight.

Brickman's work may have had much to say about the futility of pursuing happiness. But that wasn't Brickman's problem, ultimately. Those who kill themselves don't do so because they find happiness elusive, or even if they're outrageously unhappy. They do it to liberate themselves from unendurable pain. That's what Brickman was experiencing. Pain without cease. "Suicide," wrote Jamison, "is the last and best of bad possibilities."

A colleague who knew I was working on this story asked me whether there was any suicide in my own family. Yes, I told him, there is, though I'd been lucky enough to escape the curse myself: I am not one of the 4.8 percent of American adults who seriously contemplated self-slaughter in 2019 (or any other year). But two of my eight great-grandparents died by suicide, I said, one on each side of the family.

One left a note saying he had cancer and gassed himself. So there was some ambiguity there, some controversy. Maybe it wasn't crushing depression.

And the other . . . ?

The other, I said. The other, the other . . .

She had jumped off the roof of a building in Brooklyn.

Never once, at least consciously, had I thought about this while reaching out to Brickman's friends and colleagues and relations. It was my other great-grandparent who had consumed all the family-suicide mindshare. My mother and I had more than once discussed whether it was truly cancer, or whether it was depression masquerading behind an explanation my great-grandfather considered less shameful.

But my other great-grandparent: there was no ambiguity there. She jumped. It is virtually the only thing I've ever known about her.

We all find ways to study ourselves.

This month, a trio of economists published a paper that added both more nuance and more sizzle to the lottery study, showing that winners in a variety of Swedish lotteries did indeed rate their overall life satisfaction higher, even a decade later, than did a control group. Not their day-to-day happiness, the authors noted—that didn't seem to show a statistically significant change. But it affected the overall evaluation of their lives.

Maybe the playwright Richard Greenberg was right. Money doesn't buy you happiness. But it does upgrade despair.

The study was a welcome addition to the growing body of literature about the relationship between money and well-being. Brickman's paper had a comically small sample size, and its controls were lifted from a neighborhood phone book. Today, such a design would be laughed out of peer review.

But there's one part of Brickman's study—much less known, much less discussed—that I return to, stubbornly, in my head, whenever I think about him. It is this: all three groups were asked how they thought they'd feel in a couple of years. The accident victims. The lottery winners. The controls. And of those three groups, it was the accident victims who envisioned the happiest tomorrow. They carried the most optimism in their hearts.

Yet Brickman somehow couldn't do that. At the end of his life, the most pertinent lesson from his most famous study was something he could only know, not feel. I asked Coates about this. He said it was consistent with an idea he and Brickman had discussed at length. It even appears in Brickman's book.

"When depression is actually functional," he explained, "that pain *helps,* because it lowers your baseline, so that you're willing to accept alternatives you wouldn't have accepted before. And then you can start to build new relationships and new purposes and new meanings." To remarry, for instance. Or find a new job. "So that's the part that puzzles me," said Coates. "Why didn't Phil realize that eventually you get through this pain?"

That is, perhaps, the most relevant question for those who die by suicide. Why do they believe they're trapped in a permanent present of their own suffering?

If Matthew Nock or anyone else could answer it, suicide wouldn't be the tenth leading cause of death in the United States. But as it stands, the conviction that pain is irremediable remains one of the hallmarks of the suicidal. Nock cited a classic paper in psychology called "Pain Demands Attention," which is pretty straightforward: people who are in extreme pain are focused on the present, because their pain is so distracting they can think of little else. They have a much harder time remembering the past, when that awful pain did not exist.

"And in order to think about the future," Nock pointed out to me, "you have to flexibly recombine memories from the past."

So what do you do, as a clinician or loved one, when faced with such suffering? How do you see them through?

You help them generate thoughts about the future in concrete, specific detail.

You point out, using specific numbers, how many years they lived without thoughts of suicide, versus the number of years they have had such thoughts.

You help them find therapy, ideally cognitive behavioral therapy, and drag them to sessions if you have to.

You take away their ability to kill themselves.

You try to keep them safe.

These efforts may not work. Suicidal people can be determined, and they know how to dissemble, how to feign stability. But you try. Most are relieved when they fail. And approximately half of them, according to the most up-to-date research, will not be feeling suicidal at the same time the following year—or ever again.

You are six, ten, and thirteen years old. It is nighttime. A police car pulls into your driveway, and an officer pops out, rings the doorbell, and asks for your mother. But she is not home; she is at a PTA meeting. You all go to bed. Eventually, she returns, is told the news. The next morning, she tells you that your father is dead. He has jumped off a building.

The Brickman daughters were raised by one of the most psychologically oriented men of his generation. Yet after he died, BB never spoke of their father again. His name was not mentioned. His death was not discussed. The girls never visited his grave. "She was the only parent left," Sarah, the middle daughter, told me. "And she wasn't a guide to helping us know or love the father who'd just left *us*. At all."

They buried whatever questions and feelings they had beneath a separate headstone.

The oldest, Rachel, now has Parkinson's and lives with her youngest sister, Katharine. Speech for her is painful, effortful, but the one thing she communicates is that her mother went to great lengths to put the episode behind her. When BB became a nurse many years later, she told colleagues her husband died of cancer.

Katharine was probably closest to her father. She sensed his depression even as a little girl—he had that penumbra of sorrow about him—and she tried, in her own way, to help, deliberately allowing him to win at cards. For twenty-one years, she wandered the world with a misunderstanding common to children who witness their parents' distress: that her father's suicide was her fault, that she should have been able to save him. Then, at twenty-seven, she finally realized: "He wanted to leave, and it was his *right* to leave. It wasn't about me."

But it was Sarah, Brickman's middle child, who possibly paid the steepest price. In adulthood, she told me, she keeps finding men who need saving, who are broken somehow. Though she remembers little about her father, what memories she does have are grim, painful: Of her dad begging her to ask her mother whether she still loved him, for instance, because he still loved her—would she tell her mom that?

Or: Of sitting in his office at the Institute for Social Research, probably on the sixth floor, and asking what would happen if she fell out the window. The place had such big windows! She doesn't remember his response. She just knows that shortly thereafter, he went to the top of the tallest building in the city and jumped.

I asked if this meant she'd spent the last thirty-eight years secretly believing she'd seeded a family tragedy. She first told me no, not really; she knew her father had tried suicide before. "But if I did, what a jerk," she blurted. "Did he think that wouldn't negatively affect me for the rest of my life?" She cried as she said this. It was the first time, she explained, that she'd said this thought out loud.

But she is mainly protective of her father, more angry with her mother for not allowing the children to properly grieve. About ten years ago, she finally started to investigate who Philip Brickman was, hunting down former colleagues. She was stunned to discover how affected they were by his death.

"I never knew that he was such a superstar," she told me, her voice breaking again. "I was raised to think he was a nobody."

If those who die by suicide finally escape the tyranny of the awful present, they pass along that tyranny to the survivors they leave behind. The living spend years, decades, sometimes a lifetime imagining the final moments of the dead, guessing at their motives, berating themselves for the things they could have done. "Irresolvable guilt," is what Jamison calls it.

Children may have one version. Adults have another. Julie Brickman, a warm and perceptive woman (once a therapist herself, now a novelist), told me it took her a decade to stop imagining what her brother's ascent to the roof of Tower Plaza looked like. "I still want to see *how* you get up there," she said.

I asked how she coped with the plague of counterfactuals that beset most survivors: What if I'd done X, what if I'd done Y? She told me the most useful thing she'd learned after years of therapy, an elegant and simple epiphany: you have to take the suicide out of the question.

"You have to ask yourself: If life had just progressed"—meaning if your loved one *hadn't* committed suicide—"did you do anything terrible? And the answer, 95 percent of the time, is *No. I didn't.*"

She didn't. When Julie last saw her brother, she treated him with love and support and kindness.

Brickman's death reverberated in different ways with different colleagues and students. But none was left unchanged by it. Rabinowitz abandoned her work on helping and coping. Coates left academia altogether and went to work in public school administration in St. Louis and its suburbs. "What hit me was, *Gosh, if this career ended up killing him and he was so much better at it than I am, what's it going to do to me?*" Off the treadmill he hopped.

Wortman, on the other hand, leaned into her distress, learning everything she could about suicide prevention—she taught a class on it each year, dedicating it to Brickman—and focused her research exclusively on how survivors cope with the sudden loss of a family member or friend. Silver, who had already homed in on this same subject, had the unlucky-but-instructional opportunity to test her hypotheses about adversity firsthand. The death of Phil Brickman was her first true tragic event. She was twenty-six years old.

"It helped me solidify my view," she told me, "that coming to terms with a loss does not necessarily mean that one must *make sense* of that loss."

But that doesn't mean forgetting. "Every time Camille and I talked, *all* we talked about was Phil—*for years,*" she said. "I was immersed in thoughts about him: Why he did it and how he did it and did he regret it once he started to jump?"

There was a reason she thought about those questions. "They were the kinds of things that Phil and I would talk about."

It was the quintessential Philip Brickman conversation. A man lingers on the rooftop of a high-rise, despairing. He makes his way toward the edge. And after some time—seconds? minutes? hours? no one knows—he decides to jump. The ultimate commitment. Or is it really a rejection of all commitments? Whatever it is, does it feel right? Does it help? Or does it suddenly reveal, my God, that there was actually another way?

Did he make the right choice?

If only he'd been around to discuss it. It is terrible that Brickman is not here to discuss it. He would have had, you can only imagine, so very much to say.

If you are having thoughts of suicide, call the National Suicide Prevention Lifeline at 1-800-273-8255 (TALK). You can find a list of additional resources at SpeakingOfSuicide.com/resources.

LATRIA GRAHAM

Out There, Nobody Can Hear You Scream

FROM *Outside Online*

IN THE SPRING of 2019, right before I leave for my writing residency in Great Smoky Mountains National Park, my mama tries to give me a gun. A Ruger P89DC that used to belong to my daddy, it's one of the few things she kept after his death. Even though she doesn't know how to use it, she knows that I do. She's just had back surgery, and she's in no shape to come and get me if something goes wrong up in those mountains, so she tries to give me this. I turn the gun over in my hand. It's a little dusty and sorely out of use. The metal sends a chill up my arm.

Even though it is legal for me to have a gun, I cannot tell if, as a Black woman, I'd be safer with or without it. Back in 2016, I watched the aftermath of Philando Castile's killing as it was streamed on Facebook Live by his girlfriend, Diamond Reynolds. Castile was shot five times at close range by a police officer during a routine traffic stop, when he went to reach for his license, registration, and permit to carry a gun. His four-year-old daughter watched him die from the back seat. In his case, having the proper paperwork didn't matter.

I'll be in the Smokies for six weeks in early spring, the park's quiet season, staying in a cabin on my own. My local contact list will be short: the other writer who had been awarded the residency, our mentor, maybe a couple of park employees. If something happens to me, there will likely be no witnesses, no one to stream my last moments. When my mother isn't looking, I make sure the safety is on, and then I put the gun back where she got it. I leave my fate to the universe.

Before I back out of our driveway, my mama insists on saying a protective blessing over me. She has probably said some version of this prayer over my body as long as I've been able to explore on my own.

In 2018, I wrote an article for this magazine titled "We're Here. You Just Don't See Us," about my family's relationship to nature and the stereotypes and obstacles to access that Black people face in the outdoors. As a journalist, that piece opened doors for me, like the residency in Great Smoky Mountains National Park.

It also inspired people to write me.

Two years later, the messages still find me on almost every social media platform: Twitter, Instagram, even LinkedIn. They come through my Gmail. Most of them sound the same—they thank me for writing the article and tell me how much it meant to them to see a facet of the Black experience represented in a major out-door magazine. They express apprehension about venturing into new places and ask for my advice on recreating outside of their perceived safety zone. They ask what they can do to protect themselves in case they wind up in a hostile environment.

Folks have their desires and dreams tied up in the sentences they send me. They want to make room for the hope that I cautiously decided to write about in 2018.

Back then, as a realist, I didn't want my essay's ending to sound too optimistic. But I still strayed from talking about individual discrimination in the parks, often perpetrated by white visitors, like the woman who recently told an Asian American family that they "can't be in this country" as they finished their hike near Mount Tamalpais in Marin County, California, this past Fourth of July. Or the now famous "BBQ Becky" who called the police on two Black men at Lake Merritt in Oakland, California, in 2018, for using a charcoal grill in a non-charcoal-grill-designated area. Nor did I mention that when I venture into new spaces, I am always doing the math: noting the lengths of dirt roads so I know how far I have to run if I need help, taking stock of my gas gauge to ensure I have enough to get away.

I have been the target of death threats since 2015, when I started writing about race. I wasn't sure if magazine readers were ready for that level of candid conversation, so in 2018 I left that tidbit out.

There are risks to being Black in the outdoors; I am simply willing to assume them. And that's why I struggle to answer the senders of these messages, because I don't have any tips to protect them. Instead I invoke magical thinking, pretending that if I don't hit the Reply button, the communication didn't happen. Sometimes technology helps: when I let the message requests sit unaccepted in Instagram, the app deletes them after four weeks.

I deem myself a coward. I know I am a coward.

There are two messages that still haunt me.

The first is an email from a woman who wanted to know what she and her brown-skinned husband should do if they encounter another campground with a Confederate flag hanging in the check-in office. She described to me a night of unease, of worrying if they and their daughter would be safe. I filed her email so deep in my folders that I don't even think I can find it anymore. I was dying to forget that I had no salve for her suffering.

The second was even more personal. It came via Facebook Messenger, from a woman named Tish. In it she says: "I came across a read of yours when I was searching African Americans and camping. I want to rent an RV and go with my family. I live in Anderson S.C. Had a daughter that also attended SCGSAH. Is there a campground you recommend that is not too far and yes where I would feel comfortable? Thank you."

The signaling in it, of tying me to her daughter, examining my background enough to offhandedly reference the South Carolina arts high school I attended and saying, *Please, my daughter is similar to you.*

I leave her message in the unread folder.

These women have families, and they too are trying to pray a blessing over the ones they love while leaving room for them to play, grow, and learn—the same things their white peers want for their offspring. In their letters, they hang some of their hopes for a better America on me, on any advice I might be able to share.

I haven't written back because I haven't had any good advice to offer, and that is what troubles me. These letters have been a sore spot, festering, unwilling to heal.

Now, in the summer of 2020, there are bodies hanging from

trees again, and that has motivated me to pick up my pen. Our country is trying to figure out what to do about racial injustice and systemic brutality against Black people. It's time to tell those who wrote to me what I know.

Dear Tish, Alex, Susan, and everyone else:

I want to apologize for the delayed reply. It took a long time to gather my thoughts. When I wrote that article back in 2018, I was light on the risks and violence and heavy-handed on hope. I come to you now as a woman who insists we must be heavy-handed on both if we are to survive.

I write to you in the middle of the night, with the only light on the entire street emanating from my headlamp. Here in upstate South Carolina, we are in the midst of a regional blackout. My time outdoors has taught me how to sit with the darkness—how to be equipped for it. Over the years, I have found ways to work within it, or perhaps in spite of it. If there's anything I can do, maybe it's help you become more comfortable with the darkness too.

But before I tell you any more, I want you to understand that you and I are more than our pain. We are more than the human-rights moment we are fighting for.

It isn't an exaggeration to say that the *Outside* article changed my life. People paid me for speaking gigs and writing workshops. They put me on planes and flew me across the country to talk about equity, inclusion, and accountability. I know the statistics, the history, the arguments that organizations give about why they have no need to change. I call them on it.

I have to apologize for not being prepared for the heaviness of this mantle at the time. I have to admit my hesitation back then to call white supremacy and racism by their names. The unraveling of this country in the summer of 2020 has forced me to reckon with my actions, my place in the natural world, and the fact that as a Black woman writer in America, I am tasked with telling you a terrible truth: I am so sorry. I have nothing of merit to offer you as protection.

I am reluctant to inform you that while I can challenge white people to make the outdoors a nonhostile, equitable space where you can be your authentic selves, when the violence of white supremacy turns its eyes toward you, there's nothing I can give you to protect yourself from its gaze and dehumanization.

I do not wish to ask you to have to be brave in the face of inequality. This nation's diminished moral capacity for seeing Black people as human beings is not our fault. Their perception of you isn't your problem—it's theirs, the direct result of the manifest-destiny and "anybody can become anything in America" narratives they have bought into. We are made to suffer so they can slake their guilt. I want you to be unapologetically yourselves.

I check with my fellow Black outdoor friends, and they say they've gotten your email and messages too. They also waffle on what to say, telling y'all to carry pepper spray or dress in a non-threatening way. I am troubled about instructing people who have already been socially policed to death—to literal, functional death—to change the way they walk, talk, dress, or take up space in order to seem less threatening to those who are uncomfortable with seeing our brown skin.

I have no talisman that can shield you from the white imagination. The incantation "I'm calling the police" will be less potent coming from your mouth, and will not work in the same way. In the end, your utterance could backfire, causing you more pain.

I want to tell you to make sure you know wilderness first aid, to carry the ten essentials, to practice leave no trace, so no one has any right to bother you as you enjoy your day. I want to tell you to make sure you know what it means not to need, to be so prepared that you never have to ask for a shred, scrap, or ribbon of compassion from anybody.

But that is misanthropic—maybe, at its core, inhumane.

I resist the urge to pass on to you the instinct my Black fore-mothers ingrained in me to make ourselves small before the denizens of this land. I have watched this scenario play out since I was a child: my father, a tall fifty-year-old man with big hands, being called "boy" by some white person and playing along, willing to let them believe that they have more power than he does, even though I have watched him pin down a 400-pound hog on his own. I have seen my mother shrink behind her steering wheel, pulled over for going five miles above the speed limit on her way to her mom's house. She taught me and my brother the rules early: only speak when spoken to, do not ask questions, do not make eye contact, do not get out of the car, keep your hands on the wheel, comply, comply, comply, even if it costs you your agency. Never, ever show your fear. Cry in the driveway when you get to your destination alive. Those traffic stops could've ended very differently. The corpses of Samuel DuBose, Maurice Gordon, Walter Scott, and Rayshard Brooks prove that.

I will not pass on these generational curses; they were ways of compensating for anti-Black thinking. They should never have been your burden.

It would be easy to tell you to always be aware of your surroundings, to never let your guard down, to be prepared to hit Record in case you run into an Amy Cooper or if a white man points an AR-15 at you and your friends as you take a break from riding your motorcycles, hoping to make the most of a sunny almost-summer day in Virginia.

These moments—tied to a phone, always tensed in fear—are not what time in nature is supposed to be. Yet the videos seem to be the only way America at large believes us. It took an eight-minute-and-forty-six-second snuff film for the masses to wake up and challenge the unjust system our people have had to navigate for more than 400 years. They are killing us for mundane things —running, like Ahmaud Arbery; playing in the park, like Tamir Rice. They've always killed us for unexceptional reasons. But now the entire country gets to watch life leak away from Black bodies in high definition.

I started writing this on the eve of what should have been Breonna Taylor's twenty-seventh birthday. The police broke into her home while she was sleeping and killed her. I write to you during a global pandemic, during a time when Covid-19 has had disproportionate impact on Black and brown communities. I conclude my thoughts during what should have been the summer before Tamir Rice's senior year of high school. All the old protective mechanisms and safety nets Black people created for ourselves aren't working anymore. Sometimes compliance is not enough. Sometimes they kill you anyway.

Having grown up in the Deep South, I have long been aware of the threat of racial violence, of its symbolism. In middle school, many of my peers wore the Dixie Outfitters T-shirts that were in vogue in that part of the country during the late nineties. The shirts often featured collages of the Confederate flag, puppies,

and shotguns on the front, with slogans like STAND AND FIGHT
FOR SOUTHERN RIGHTS and PRESERVING SOUTHERN HERITAGE
SINCE 1861 printed on the back.

I was eleven years old, and these kids—and their commitment
to a symbol from a long-lost war—signaled that they believed I
shouldn't be in the same classroom with them, that I didn't belong
in their world.

But that was nothing compared with the routine brutality perpetrated upon Black people in my home state. In 2010, years before the deaths of Trayvon Martin, Michael Brown, and Sandra
Bland, there was the killing of Anthony Hill. Gregory Collins, a
white worker at a local poultry plant not far from my family farm,
shot and killed Hill, his Black co-worker. He dragged Hill's body
behind his pickup truck for 10 miles along the highways near my
grandmother's house, leaving a trail of blood and tendons. Abandoned on the road, the corpse was found with a single gunshot
wound to the head and a rope tied around what remained of the
body. Collins was sentenced for manslaughter. Five years ago, a
radicalized white supremacist murdered nine Black parishioners
as they prayed in Mother Emanuel African Methodist Episcopal
Church in Charleston. South Carolina is one of three states that
still does not have a hate-crime law.

Before my writing residency, I did not own a range map. Traditionally, these are used to depict plant and animal habitats and
indicate where certain species thrive. Ranges are often defined by
climate, food sources, water availability, the presence of predators,
and a species' ability to adapt.

My friend J. Drew Lanham taught me I could apply this sort of
logic to myself. A Black ornithologist and professor of wildlife ecology, he was unfazed by what happened to bird-watcher Christian
Cooper in Central Park—he's had his own encounters with white
people who can't understand why he might be standing in a field
with binoculars in his hand. Several years ago he wrote a piece for
Orion magazine called "9 Rules for the Black Birdwatcher."

"Carry your binoculars—and three forms of identification—
at all times," he wrote. "You'll need the binoculars to pick that
tufted duck out of the flock of scaup and ring-necks. You'll need
the photo ID to convince the cops, FBI, Homeland Security, and
the flashlight-toting security guard that you're not a terrorist or

escaped convict." Drew frequently checks the Southern Poverty Law Center's hate-group map and the Equal Justice Initiative's "Lynching in America" map and overlays them. The blank spaces are those he might travel to.

I never thought to lay out the data like that until the day I went to Abrams Creek.

Three weeks into my residency, I made an early-afternoon visit to the national-park archives. I needed to know what information they had on Black people. I left with one sheet of paper—a slave schedule that listed the age, sex, and race ("black" or "mulatto") of bodies held in captivity. There were no names. There were no pictures. I remember chiding myself for believing there might be.

Emotionally wrought and with a couple of hours of sunlight ahead of me, I decided to go for a drive to clear my mind. I came to the Smokies with dreams of writing about the natural world. I wanted to talk about the enigmatic Walker sisters, the park's brook trout restoration efforts, and the groundbreaking agreement that the National Park Service reached with the Eastern Band of Cherokee Indians about their right to sustainably harvest the edible sochan plant on their ancestral lands. My Blackness, and curiosity about the Black people living in this region, was not at the front of my mind. I naively figured I would learn about them in the historical panels of the visitor's center, along with the former white inhabitants and the Cherokee. I thought there would be a book or a guide about them.

There was nothing.

Vacations are meant to be methods of escapism. Believing this idyllic wilderness to be free of struggle, of complicated emotions, allows visitors to enjoy their day hikes. Many tourists to Great Smoky Mountains National Park see what they believe it has always been: rainbow-emitting waterfalls, cathedrals of green, carpets of yellow trillium in the spring. The majority never venture more than a couple of miles off the main road. They haven't trained their eyes to look for the overgrown homesites of the park's former inhabitants through the thick underbrush. Using the park as a side trip from the popular tourist destinations like Dollywood and Ripley's Believe It or Not, they aren't hiking the trails that pass by cemeteries where entire communities of white, enslaved, and emancipated people lived, loved, worked, died, and were buried, some, without ever being paid a living wage. Slavery here was arguably more intimate. An owner had four slaves, not four hundred. But it happened.

There is a revisionist fantasy that Americans cling to about the people in this region of North Carolina and Tennessee: that they were dirt-poor, struggled to survive, and wrestled the mountains into submission with their own brute strength. In reality, many families hired their sharecropping neighbors, along with Black convicts on chain gangs, to do the hard labor for them.

These corrections of history aren't conversations most people are interested in having.

After a fruitless stop at Fontana Dam, the site of a former African American settlement where I find precious little to see, I try to navigate back to where I'm staying. Cell service is spotty. My phone's GPS takes me on a new route along the edge of the park, through Happy Valley, which you can assume from the moniker is less than happy.

Early spring in the mountains is not as beautiful as you might believe. The trees are bare, and you can see the Confederate and Gadsden flags, the latter with their coiled rattlesnakes, flapping in the wind, so they do not take you by surprise. At home after home, I see flag after flag. The banners tell me that down in this valley I am on my own, as do the corpses of Jonathan A. Ferrell and Renisha McBride, Black people who knocked on the doors of white homeowners asking for help and were shot in response.

In the middle of this drive back to the part of the park where I belong, I round a corner to see a man burning a big pile of lumber, the flames taller than my car.

I am convinced that pyrophobia is embedded in my genes. The Ku Klux Klan was notorious for cross burnings and a willingness to torch homes. The fire over my shoulder is large enough to burn up any evidence that I ever existed. There is a man standing in his yard wearing a baseball cap and holding a drink, watching me as my white rental car creeps by. I want to ask him how to get out of here. I think of my mama's frantic phone calls going straight to voicemail. I stay in the car.

Farther down the road, another man is burning a big pile of lumber. I know it's just coincidence, that these bundles of timber were stacked before I set off down this path, but the symbolism unnerves me.

I round a bend and a familiar sign appears—a national-park placard with the words ABRAMS CREEK CAMPGROUND RANGER STATION in white letters. Believing some fresh air might settle my stomach and strengthen my nerves, I decide to enter that section of the park. The road I drive is the border between someone's property and the park. Uneven, it forces me to go slowly.

The dog is at my car before I recognize what is happening. It materializes as a strawberry blond streak bumping up against my driver-side door. Tall enough to reach my face, it is gnashing at my side mirror, trying to bite my reflection.

I'm not scared of dogs, but this one, with its explicit hostility, gives me pause.

Before emancipation, dogs hunted runaway slaves by scent, often maiming the quarry to keep them in place until their owner could arrive. During the civil rights movement, dogs were weaponized by police. In the modern era, use of K-9 units to intimidate and attack is so common that police have referred to Black people as "dog biscuits."

I force myself to keep driving.

When I reach the ranger station, the building is dark: closed for the season. I see a trail inviting me to walk between two short-leaf pines, but I decline. There is something in me that is more wound up than it has a right to be. No one knows my whereabouts. Despite making up 13 percent of the population, more than 30 percent of all missing persons in the United States in 2019 were Black. A significant portion of these cases are never covered by the news. The chances of me disappearing without a mention are higher than I'd like.

There are three cars in the little gravel parking lot. A pair of men, both bigger than me, are illegally flying drones around the clearing, and there is palpable apprehension around my presence. They don't acknowledge me, and I can't think of what I'm supposed to say to convince them I'm not a threat. I have no idea who the third car belongs to—they are somewhere in my periphery, real and not real, an ancillary portion of my calculation.

I take photos of the clearing, including the cars, just in case I don't make it out. It is the only thing I know to do.

I run my odds. No one in an official capacity to enforce the rules, no cell service to call for help, little knowledge of the area.

I leave. Later, my residency mentor gently suggests that maybe I don't visit that section of the park alone anymore.

I promise that there are parts of this park, and by extension the outdoors as a whole, that make visiting worth it. Time in nature is integral to my physical, spiritual, and mental health. I chase the radiant moments, because as a person who struggles with chronic depression, the times I am enthusiastically happy are few and far between. Most of them happen outside.

I relish the moments right before sunrise up at Purchase Knob in the North Carolina section of the Smokies. The world is quiet, my mind is still, and the birds, chattering to one another, do not mind my presence. I believe this is what Eden must have been like. I still live for the nights where I sink into my sleeping pad while I cowboy-camp, with nothing in or above my head except the stars. I believe in the healing power of hiking, the days when I am strong, capable, at home in my body.

The fear, on some level, will always exist. I say this to myself all the time: I know you're scared. Do it anyway.

Toward the end of my writing residency, the road to Clingmans Dome opens. At 6,643 feet, Clingmans is the highest point in Tennessee and in the park. About two days before I'm scheduled to leave, I go to see what this peak holds for me.

There is a paved trail leading to the observatory at the summit. It isn't long, just steep. Maybe it's the elevation; I have to do the hike twenty steps at a time, putting one foot in front of the other until I get to twenty, then starting over again. I catch my breath in ragged clips, and there are moments when I can feel my heartbeat throbbing in my fingertips. I'd planned to be at the top for sunset, but I realize the sun might be gone when I get there. I continue anyhow. I'm slow but stubborn.

If there's anything I appreciate about the crucible we're living in, it's the role of social media in creating a place for us when others won't. We're no longer waiting for outdoors companies to find the budget for diversity, equity, and inclusion initiatives. With the creation of a hashtag, a social media movement, suddenly we are hyper-visible, proud, and unyielding.

As I make my way up the ramp toward its intersection with the Appalachian Trail, I think about Will Robinson (@akunahikes on

Instagram), the first documented African American man to complete the triple crown of hiking: the Appalachian, Pacific Crest, and Continental Divide Trails. I understand that I'm following in Robinson's footsteps, and those of other Black explorers like writer Rahawa Haile (@rahawahaile) and long-haul hiker Daniel White (@theblackalachian)—people who passed this way while completing their AT through-hikes and whom I now call friends, thanks to the Internet. I smile and think of them as the trail meets the pavement, and stop for a moment. We have all seen this junction.

Their stories, videos, and photographs tell me what they know of the world I'm still learning to navigate. They are the adventurers I've been rooting for since the very beginning, and now I know they're also rooting for me.

It's our turn to wish for good things for you.

When I get to the summit the world is tinged in blue, and with minimal cloud cover I can see the borders of seven states. There is nothing around me now but heaven. I'm grateful I didn't quit.

My daddy had a saying that I hated as a child: "The man on top of the mountain didn't fall there." It's a quote by NFL coach Vince Lombardi, who during the fifties and sixties refused to give in to the racial pressures of the time and segregate his Green Bay Packers. It took me decades to understand what those two were trying to tell me, but standing at the top of Clingmans Dome, I get it. The trick is that there is no trick. You learn to eat fire by eating fire.

But none of us has to do it alone.

America is a vast place, and we often feel isolated because of its geography. But there are organizations around the country that have our backs: Black Outside, Inc., Color Outside, WeGotNext, Outdoor Afro, Black Folks Camp Too, Blackpackers, Melanin Base Camp, and others.

The honest discussions must happen now. I acknowledge that I am the descendant of enslaved people—folks who someone else kidnapped from their homeland and held captive in this one.

We were more than bodies then.

We are more than bodies now.

We have survived fierce things.

My ancestors survived genocide, the centuries-long hostage situation they were born into, and the tortures that followed when they called for freedom and equality. They witnessed murder. They endured as their wages and dreams were taken from them by systemic policies and physical force. And yet, because of their drive to survive, I am here.

I stand in the stream of a legacy started by my ancestors and populated by present-day Black trailblazers like outdoors journalist James Edward Mills, environmental-justice activist Teresa Baker, and conservationists Audrey and Frank Peterman. Remembering them—their struggles and triumphs—allows me to center myself in this scenery, as part of this landscape, and claim it as my history. This might be the closest thing to reparations that this country, founded on lofty ideals from morally bankrupt slaveholders, will ever give me.

I promised you at the beginning that I would be candid about the violence and even-keeled about the hope. I still have hope—I consider it essential for navigating these spaces, for being critical of America. I wouldn't be this way if I didn't know there was a better day coming for this country.

Even when hope doesn't reside within me—those days happen too—I know that it is safely in the hands of fellow Black adventurers to hold until I am ready to reclaim my share of it. I pray almost unceasingly for your ability to understand how powerful you are. If you weren't, they wouldn't be trying to keep you out, to make sure they keep the beauty and understanding of this vast world to themselves. If we weren't rewriting the story about who belongs in these places, they wouldn't be so focused on silencing us with their physical intimidation and calls for murder.

The more we see, the more we document, the more we share, the better we can empower those who come after us. I've learned during all my years of historical research that even when white guilt, complacency, and intentional neglect try to erase our presence, there is always a trace. Now there are hundreds of us, if not thousands, intent on blazing a trail.

It is true: I cannot protect you. But there is one thing I can continue to do: let you know that you are not alone in doing this big, monumental thing. You deserve a life of adventure, of joy, of enlightenment. The outdoors are part of our inheritance. So I will keep writing, posting photos, and doing my own signaling. For every new place I visit, and the old ones I return to, my message to you is that you belong here too.

BATHSHEBA DEMUTH

The Empty Space Where Normal Once Lived

FROM *The Atlantic*

ON THE FIRST day of summer, Siberia and I were the same temperature. In Verkhoyansk, roughly 3,000 miles northeast of Moscow, a searing week ended in an afternoon hotter than any before recorded north of the Arctic Circle. Half a planet away in New England, a thermometer under my tongue gave the same reading: 100.4 degrees Fahrenheit. In a human, this is the clinical threshold for a fever.

We had also been too hot for too long, Siberia and I. Most days since mid-May, I have lived in a body a degree or two and occasionally three above normal. Siberia is even more pyretic, averaging more than nine degrees above the twentieth-century norm from January onward. Russian friends sent photos of berries flushed ripe a month early. Dark clots of mosquitoes stuck on window screens. In places, the land itself is wobbly and out of joint, as melting permafrost opens large slump pits and gullies. By May, tundra peat was burning. The boreal forests broke into wildfires. These conflagrations are now the worst that Russia has seen, blazing on toward autumn.

The heat across northern Eurasia is uncanny but not mysterious. Our atmosphere, saturated with the carbon of burned fossil fuels, is becoming that of a prehuman time, one last present on Earth 3 million years ago. The poles are warming at about twice the rate of temperate regions, making Siberia's current climate anomalies the future of those regions too. Such transformations crack open ecological worlds and the lives within them. Given the scale and implication of events in northern Eurasia, calling this season the Summer of Siberia would not be hyperbole.

Yet it is my malady, not Siberia's, that rules conversations and headlines. I am too hot because of a persistent case of Covid-19, what its sufferers have begun calling "long Covid." Mine is one case among millions, a pace of infection that, like distant wildfires, will roar into fall. So perhaps it is no wonder that, from America, Eurasia's heat feels like an abstraction. Siberia and its inhabitants are far; much suffering is close. How do we take in the ruptures of a burning world when our own bodies are alight?

My body has been alight for months now. From within this illness, I have come to think that Siberia and I endure more than a coincidence in temperature. Our fevers are stoked by related patterns of economic production, patterns both relatively new and seemingly inevitable. And my corporeal fire says something about how a continental fire can go unseen, offering a lesson in the implications of duration: how as a condition lingers, its origins or significance grow harder to see. Long Covid and climate change are alike in this: live ill for long enough, and the absence of health threatens to become normal.

Two summers ago, I was in Russia, on a comma of rocky Bering Sea beach called Napkum Spit. It was August, the turquoise water free of ice and full of spotted seals. Far off, a gray whale spouted, its breath tracking a shimmering mist against the horizon.

North and south from this place, Indigenous kin and culture are nourished by hunts of gray and bowhead whales. For Yupik and Chukchi, peoples whose ancestors have lived along the Bering Sea for thousands of years, whales' flesh is food, their beings woven into the necessities and ceremonies of daily life. I was on Napkum Spit because of a different kind of whaling. My work as a historian had led me to the logbooks of the New England fleet, which began killing Bering Sea cetaceans in the 1840s. These sailors hunted whales for oil and baleen, to light homes and brace corsets where I now live, in Rhode Island, and all along the Eastern Seaboard of the United States. I wanted to see what marks remained in this whaling ground.

I expected something tangible, even monumental. In fifty years, the Yankee fleet killed tens of thousands of whales. Around that loss, the Bering Sea ecosystem transformed, likely feeding more squid and fish in the ecological spaces once home to bowheads and grays. But these species were inaccessible to Yupik and Chukchi. By the 1880s, famine claimed families, then whole villages, many also suffering from epidemic diseases transported north by wooden ships.

Had I lived in Rhode Island then, I would have lit whale-oil lamps at dusk, with baleen cinching my ribs, and seen nothing of that suffering. A hundred and twenty years later, to one recently arrived on Napkum Spit from New England, the traces of commercial whaling were imperceptible still. There were no hulking shipwrecks, or graves, or mounds of whale skulls, only that single whale spout on the horizon. The memorial to market killing is absence. The Yankee fleet ceased harrowing the waters off Siberia by 1900, yet bowhead and gray whale populations are still shy of their former plenty. The only Bering Sea I have ever seen, the only one I can experience, would have seemed eerily bereft in 1840.

That same year, I visited an abandoned mining town to the southwest. Shakhtyorsky was a Soviet creation, wrested from the tundra's green sedges. A vein of lignite coal ran under the hills; miners, or *shakhtyory,* came here to peel away dirt, permafrost, and stone, then haul out fuel by the black, dusty ton. At its peak, the town was coated in a thin layer of grime, exhaled from the mines and the generators that powered heavy equipment. Coal dust stunted plant growth; coal heaps leached acid into streams. All around Shakhtyorsky, the process of extraction left the earth hollow and pocked. In miniature, it did the same to human lungs. Years of breathing sulphur and black dust caused coal workers' pneumoconiosis, or black lung; in severe cases, lesions in the lung tissues necrotized, leaving empty, dead cavities. Yet had I flipped a switch powered by Shakhtyorsky coal in 1970, or 1980, I would have seen no sign of the scarred lands or bodies behind the light.

Napkum Spit and Shakhtyorsky are alike in this: they are monuments to how, in the nineteenth and twentieth centuries, people of comparative wealth could consume parts of the Arctic—as they consumed Indian cotton, Caribbean sugar, Middle Eastern oil, South American bananas, and dozens of other products from distant parts of the world—at a remove from the costs of their manufacture. Long before Covid-19 turned grocery stocking and Amazon delivery into dangerous work, consumption was healthier at a distance. And that severed use from consequence. If most Americans now pay little heed to Siberia's burn, perhaps it is because recent history has made material plenty and heedlessness coincident. Wealth is freedom not just from bearing the consequence of using up the world, but from paying attention to it.

That might have worked 200 years, ago, a century ago, even a lifetime ago. Today, the speed and intensity of twenty-first-century life erodes the space between the costs of production and the benefits of consumption. What starts far off in the Arctic—or in China, or anywhere—does not remain there. Or, put another way, the same dynamics that warm Siberia also warm me.

There are many examples of this, how burning trees and fossil fuels alter the composition of the atmosphere while moving people and, with them, viruses. Modern agriculture, which turns petroleum into fertilizer, concentrates sites of possible infection and transmission between livestock and humans. Industrialization replaces animal homes with human ones, and with markets for fauna such as the bats that gave us this coronavirus. As loggers turn forests into furniture, they push more species into new intimacy with people. Deforestation also emits billions of tons of carbon each year. That carbon warms the planet more; a warmed planet forces more animals to move, which makes viral transfer more likely.

Siberia's wildfires are deforestation at an immense and terribly efficient scale. This year, about 50 million acres of forest and grasslands have already burned, more than a Greece's worth of plant life blown into a pall of smoke so massive it now sits over Alaska and Washington State. A month of such burning releases as much carbon dioxide as a small country—Portugal, or Sweden—does in a year. No summer on record has seen less ice in the Arctic Ocean; the greatest losses are north of Russia's baking landmass, in the Barents and Laptev Seas. Ice at the poles anchors the stability of our climate. Even if we pay it no heed, this hot summer in Siberia is shifting the terms of what normal is out from under us all.

Playing host to the coronavirus for three months has made me think about normalcy—its shifty character, how it plays with my sense of time—and the drive to pretend that things are at stasis, despite all evidence indicating turmoil. My case of Covid-19 was never acute; I was not on a ventilator or even close, nor do I have the harsher ills of many long-haulers, who report roving pain, memory loss, tachycardia. My experience of the virus has not been an event so much as a shift, erosion rather than earthquake. The most enduring symptom is a corporeal heat wave that shows no more sign of fully lifting than the warmth in northern Eurasia. As the weeks drag on, the hale clarity of my normal self is receding. Perhaps this is just what I am now: weaker, wan, soggy-brained.

An amazing and terrible thing about being human is how quickly we adapt to circumstances unthinkable just years, or months, or weeks in the past. The marine ecologist Daniel Pauly calls this the problem of "shifting baselines": assuming that observations this year or this decade represent life at its most flourishing. A whaler fresh to the Bering Strait in 1850 saw thousands of bowheads; fifty years later, a new sailor might have assumed that the species was naturally scarce. A miner who came to Shakhtyorsky in 1980 would never breathe air free of coal dust. In April I assumed I could wake each morning and work 'til evening; now I route my days around my body's weather. People born in Siberia early this century have watched summers warm dramatically. Their children may never know it otherwise; unless carbon emissions halt, this year's average temperature in Siberia will likely be the norm at the century's end.

The danger of acceptance is in how ill it leaves bodies and the places they live. I am not well, at 99.8 degrees, or 99.2, or 100.1, even if those are the temperatures I experience more days than not. The swaths of Siberia choking in smoke are not well, nor are their people. But the very slowness, the week-in, week-out constancy of climate change or enduring infection, is lulling. It is tedious to tell people I am still sick. Sustaining alarm at a thousand people dying in a day is more difficult in August than it was in April. Siberia is too hot, still, but it has not exceeded the record of 100.4 degrees Fahrenheit set in July. A 98-degree afternoon in Verkhoyansk is now not an event; it is just a day. The phenomenal becomes mere background.

As summer tips toward autumn—an autumn in which there will be too little sea ice and too much virus—I do not want to forget the possibilities of my April self, or of Siberia without fire, or of whales by the tens of thousands. The need to build a society that cares for all, that does not let some hide in the safety of distance, has never been more acute. The habits of wealth need reconditioning to account for the real costs of consumption. These are forward-looking projects. My experience of this virus makes me think, however, that we should not forget a longer view, one able to see how the conditions of 2020 are not inevitable. The line of heat that connects my body and Siberia has existed for only a few centuries. It is not inevitable. Thinking past it, as this summer of our many discontents moves into fall, requires a kind of split imagination: to conjure moments of past flourishing, and a future where we might flourish again.

EMILY RABOTEAU

This Is How We Live Now

FROM *New York Magazine/The Cut*

SOME SCIENTISTS SAY the best way to combat climate change is to talk about it among friends and family—to make private anxieties public concerns. For 2019, my New Year's resolution was to do just that, as often as possible, at the risk of spoiling dinner. I would ask about the crisis at parent-association meetings, in classrooms, at conferences, on the subway, in bodegas, at dinner parties, while overseas, and when online; I would break climate silence as a woman of color, as a mother raising Black children in a global city, as a professor at a public university, and as a travel writer—in all of those places, as all of those people. I would force those conversations if I needed to. But, it turned out, people wanted to talk about it. Nobody was silent. I listened to their answers. I noticed the echoes. I wrote them all down.

January

Tuesday, January 1

At last night's New Year's Eve party, we served hoppin' John. Nim said that when he used to visit relatives in Israel, he could see the Dead Sea from the side of the road, but on his most recent trip, he could not. It was a lengthy walk to reach the water, which is evaporating.

Chris responded that the beaches are eroding in her native Jamaica, most egregiously where the resorts have raked away the seaweed to beautify the shore for tourists.

Wednesday, January 2

After losing her home in Staten Island to Hurricane Sandy, Lissette bought an RV with solar panels and has been living off the grid, conscious of how much water it takes to flush her toilet and to take a shower, I learned at Angie's house party.

Monday, January 14

At tonight's dinner party, Marguerite said that in Trinidad, where they find a way to joke about everything, including coups, people aren't laughing about the flooding.

Wednesday, January 16

On this evening's trip on the boat Walter built, he claimed with enthusiasm that we might extract enough renewable energy from the Gulf Stream via underwater turbines to power the entire East Coast.

Moreover, Walter predicted with the confidence of a Swiss watch, no intelligent businessman will invest another dime in coal when there is more profit to be made in wind, solar, and hydrokinetic energy. Economic forces will dictate a turnaround in the next ten years, he said.

Monday, January 21

After Hurricane Irma wrecked her home in Key West, Kristina, a triathlete librarian, moved onto a boat and published a dystopian novel titled *Knowing When to Leave,* I learned over lobster tail.

February

Tuesday, February 12

We ate vegetable quiche at Ayana and Christina's housewarming party, where Christina described the Vancouver sun through the haze of forest-fire smoke and smog as looking more like the moon.

Monday, February 18

In the basement of Our Saviour's Atonement this afternoon, Pastor John said he's been preaching once a month about climate change, despite his wife's discomfort, and recently traveled to Albany to lobby for the Community and Climate Protection Act.

Saturday, February 23

When I see those brown recycling bins coming to the neighborhood, said a student in Amir's class at City College in Harlem, it tells me gentrification is here and our time is running out.

Thursday, February 28
Just between us, Mik said over drinks at Shade Bar in Greenwich
Village, it scares me that white people are becoming afraid of what
they might lose. History tells us they gonna get violent.

March

Sunday, March 17
On St. Patrick's Day, Kathy, who'd prepared the traditional corned
beef and cabbage, conversed about the guest from the botanical
garden in her master gardening class, who lectured on shifting
growing zones, altering what could be planted in central New Jersey, and when.

Tuesday, March 19
Sheila, who brought weed coquito to the tipsy tea party, said that
when people ask her, "What are you Hondurans, and why are you
at the border?" she says, "Americans are just future Hondurans."

Monday, March 25
Mat recalled vultures in the trees of Sugar Land, Texas, hunting dead animals that had drowned in Hurricane Harvey, during which he'd had difficulty fording flooded streets to reach his
mother's nursing home.

April

Tuesday, April 16
After a bite of roasted-beet salad in the Trask mansion's dining
room, Hilary spoke of the historic spring flooding in her home
state of Iowa, where the economic impact was projected to reach
$2 billion.

Thursday, April 18
Carolyn warned me at the breakfast table, where I picked up my
grapefruit spoon, that I may have to get used to an inhaler to be
able to breathe in spring going forward, as the pollen count continues to rise with the warming world. My wheezing concerned her,
and when she brought me to urgent care, a sign at the check-in
desk advised, DON'T ASK US FOR ANTIBIOTICS. Valerie, the doc-

tor who nebulized me with albuterol, explained that patients were overusing antibiotics in the longer tick season for fear of Lyme.

Tuesday, April 23

On his second helping of vegetable risotto, Antonius reflected that in Vietnam, where his parents are from, the rate of migration from the Mekong Delta, with its sea-spoiled crops, is staggering.

Sunday, April 28

Due to Cyclone Fani, Ranjit said he was canceling plans to visit Kerala and heading straight back to Goa, where he would be available for gigs, lessons, jam sessions, meals.

Michael said that beef prices were up after the loss of so much livestock in this spring's midwestern flooding, and so he'd prepared pork tacos instead.

May

Friday, May 3

At the head of the table where we sat eating bagels, Aurash said we won't solve this problem until we obsess over it, as he had obsessed over Michael Jordan and the Lamborghini Countach as a kid.

He added that, just as his parents weren't responsible for the specific reasons they had to leave Afghanistan, in general the communities most impacted by climate change are least responsible for it.

Balancing an empty plate in his lap, Karthik said that New York City (an archipelago of thirty-odd islands), with all its hubris, should be looking to Sri Lanka, another vulnerable island community, for lessons in resilience.

We have more in common, he went on, with the effective stresses of low-lying small-island coastal regions such as the Maldives, the Seychelles, Cape Verde, Malaysia, Hong Kong, and the Caribbean than with a place like Champaign, Illinois—

"I'm from Champaign!" Pamela interrupted, her mouth full. "It's in a floodplain too!" she cried. We're all sitting at this table now.

Tuesday, May 7

"Personally, I'm not that into the future," said Centime, who had a different sense of mortality having survived two bouts of breast cancer. She uncorked the fourth bottle of wine. We'd gathered over

Indian takeout for an editorial meeting to comb through submissions to a transnational feminist journal centering on women of color. "But I can respect your impulse to document our extinction."

Sunday, May 19
Eating a slice of pizza at a kid's birthday party in a noisy arcade, Adam reminisced about the chirping of frogs at dusk in northern Long Island—the soundtrack to his childhood, now silent for a decade.

"Sad to say," he mused, "among the 9 million meaningless things I've Googled, this wasn't one. It's like a postapocalypse version of my life: 'Well, once the frogs all died, we shoulda known.' Then I strap on a breather and head into a sandstorm to harvest sand fleas for soup."

June

Friday, June 7
Hiral, scoffing at what passes for authentic Punjabi food here in New York, was worried about her family in Gandhinagar and the trees of that green city, where the temperature is hovering around 110 degrees Fahrenheit weeks before monsoons will bring relief.

Sunday, June 9
After T-ball practice at Dyckman Fields, while the Golden Tigers ate a snack of clementines and Goldfish crackers, Adeline's dad, an engineer for the Department of Environmental Protection, spoke uneasily of the added strain upon the sewage system from storms.

Saturday, June 15
Jeff, who'd changed his unhealthy eating habits after a heart attack, said, "We are running out of language to describe our devastation of the world."

Lacy agreed, adding, "We need new metaphors and new containers with which to imagine time."

Sunday, June 16
Keith confessed that he was seriously losing hope of any way out of this death spiral.

Tuesday, June 18
We sipped rosé, listening to Javier read a poem about bright-orange crabs in the roots of the mangrove trees of Estero de Jalte-

peque in his native El Salvador, where the legislative assembly had just recognized natural forests as living entities.

The historic move protects the rights of trees, without which our planet cannot support us. Meanwhile, Javier discussed the lack of rights of migrants at the border, recalling the journey he made at age nine, unaccompanied, in a caravan surveilled by helicopters.

In Sudan, where Dalia (who read after Javier) is from, youth in Khartoum wish to restore the ecosystem through reforestation using drones to cast seedpods in the western Darfur region, hoping to stymie disasters such as huge sandstorms called *haboob*.

Owing to this month's massacre, one of Dalia's poems proved too difficult for her to share. "I'd be reading a memorial," she said.

I strained to hear the unspoken rhyme between the rising sandstorms and the dying mangroves, hemispheres apart.

Wednesday, June 19

Salar wrote to me about the call of the watermelon man this morning in Tehran where groundwater loss, overirrigation, and drought have led to land subsidence. Parts of the capitol are sinking, causing fissures, sinkholes, ditches, cracks.

The damage was most evident to him in the southern neighborhood of Yaftabad, by the wells and farmland at the city's edge. There, ruptures in water pipes, walls, and roads have folks fearing the collapse of shoddier buildings. The ground beneath the airport, too, is giving way.

Thursday, June 20

"Our airport's sinking too!" mused Catherine, who'd flown in from San Francisco for this evening of scene readings at the National Arts Club, followed by a wine-and-cheese reception.

Friday, June 21

"It's not true that we're all seated at the same table," argued David, a translator from Guatemala, where erratic weather patterns have made it nearly impossible to grow maize and potatoes.

Retha, David's associate, quoted the poem "Luck," by Langston Hughes:

Sometimes a crumb falls
From the tables of joy,

Then we went out looking for the Korean barbecue truck.

Saturday, June 22

"Say what you will about the Mormons," said Paisley, who lives in Utah, "but they've been stockpiling for the end of days for so long that they're better prepared."

Sunday, June 23

At the Stone Barns farm, where tiara cabbages, garlic scapes, snow peas, red ace beets, zucchini flowers, and baby lambs were being harvested for the Blue Hill restaurant's summer menu, Laura spoke hopefully of carbon sequestration in the soil.

Edgily, Lisa argued, "There's not a single American living a sustainable lifestyle. Those who come close are either homeless or are spending most of their time growing food and chopping wood."

Tuesday, June 25

S.J. said their car as well as eight of their neighbors' cars, including a freaking Escalade, got totaled by a flash flood in the middle of the night in Charleston without warning. Living in a sea-level coastal city is becoming more terrifying by the day, said S.J. They now check the radar before parking.

Thursday, June 27

Magda turned philosophical before returning to Tepoztlán, Mexico. What is the future of memory and the memory of the future? she pondered. We were eating raw sugar-snap peas, remarkable for their sweetness, out of a clear plastic bag.

Her eyes, too, were startlingly clear. "My daughter's twenty-seven now," she said. "By midcentury, I'll be dead. I can't imagine her future or recall a historical precedent for guidance . . ." Magda lost her thread.

Meanwhile, Roy had been pointing out the slowness of the disaster; not some future apocalypse, but rather our present reality —a world's end we may look to culturally endure with lessons from *Gilgamesh, The Aeneid,* the Torah, and the Crow.

Friday, June 28

The other Adam sent word from Pearl River at breakfast: "Today's temps at camp are going to reach 100. It will feel hotter than that. We'll be taking it slower and spending more time in the shade. Don't forget sunscreen, water bottles, and hats; they're critical to keeping your kids safe."

There was no shade at the bus stop in front of the Starbucks on 181st Street. "Why wasn't climate change the center of last night's Democratic presidential debate?" asked Ezra, a rabbi.

"They didn't talk about it at all in 2016," pointed out Rhea's mom, who preferred to see the glass as half-full. "This is progress!" she cheerfully exclaimed.

Sunday, June 30

Ryan, Albert's head nurse on the cardiac unit, feared the hospital was understaffed to deal with the upswing of heat-induced diseases. Delicately moving the untouched food tray to rearrange the IV tube, he said, "It's hard on the heart."

July

Tuesday, July 2

"My homeland may not exist in its current state, a bewildering, terrifying thought I suffer daily," Tanaïs said of Bangladesh. "Every time I go to the coast, there's less and less land and now a sprawling refugee camp. Every visit feels closer to our end."

Wednesday, July 3

"Let's lay off the subject tonight," suggested Victor, as he prepared the asparagus salad for dinner with Carrie and Andy, who were back in town for the music festival.

Thursday, July 4

Holding court over waffles this morning in the stately dining room of the Black-owned Akwaaba Bed and Breakfast in Philadelphia, Ulysses, who works to diversify the U.S. Geological Survey, said, "We need representation. Earthquakes affect us too. Volcanoes affect us too. Climate change affects us too."

Charlie stirred the gumbo pot. He speculated that his girls' public school had closed early this year because its sweltering classrooms lacked air-conditioning to manage the heat wave. "Our seasons are changing," he said, regarding the prolonged summer break.

While Lucy distributed glow necklaces to her little cousins on the Fourth of July, her aunt learned the fireworks display had been canceled by the Anchorage Fire Department owing to extreme dry weather conditions. Alaska was burning.

Cyrus yanked off his headphones with bewilderment and looked up from his iPad toward his mom. "It says there's a tornado warning," he cried. All through the airport, our cell phones were sounding emergency alarms, warning us to take shelter. A siren sounded.

"Take shelter where?" begged his mother in confusion. She clutched a paper Smashburger bag with a grease spot at the bottom corner. The aircraft was barely visible through the gray wash of rain at the wall of windows rattling with wind.

Sunday, July 7

Nadia, a flight attendant in a smart yellow neck scarf, served us Würfel vom Hahnchenkeulen in Pilzsauce on the delayed seven-hour red-eye from Philly to Frankfurt, on which each passenger's carbon footprint measured 3.4 metric tons.

Monday, July 8

Owing to a huge toxic algae bloom, all twenty-one of the beaches were closed in Mississippi, where Jan was getting ready to start her fellowship, I learned before tonight's dinner at the Abuja Hilton.

Jan ordered a steak, well done, and swallowed a malaria pill with a sip of South African wine. She referred to Joy Harjo's poem "Perhaps the World Ends Here," which starts:

The world begins at a kitchen table.
No matter what, we must eat to live.

Wednesday, July 10

Eating chicken suya in the mansion of the chargé d'affaires, Chinelo spoke quietly of the flooding in Kogi state at the confluence of the Niger and Benue Rivers.

Few Nigerians realize, Buchi said, that the longevity of Boko Haram in the Northeast, the banditry in the Northwest, and the herder-farmer crises in the North Central are a result of rapid desertification and loss of arable land even as the country's population keeps exploding.

Thursday, July 11

Jide, a confident and fashionable hustler, slipped me a business card claiming his sneaker line was the first innovative, socially conscious, sustainable footwear brand in all of Africa. His enviable red-laced kicks said, "We're going to Mars with a space girl, two cats, and a missionary."

Stacey, a science officer for the CDC, was geeking out about the data samples that would help control the spread of vector-borne diseases like yellow fever and dengue when the waiter interrupted her epidemiological account with a red-velvet cake for my forty-third birthday.

"*Nel mezzo del cammin di nostra vita / Mi ritrovai per una selva oscura / Ché la diritta via era smarrita!*" shouted Nicole, my college roommate from half a lifetime ago, before we had kids, before she went blind. We had memorized the opening lines of *The Inferno,* had crushes on the Dante professor, and knew nothing yet of pain.

Tuesday, July 16

Naheed said, "The southwest monsoon is failing in Nagpur. For the first time in history, the municipal corporation will only provide water on alternate days. There will be no water on Wednesday, Friday, nor Sunday in the entire city for two weeks."

Chido told us that in Harare, she was one of the lucky ones on municipal rotation getting running water five days out of the week, until fecal sludge appeared, typhoid cases cropped up, and the taps were shut off entirely. "They are killing us," she said.

Friday, July 19

Kate said the back roads of Salisbury, Vermont, were slippery with the squashed guts and body fluids of the hundreds of thousands of northern leopard frogs—metamorphosing from tadpoles in explosive numbers—run over by cars.

Centime sent a picture of a memorial for Okjökull, the first Icelandic glacier to lose its status as a glacier. "For your time capsule," she offered. The plaque read, THIS MONUMENT IS TO ACKNOWLEDGE THAT WE KNOW WHAT IS HAPPENING AND WHAT NEEDS TO BE DONE.

Posed as a letter to the future, the message ended, ONLY YOU KNOW IF WE DID IT.

"What would you do if the power went out and you were stuck underground in a subway tunnel?" Lissette drilled, showing me the prepper items in her crowded backpack, heavy as a mother's diaper bag: water, protein bars, flashlight, battery, filter, knife . . .

Saturday, July 20

"Bobby was stuck underground on the 1 train during last night's commute for forty-five minutes," said his wife, Angela, describing the clusterfuck of six suspended subway lines. "And in this heat wave too," she griped. "Folks were bugging *out!*—ten more minutes and there woulda been a riot."

Monday, July 22

Morgan wasn't the only one to observe it was the poorer neighborhoods in Brooklyn that had power cut off in yesterday's rolling blackout. The powerless scrambled to eat whatever food was in their fridges before it spoiled. Wealthier hoods were just fine.

Tuesday, July 23

"Can you rummage in my mind and take out the fire thoughts and eat them?" asked eight-year-old Geronimo at bedtime. This was the ritual. He felt safer with his anxieties in my stomach than in his brain.

Just back in L.A. from an empowering trek to Sicily where she'd visited the Shrine to the Black Madonna despite sizzling temperatures, Nichelle shared her two rules for dealing with the global heat wave: "(1) Drink lots of water. (2) Watch how you talk to me."

Wednesday, July 24

Marking the fiftieth anniversary of the moon landing, the Reverend John sermonized, "You'd think after seeing the Earth from afar, we would do anything to protect this planet, this home. You'd think wrong."

"We've become drunk on the oil and gas poisoning the waters that give us life," he preached. "And we have vomited that drunkenness into the atmosphere. Truly, the prophet is right," he said, quoting Isaiah 24:4. "The Earth dries up and withers. The world languishes and withers. The heavens languish with the Earth."

"We have broken the everlasting covenant," reasoned the Reverend John. "Nevertheless, the Bible tells us that God loves this world."

Thursday, July 25

At last night's "Intimate Dilemmas in the Climate Crisis" gathering at a software company on Madison Avenue, we were told to write our hopes and fears for the future on name tags as a silent icebreaker, then to stick these messages to our chests and walk about the room. Sebastian's was only one word: war.

Mary, who left the event early, said she worried about her aging mother down South. "I'm the first person in my family born after Jim Crow. They fought battles so I could live the dreams my mother couldn't. How can I talk to her about this existential grief of mine when she's already been through so much?"

"Having one less child reduces one's carbon footprint 64.6 U.S. tons per year," Josephine from Conceivable Future informed us.

"Why is it so easy to police reproductive rights of poor women and so hard to tell the fossil-fuel industry to stop killing us?" asked Jade, a Diné and Tesuque Pueblo activist in New Mexico, whose shade of red lipstick I coveted.

Friday, July 26

Ciarán set down our shepherd's pie and Guinness on a nicked table at Le Chéile. On one of the many drunken crayon drawings taped to the walls of that pub were scrawled these lines from Yeats:

All changed, changed utterly:
A terrible beauty is born.

Protesters from Extinction Rebellion Ireland staged a die-in at the Natural History Museum in Dublin, where Ciarán's family is from, arranging their inert bodies on the floor among silent stuffed "Mammals of the World."

Tuesday, July 30

Ari cooked lamb shoulder chops with eggplant and cilantro puree, a family recipe from Yemen, where swarms of desert locusts, whose summer breeding was ramped up by extraordinary rainfall, are invading crops, attacking farms, and eating trees.

Meanwhile, Yemeni villagers are eating the locusts, shared Wajeeh, catching them in scarves at nightfall, eating them with rice in place of vegetables, carting sacks of them to Sanaa and selling them, grilled, near the Great Mosque.

Wednesday, July 31

When Nelly and I chewed khat with Centime in Addis Ababa a decade ago, discussing creation myths at the New Flower Lounge while high as three kites, we never imagined that Ethiopia would plant 350 million trees in one day, as they did today.

Eric distributed Wednesday's fruit share under a canopy in Sugar Hill, Harlem. I took note of the Baldwin quote on the back of his sweat-soaked T-shirt when he bent to lift a cantaloupe crate:

THE MOMENT WE BREAK FAITH WITH ONE ANOTHER, THE SEA ENGULFS US AND THE LIGHT GOES OUT.

August

Thursday, August 1

Off the rugged coast of Devon, where Jane grew up picking wild blackberries, the Cloud Appreciation Society gathered to slow down and gaze up at the sky in gratitude and wonder. Nobody spoke of the modeled scenario released by scientists of a cloudless atmosphere.

"In the beginning," said Elizabeth, who lives in Pass Christian, a block from the Mississippi shore, "before they closed the beaches, I saw the death with my own eyes. Dead gulf redfish, dead fresh-water catfish dumped from the river. Thousands. I saw a dead dolphin in the sand."

Friday, August 2

"I'm always so pissed at plastic bags and idling cars, but I feel like there's no point in caring anymore," said Shasta upon learning that between yesterday and today, more than 12 billion tons of water will have melted from the Greenland ice sheet.

Saturday, August 3

Meera grew disoriented when she returned to the Houston area to finish packing up the house that her family had left behind and could not sell; it was languishing on the market for a year as if cursed.

Sunday, August 4

Because he dearly loved taking his boys camping in the Mojave Desert, Leonard felt depressed about the likely eventual extinction of the otherworldly trees in Joshua Tree National Park.

Monday, August 5

The El Paso shooter's manifesto said, "My whole life I have been preparing for a future that currently doesn't exist . . . If we can get rid of enough people, then our way of life can become more sustainable."

In her kitchen, Angie nearly burned the platanos frying in oil on her stovetop. "That ecofascist targeted Mexicans," she said, swatting at the smoke with a dish towel. "He called us invaders."

Wednesday, August 7

"In the Black Forest," said Daniel, "there are mainly firs and spruces. Many of them die because it is too dry. We used to have something called land-rain. That was light rain for days. It's gone. When it rains (like now) it feels like an Indian monsoon. What I really want to say to you about *Waldersterben* (dying forest): come now, as long as the Black Forest exists."

Friday, August 9

Claire, a former Colorado farmer, spoke of intensifying forest fires. "The mountains are full of burn scars like this," she said, sharing a shot from a blaze near Breckenridge.

None of us will be able to say later that we didn't know we were doing this to the Earth.

Thursday, August 15

Isobel stopped planning our twenty-fifth high school reunion to study the weakening of global ocean circulation and the tanking of the stock market when the Dow dropped 800 points today. Back to back, she traced with a painted fingernail the lines of the inverted yield curve and the slowing Gulf Stream.

Friday, August 16

Zulema wasn't surprised when Pacific Gas & Electric went bankrupt from the billions of dollars in liability it faced from two years of raging California wildfires, though it wasn't a downed power line that ignited the Detwiler fire she fled. It was a discharged gun.

On being evacuated from Mariposa for six days by that fire, whose smoke reached Idaho as it burned 80,000 acres of trees dried into tinder by bark beetles and drought, she said over soup dumplings: "I almost lost my house. It's surrounded by charred forest now. We're like those frogs in the boiling pot."

Sunday, August 18

"The developers don't live here, so they don't care," said Jimmy, the tuxedoed waiter who served me linguini with clam sauce for lunch at Gargiulio's on Coney Island, where the new Ocean Dreams luxury apartment towers are topping out despite sea-level rise. "All they care about is making a buck."

Monday, August 19

Manreet said she felt anxious. Yesterday in Delhi, where her sister-in-law lives, the government sounded a flood alert as the Yamuna River swelled to breach its danger mark.

"Punjab, where I come from, means 'The Land of Five Rivers,'" she explained. "It's India's granary. After a severe summer left the fields parched, the brimming rivers are now flooding them. It's worse and worse each year. I feel weirdly resigned."

September

Tuesday, September 3

Although the sky directly above her wasn't blackened by smoke from the burning Amazon rain forest, Graduada Franjinha saw protests along the road to a capoeira competition in Rio. "It's so

sad to see how humankind destroys the lungs of the earth that
gives us breath," she said.

Saddened by the loss of twenty-eight wild horses in Pamlico
Sound to a mini-tsunami, Chastity remembered seeing them as a
kid and swearing to commit them to her forever memory. "You
don't see beautiful things like that and question whether there's a
higher being," she said. "You just don't."

Wednesday, September 4
Chaitali said she can't stop thinking about Grand Bahama after
learning that 70 percent of it is now underwater. "Where are all
those people going to go?" she asked, mystified and horror-struck.

It is an unprecedented disaster, said Christian, struggling to
control his voice. He had cut his hair since last I saw him. Dorian
was still hovering over his birthplace of Grand Bahama. "Natural
and unnatural storms reveal how those most vulnerable are disproportionately affected," he said.

Friday, September 6
At last night's party, Jamilah, a Trini-Nigerian Toronto-based sound
artist and former member of the band Abstract Random, took a
bite of pastelito and said she'd like to get to the Seychelles before
they drop into the Indian Ocean.

Saturday, September 7
"Eat the fucking rich," said Jessica, in reply to a quarterly investment report on how to stay financially stable when the world may
be falling apart.

Thursday, September 9
Arwa feared that the plight of 119 Bahamian evacuees thrown off
a ferryboat to Florida for being without visas they did not legally
need was a sign of climate apartheid.

Wednesday, September 11
"Ma'am, I *am* the heat," Maurice replied to the woman in New Or-
leans's Jackson Square who warned him against jogging outdoors
because of the heat advisory in effect.

Thursday, September 12
Maya, proud owner of a Chihuahua–pit bull–mini-pin mix in
Montclair, was saddened to learn that nearly 300 animals had
drowned at a Humane Society shelter in Freeport during the hurricane.

Melissa, incensed, asked why they didn't let the animals out of
their damn crates.

"Well, if it's any consolation, a shit ton of *people* died too," argued Sanaa.

Tons of babies, tons of elderly and infirm people, even perfectly healthy people died too. Over 2,500 people are still missing, and 70,000 now homeless.

"Did you not see the videos of people trapped in their attics with the waves crashing over their houses? Y'all sound fucking stupid," Sanaa fumed.

Friday, September 13

"Did you hear the NYC Department of Education approved absences from school for the youth climate strike next Friday?" Elyssa asked during the Shabbat Schmooze while the children swarmed around a folding table tearing off hunks of challah and dunking them in Dixie cups of grape juice.

"I'd rather go to school," said Jacob. His dislike of large crowds outweighed his dislike of third grade.

Wednesday, September 18

Amanda, whom I last saw at Raoul's, where we ate steak au poivre and pommes frites, said she had to sell off half the herd on her family's Texas cattle ranch after a drought left the tanks dry, the lake depleted, and the hayfield shriveled.

She mentioned, almost as an aside, that they'd lost half the honeybees in their hives to colony-collapse disorder in the past five years too.

"Everyone here is linked to someone who works in oil," she said. "It's the center of the damage, and all that industry makes my efforts feel small. Sailing in Galveston Bay after a tanker spill, I wondered if my soaking-wet clothes were flammable."

Thursday, September 19

TaRessa, from Atlanta, said, "I have always loved awakening to birdsong. This year, for the first time, I hear none." A third of North American birds had vanished from the sky in the span of her lifetime.

Friday, September 20

"I'm here to sign out my child for the climate strike," said a dad to Consuelo, the parent coordinator in the main office at Dos Puentes Elementary.

"By the time they're our age, they won't have air to breathe," worried Consuelo. "They'll be wearing those things on their faces — *mascarillas respiratorias*."

Ben's sign said, I'M MISSING SCIENCE CLASS FOR THIS. He was six, in the first grade, and studying varieties of apples, of which he knew there were thousands. He'd also heard that as many as 200 species were going extinct every day.

Shawna told her daughter on the packed A train down to Chambers Street that a teenage girl had done this, had started protesting alone until kids all over the world joined her to tell the grown-ups to do better, had sailed across the ocean to demand it.

Along Worth Street toward Foley Square, the signs said:

SHIT'S ON FIRE, YO
COMPOST THE RICH
THIS IS ALL WE HAVE
I WANT MY KID TO SEE A POLAR BEAR
SEAS ARE RISING AND SO ARE WE
MAKE EARTH GREAT AGAIN
SAVE OUR HOME
PLEASE HELP

In yellow pinafores, Grannies for Peace sang "The Battle Hymn of the Republic" while a nearby police officer forced a protester to the ground for refusing to move off the crowded street to the sidewalk. "*Shame!*" chanted the massive crowd in lower Manhattan.

"When our leaders act like kids, then we, the kids, will lead!" shouted a gaggle of outraged preteen girls in Catholic-school uniforms. Their voices grew hoarse, though the march had not yet begun.

Saturday, September 21
Humera's Sufi spiritual guide, Fatima, said, "*Alhamdulillah!* Let's offer a Fatiha for the young generations who are inheriting a heavy, sad burden left by their predecessors but who are in process of finding their own voice of goodness. This is a movement of consciousness."

Thursday, September 26
"You need to use an AeroChamber that goes over his nose with the pump so he gets all the asthma medicine," La Tonya, the school nurse, instructed me. Her office was full of brown boys like our son, lined up for the first puff of the day.

Friday, September 27
"The point of the shofar is to wake us up," Reb Ezra said, lifting the ram's horn to his mouth. He blasted it three times with all he

had. "Shana tova!" he shouted. The table was dressed for the New Year with apples and honey.

"Who shall perish by water and who by fire?" went a line in the Rosh Hashanah service as we were asked to think about atonement. So began the Days of Awe.

Sunday, September 29

Namutebi said at Andrew's memorial service that in the twenty-five years since that picture of him holding his son in Kampala was taken, Uganda has lost 63 percent of its trees.

Monday, September 30

"The Rollerblades are $5," said Abby, who sold books, clothes, toys, puzzles, and games she'd outgrown, spread over a blanket on the sidewalk leading to the Medieval Festival, to make money to fight climate change.

October

Tuesday, October 8

Danielle made risotto in the pressure cooker for dinner tonight in Marin County to feed her ninety-one-year-old grandparents, who are staying over because they lost power in Sonoma as part of the huge, wildfire-driven blackout.

"I'm almost scared they aren't turning off our power and we're going to end up engulfed in flames," said Danielle. "My grandfather keeps asking when the storm is coming, and I keep trying to explain to him that this isn't like a hurricane."

She was curious about how the rest of America sees this— 800,000 people without power as risk mitigation by the gas-and-electric company against wildfires during high winds. She asked, "Do they know this is how we live now?"

Wednesday, October 9

"We are okay, but it is starting to get smoky, and we are sorry about our friends closer to the fire," Zulema alerted us. The Briceburg fire was 4,000 acres and 10 percent contained. "PG&E will cut power to the northern part of the county," she said.

Friday, October 11

"You're going to feel some discomfort," Dr. Marianne warned me at yesterday's annual gynecological checkup. She inserted the speculum. I stared at the wall with a picture of her taken five years prior on the white peak of Kilimanjaro.

"Are you in pain?" the doctor asked, discomfited by my tears. The glaciers that ring the mountain's higher slopes were evaporating from solid to gas, the wondrous white ice cap towering above the plains of Tanzania for as long as anyone can remember disappearing before our eyes.

Saturday, October 12

In the highlands of Tanzania, Kenya, Ethiopia, and Uganda— where Damali is from—the climate is no longer hospitable for growing coffee. Damali will likely serve hot milky spiced tea at the family gathering she invited us to with a proper note card through the mail.

Baby Kazuki's mother feared her breast milk had sickened him after she reintroduced eggs into her diet. And she feared for the 8 million people ordered to evacuate their homes, as Typhoon Hagibis flayed Tokyo, including the house where her father was born.

Sunday, October 13

In the park this morning, Ana said her Realtor had advised against the offer she wished to make on purchasing her first home through the subsidized Teacher Next Door program. The house she'd fallen in love with was in a flood zone.

Tuesday, October 15

Romy sent us video of the churches in Damour ringing bells before sunrise to warn people of the raging wildfires. "Lebanon is burning," Romy said. "Probably the biggest fire this country has seen. Please send help."

Amaris said, "Mount Lebanon, the refuge of persecuted native minorities and their history in the Middle East, is on fire. For a place that represents holy land for us, I'm not joking when I say I feel my soul has been set aflame."

And then, as if by listing the scorched villages, she could turn them verdant again, she mourned their names: "Mechref, Dibbeyye, Damour, Daqqoun, Kfar Matta, Yahchouh, Mazraat Yachoua, Qournet El Hamra, Baawarta, Al Naameh . . ."

Wednesday, October 16

Yahdon, bred in Bed-Stuy, bought his gold Maison Martin Margiela designer sneakers secondhand to stay sustainably fly, he said.

Tuesday, October 22

Amelia posted a picture of the view from her kitchen window in Quito last week. "*Gracias a Dios,* we escaped the fire and the house is still standing!" she said amid nationwide civil unrest, wherein protesters clashed with riot police and a state of emergency was declared.

"Fossil-fuel subsidies were reinstated to stop the protests in Ecuador, a petrostate where the price of an unstable, fossil-fuel-dependent economy is paid by the poor. It's been a tough week," said Amelia, following up with a picture of a chocolate cupcake. "We all need a treat sometimes."

"What's your position on public nudity?" slurred Elliott, my seatmate on this morning's flight to San Francisco. In Melbourne, where he's from, Extinction Rebellion activists had stripped for a nudie parade down Exhibition Street.

Thursday, October 24

"Are we under the ocean or in the clouds?" asked Geronimo, looking up at the illusion of undulating blue waves made by a trick of laser light and fog machines at tonight's *Waterlicht* show, both dream landscape and flood.

"Anyone else have their fire go-bag ready just in case?" asked Lizz, who paints wrought iron in San Diego and writes about *brujas*. Six hundred fires had burned in California in the past three days.

"For me as a parent, knowing that my ancestors have overcome the brutality of colonialism gives me hope for the future," said Waubgeshig, originally from the Wasauksing First Nation near Parry Sound, Ontario. "My people have seen the end before."

Tuesday, October 29

Salar, just back from Beirut, described a contrast between streets of festering trash and citizens forming a human chain, across sect, at the start of revolution. "It's like we forgot the planet was our house until it grew so dirty we had to wake up," he said.

Wednesday, October 30

Felicia, Mark, Dean, Robin, Dara, Kellen, Alexandra, Roxane, Alethea, Susan, David, and Roy all marked themselves safe in Los Angeles during the Getty fire, which started near I-405 and Getty Center Drive, destroying twelve homes and threatening 7,000 more.

No word as yet on the safety of Samara, Marisa, Nkechi, Josh, Kelela, Anika, or Laila.

Thursday, October 31
"It's because of global warming," said Geronimo, dressed as a wizard, when his father recalled having to wear a winter coat over Halloween costumes during his own New York City childhood. The jack-o'-lanterns were decaying. It was 71 degrees when we walked to the parade.

November

Friday, November 1
Naheed brought us back a painting of Lord Shiva, the Destroyer, and his wife Parvati, from the Dilli Haat handicraft bazaar in New Delhi, where schools have closed because of the dirty, toxic air.

Tuesday, November 5
"I feel guilty," said Alejandra, a City College student, at the first Extinction Rebellion meeting held on campus, the same day 11,000 scientists declared a global climate emergency.

"Is there going to be food at this meeting?" Hector asked, poking his head in the door of the nearly empty classroom with mismatched, broken chairs. Down the hall was a food pantry. "You'd get more students to act if you offered food," Hector said, then left.

"Our aim is to save humanity from extinction," said Tom, an Iowa native. He'd volunteered to give the presentation, having joined the protest back in August. The slideshow included a picture of him drenched in fake blood at the feet of the Wall Street bull.

"This is a decentralized movement. Our nonviolent civil-disobedience actions are theatrical. We disrupt the status quo by occupying space. This was my first time getting arrested," Tom said. "You can do this too."

"Not me," said Cedric, referencing the obstacles to his participation, as a Black man. "If I get arrested, will it go on my record? Who pays my bail?"

Valentin, a full-time rebel since graduating with a degree in architecture, said we could address the criticism of the rebellion as a white movement that fetishizes arrests at our next house meeting. Demanding divestment, he added, should be on the agenda.

Wednesday, November 6

"Back home in Ontario, the backyard rinks are gone," lamented Michael, the man we met playing solo street hockey in the school-yard of PS 187. He showed my boy, wobbling on new inline skates, how to balance himself with a hockey stick, how to gracefully sweep the puck across concrete.

Sunday, November 10

At Václav's baby shower, Yana, who'd ordered the usual Mediter-ranean platter, told him to just rip the wrapping paper off the gift. That's how Americans do it, she said. Václav held up the bibs, boo-ties, and dresses she'd bought for his baby, due in five weeks.

"Is it just me or does it feel like this is the last baby we will pro-duce?" whispered Renata, depressed by our aging and shrinking department in an age of endless austerity with several retirements on the horizon but no new hires. "It feels like *Children of Men*."

Monday, November 11

Geronimo climbed into our bed with *The Children's Book of Mythical Beasts and Magical Monsters* open to a page of flood stories, floods delivered by vengeful gods: Utnapishtim, Viracocha, Zeus, Vishnu, Noah, and Chalchiuhtlicue.

"'The Mexican goddess of rivers and lakes once flooded the whole world to get rid of all those who are evil, but those who were good were turned into fish and were saved,'" he read. "Will I be saved?"

"You will be safe because we are privileged, not because we are good," I said, torn between wishing to comfort him and wanting to tell him the truth. "Those who are less safe aren't drowning be-cause they are bad but because they are poor."

Thursday, November 14

"Samantha's got serious respiratory issues now too," said her mother, as we waited for the school bus to drop off our kids out-side our building around the corner from a busy bus terminal in a neighborhood at the nexus of three major highways and the most heavily trafficked bridge in the world.

Friday, November 22

"Are we rebels or are we not?" asked Lena, a French international student studying environmental biotechnology. "The best way to make people know the movement is to plan an action and make demands," she said.

Saturday, November 23
"Wow, and here I thought it was going to be just another game," said Aaron, class of '98, after student activists from both schools disrupted today's Harvard-Yale football game, rushing the field to demand fossil-fuel divestment. "I guess I should have gone in to bear witness instead of hanging out at the tailgates."

Friday, November 29
Next to me at Kathy's Thanksgiving table sat her eldest son, who'd driven up for the holiday from Virginia, where he said his neighbors in the coalfields knew their industry was dead and were understandably fearful of the transition into new lines of work.

December

Sunday, December 8
The Ghost of Christmas Present encouraged Ebenezer Scrooge to do the most he could with the time he had left, in the Harlem Repertory Theater's opening-night production of *A Christmas Carol*. The last ghost waited in the wings.

Monday, December 9
Sujatha said it was getting harder to see outside in Sydney, but the failure of state and federal government action was clear: No mitigation policy. No adaptation policy. No energy-transition policy. No response equal to the task of this state of climate emergency.

"I am worried," she said, as ferries, school days, and sports were canceled because of air quality eleven times the hazardous levels. Mike bought air filters for the house, face masks for their two kids. Shaad had asked her, "Will this be the future?"

Friday, December 13
The other Ben had been at the UN climate conference in Madrid all week and felt depressed about our chances of getting through this century "if it wasn't for these kids," he said, sharing a picture of teens with eyes drawn on the palms of their upheld hands. "They are watching and awake."

"We're not here for your entertainment. The youth activists are not animals at a zoo to look at and go, *Awww, now we have hope for the future.* If you want hope for the future, you have to act," said Vega, a Swedish Fridays for Future leader.

Wednesday, December 18

"You know it's bad when the sun looks red and there's ash on every windshield," said Sarah from Sacramento, who could feel it constricting her lungs.

"What's the right balance of hope and despair?" asked the other Laura.

Friday, December 20

In the Netherlands, where Nina just submitted her doctoral dissertation proposal to the University of Amsterdam, the Dutch Supreme Court ruled that the government must protect the human rights of its citizens against climate change by cutting carbon emissions.

"Everyone not from Australia, *I'm begging you*," said Styli in Sydney. She feared international ignorance due to the lack of celebrity and location. "The truth is, our country is burning alive," she said, on the nation's hottest day on record, one day after its prior record.

Sunday, December 22

"It looks like an alligator's head," said Ben from the back seat on the drive to Nana's for Christmas. "No, a hydra," said Geronimo. Billowing smoke from the towers of the oil refinery and petrochemical plant to the side of the New Jersey Turnpike at Linden took shifting monstrous shapes.

Monday, December 23

"It's always the women who pick up the mess at the end of the meal," sighed Angie, doing the dishes at the kitchen sink in a pink T-shirt that said, SIN MUJERES NO HAY REVOLUCIÓN.

Tuesday, December 24

Though it was the third night of Hanukkah, Rebecca was still preoccupied by the Parshas Noach she'd heard weeks before, admonishing her to be like Noah, who organized his life around saving his family despite the part of him that couldn't fathom the flood.

The hardest pill for her to swallow was this: knowing that a single transatlantic flight for one person, one way, is equivalent to commuting by car for an entire year, she now feels flying to Uruguay to see friends and family for the holidays is a kind of violence.

Friday, December 27

Home in Bulawayo for the holidays during Zimbabwe's worst drought of the century, NoViolet described hydropower failure at Kariba Dam. Downstream from Victoria Falls, shrunken to a trickle, the Zambezi River water flow was too anemic to power the dam's plants, and so, NoViolet said, there was no running water three to four days a week, and power only at night. "A terrible living experience."

"The time of the month can be a nightmare for women and girls. Showers are a luxury. Those who can afford to turn to generators and solar power, but for the poor, it means adapting to a maddening and restricted life," she said.

Saturday, December 28

"Mom!" called Geronimo from the bath. "I can't breathe."

Sunday, December 29

Ben was disturbed by the dioramas on our visit to the American Museum of Natural History. "Who killed all these animals?" he demanded. "Don't they know this is their world too?"

"I learned to fish at my grandparents' house on the beach, and now my kids enjoy its calm waters," said Trever from Honolulu. "Every year, the ocean inches higher. We will sell the house next year."

Monday, December 30

From Gomeroi Country, Alison wrote, "Even away from the fires, we saw a mass cockatoo heat-kill on the Kamilaroi Highway near Gunnedah. Willy-willy after willy-willy followed us home down that road. I can't find it in me to be reflective about the decade right now. Love to everyone as you survive this, our night."

"The worst part is feeling helpless, held hostage at the whim of an abusive, inconsistent parent who wreaks havoc, then metes out arbitrary punishment in the name of protecting us," said Namwali from Zambia, about the failing of the hydroelectric company and the failures of those in power. "In a word, capitalism."

Tuesday, December 31

Another New Year's Eve. In distant parts of the planet, it was already tomorrow. The future was there and almost here. We drank prosecco at Angie's party, awaiting the countdown while thousands of people in the Land Down Under fled from the raging bushfires and headed for the beach, prepared to enter the water to save their lives on New Year's Day.

The screen of my phone scrolled orange, red, gray, black—fire, blaze, smoke, ash. A window into hell on earth. I shut it away to be present for the party and the people I loved. Before he kissed me, Victor said, "Here's to a better 2020 for our country and the whole world."

One hundred forty blocks to the south of us in Times Square, the ball is about to drop.

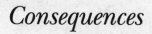

Consequences

MEEHAN CRIST

What the Coronavirus Means for Climate Change

FROM *The New York Times*

SOMETHING STRANGE IS happening. Not just the illness and death sweeping the planet. Not just the closing of borders and bars and schools, the hoarding of wipes and sanitizer, the orders— unimaginable to most Americans weeks ago—to "shelter in place." Something else is afoot. In China and Italy, the air is now strikingly clean. Venice's Grand Canal, normally fouled by boat traffic, is running clear. In Seattle, New York, Los Angeles, Chicago, and Atlanta, the fog of pollution has lifted. Even global carbon emissions have fallen.

Coronavirus has led to an astonishing shutdown of economic activity and a drastic reduction in the use of fossil fuels. In China, measures to contain the virus in February alone caused a drop in carbon emissions of an estimated 25 percent. The Center for Research on Energy and Clean Air estimates that this is equivalent to 200 million tons of carbon dioxide—more than half the annual emissions of Britain. In the short term, response to the pandemic seems to be having a positive effect on emissions. But in the longer term, will the virus help or harm the climate?

To be clear, the coronavirus pandemic is a tragedy—a human nightmare unspooling in overloaded hospitals and unemployment offices with unnerving speed, barreling toward a horizon darkened by economic disaster and crowded with portents of suffering to come. But this global crisis is also an inflection point for that *other* global crisis, the slower one with even higher stakes, which remains the backdrop against which modernity now plays out. As the United Nations' secretary general recently noted, the threat from coronavirus is temporary whereas the threat from heat waves, floods, and extreme storms resulting in the loss of human life will remain with us for years.

Our response to this health crisis will shape the climate crisis for decades to come. The efforts to revive economic activity—the stimulus plans, bailouts, and back-to-work programs being developed now—will help determine the shape of our economies and our lives for the foreseeable future, and they will have effects on carbon emissions that reverberate across the planet for thousands of years.

How hopeful you feel about the direction this response is taking may depend on how long ago you refreshed your news feed. Just last week (which feels like a hundred years ago), a friend suggested that there may be a sort of Freudian transference from coronavirus to climate—that the fear and sense of urgency will be lifted from the faster-moving crisis and settle on the slower one, becoming a catalyst for much-needed action. So far, it seems any transference is working in the opposite direction: lockdowns and social distancing are providing a litany of necessary actions ripe for the transferal of nebulous climate anxieties and fears. In this context, consumerism perversely provides some relief—you can finally go buy dry goods to prepare for the apocalypse.

But personal consumption and travel habits are, in fact, changing, which has some people wondering if this might be the beginning of a meaningful shift. Maybe, as you hunker down with cabinets full of essentials, your sense of what consumer goods you *need* will shrink. Maybe, even after the acute phase of the coronavirus crisis has passed, you will be more likely to telecommute. Lifestyles that include, for example, frequent long-distance travel already seem ethically questionable in light of the climate crisis, and, in an age irrevocably scarred by pandemic, these lifestyles may come to be seen as grossly irresponsible. Maybe among the relatively wealthy, jumping on a plane for a weekend away or for a destination wedding will come to seem unthinkable.

Sweeping changes in individual habits—particularly in wealthy countries with high per capita consumption—could lead to lower emissions, which would be an unequivocal good. But personal habits may matter less because of direct reductions in carbon emissions and more because of "behavioral contagion," a term from the social sciences that refers to the way ideas and behaviors spread through a population and can, in terms of climate action, lead to changes in voting and even policy.

Which is to say, in order to be meaningful for global emissions, changes in consumption habits as a result of the virus would need to extend beyond individuals to the larger structures that shape our lives. In China, it wasn't telecommuting or grounded planes that led to the 25 percent drop in emissions. It was the abrupt halt of industrial manufacturing. (The concept of the "personal carbon footprint" was popularized by BP in a 2005 media campaign costing over $100 million—a campaign that, research has indicated, deflected responsibility for climate change away from the corporation and onto the individual consumer.) This is not to say that personal consumption is meaningless—a significant reduction in air travel could decrease aviation emissions. But aviation accounts for only about 2.5 percent of global emissions, an amount that looks downright puny in the shadow cast by heavy industry.

If anything, the short-term positive effects on the climate that we're seeing today serve as a dramatic reminder that changing personal consumption habits will mean very little going forward if we also fail to decarbonize the global economy.

Of course, there's good reason for concern that despite the clean air and canals of the past three weeks, coronavirus will be a disaster for the climate.

According to the oil-trading firm Trafigura, coronavirus may lead to global oil demand seeing its biggest contraction in history, perhaps by more than 10 million barrels per day. While this may be good news for carbon emissions now, it signals a human disaster of epic proportions without any guarantee that emissions will remain low.

Yes, we could see a sustained emissions drop as economies stagnate and people struggle with the harsh daily realities of a global recession. But there were also dips in emissions during the 2008 financial crisis and the oil shocks of the 1970s, and emissions bounced back as economies recovered. The current crisis is different, to be sure, but after the acute phase passes, industrial production and carbon emissions are likely to ramp back up.

A global recession as a result of coronavirus shutdowns could also slow or stall the shift to clean energy. If capital markets lock

up, it will become difficult for companies to secure financing for
planned solar, wind, and electric grid projects, and it could tank
proposals for new projects; renewable energy projects around the
world are already stumbling because of disruptions to the global
supply chain. (A huge share of the world's solar panels, wind turbines, and lithium-ion batteries are produced in China.) Going
forward, a shutdown of trade between China and the United States
—for economic or political reasons—would also hit these projects
hard. The clean energy analyst BloombergNEF has already down-
graded its 2020 expectations for the solar, battery, and electric-
vehicle markets, signaling a slowdown in the clean energy transition when we urgently need to speed it up.

If oil prices stay low, that could be bad news for the climate too.
Falling demand has converged with skittish investors spooked by
the pandemic and with an oil price and production war between
Russia and Saudi Arabia. Cheaper energy often leads consumers to
use it less efficiently. Low prices could help depress electric-vehicle
sales and make people less inclined toward projects like retrofit-
ting homes and offices to save energy.

Coronavirus is bad for the climate even on the most macro levels. Lockdowns and social distancing have slowed climate research
around the world or ground it to a halt. NASA is on mandatory
telework. Research flights to the Arctic have been stopped, and
fieldwork everywhere is being canceled. No one knows how much
climate data will never be collected as a result, or when research
might be able to start up again.

Gatherings of world leaders to address the climate crisis also
have been delayed or canceled, and the COP26 climate summit
in Glasgow planned for November could be next, meaning that
the pandemic will very likely slow already sluggish international
action. This could derail climate talks at a time when, under the
Paris Agreement, countries are supposed to announce new pledges
to reduce emissions. Such a derailment would make it even more
likely that countries would blow past warming-limit goals. Going
forward, public attention is likely to be diverted from the climate
by ballooning fears over health and finances, and climate activism
that depends on large public protests is being forced indoors and
online.

There is a world in which stimulus measures could outweigh
short-term impacts on energy and emissions, driving emissions

up over the long term. This is what happened in China after the 2008 global economic crisis. Already, China is indicating that it will relax environmental supervision of companies to stimulate its economy in response to coronavirus shutdowns, which means that astonishing 25 percent cut in carbon emissions could evaporate, followed by even more emissions than before.

In the United States, we could see similarly shortsighted recovery packages aiming to ramp up the economy to prepandemic levels that double down on soaring carbon emissions. So far, the American government's aid legislation has failed to address clean energy or the climate. The $2 trillion stimulus bill passed by Congress this week, the largest fiscal stimulus package in modern American history, includes direct payments to individuals, expanded and extended unemployment benefits, and $500 billion in loans to bail out affected industries. It does not include relief for renewables, such as crucial tax credit extensions for solar and wind.

This isn't likely to be the last stimulus. Already, there is talk of the next phase of economic relief, and climate and clean energy advocates are looking to future legislation that might aim to relieve specific industries.

The two biggest wild cards for climate going forward are how policymakers respond to the threat of a global recession and how the pandemic changes political will for climate action around the world. Prime Minister Andrej Babiš of the Czech Republic has already said that the European Green Deal, a new policy package that commits European Union member states to zero emissions by 2050, should be set aside so that countries can focus on fighting the pandemic.

This week has seen a chilling shift in conservative rhetoric around the virus that echoes all-too-familiar patterns of climate denialism, suggesting that a more dangerous sort of transference is taking place. As the climate scientist Katharine Hayhoe wrote on Twitter, "The six stages of climate denial are: It's not real. It's not us. It's not that bad. It's too expensive to fix. Aha, here's a great solution (that actually does nothing). And oh no! Now it's too late. You really should have warned us earlier."

There is another world in which policymakers and politicians planning for economic recovery decide to make building a carbon-neutral society a priority. Because while the new reality could easily drain political will and funding from efforts to address the climate crisis, it could also inject a sense of urgency at a time when politicians are suddenly willing to spend vast sums of money. In this world, governments would create meaningful jobs in areas such as education, medical care, housing and clean energy, with an emphasis on "shovel-ready" projects that put people to work immediately.

The U.S. government, for example, could continue to provide jobs as needed—the program would expand during recession and contract when the economy recovered and people could find work elsewhere. As Kate Aronoff writes in the *New Republic*, "One possible benefit to such a program is that it could provide an alternative to low-paid work bound up in carbon-intensive supply chains like those at McDonald's and Walmart—currently the only employment on offer in many communities around the country." This approach would address the climate crisis with the urgency it demands while also addressing the immediate needs of workers who will be laid off or have hours reduced because of shutdowns.

Rather than seeing the clean energy transition stall, such an approach could jump-start it, while also stimulating the economy. Governments drive more than 70 percent of global energy invest-ments, and recovery plans could shift those investments as well as include new large-scale investments to turbocharge the develop-ment, deployment, and integration of clean energy technologies. As Fatih Birol, the executive director of the International Energy Agency, recently pointed out, the drop in oil prices also offers an opportunity for countries around the world to lower or remove subsidies for fossil-fuel consumption, which disproportionally line the pockets of wealthy individuals and corporations with money that could go to education, health care, or clean energy projects.

There are, of course, more radical policy interventions that could improve the health of the planet, our communities, and our lives. Adopting a thirty-two-hour workweek in the United States could lower emissions and vastly improve the quality of American life. It's unlikely we will see a four-day workweek anytime soon, but the profound disruptions of the pandemic provide a rare oppor-tunity, even in the midst of great suffering, for rewiring our sense of what is possible in American society. Maybe the rupture caused by "shelter in place" orders provides a glimpse of what work is "essential" to society—care work, education, and food distribution. Maybe it offers a glimpse, distorted though it may be, of what life might be like if we all went to work a little less.

A best-case outcome might include a rethinking of the social contract that helps protect and provide for the most vulnerable members of society at a time of increasing risk. We need to ask: What does a government owe to its people? The climate crisis has already demonstrated that the way our societies and economies are organized is unsustainable on a planet of finite resources. And as people face increasing and unevenly distributed climate risk, it is reasonable to wonder what sort of support we can expect from our government. When your community is in crisis, how will your government respond? The pandemic is a gut-wrenching reality check.

The crushing blows of the coronavirus pandemic, like those of the climate crisis, will be felt hardest by our most vulnerable populations—the poor, the elderly, the homeless, the stateless, the incarcerated, and the precariously employed—while international corporations driven by the logics of profit and endless growth to seek new markets, cheap labor, and what the sociologist Jason Moore has called "cheap nature," thereby connecting the world and helping create the conditions for crisis, will most likely remain relatively protected.

The new coronavirus spread through the activity of global markets, and it remains to be seen whether we can respond to this crisis without relying on and reinforcing the same market logics that got us into this mess. Rather, to face the profound challenges of pandemics—of which this coronavirus will not be the last—as well as the threat of climate change, to survive and even flourish on this interconnected planet, we have to learn to subordinate the needs of the market to our own needs.

It is tempting to say that humans are a pox on the Earth. That where we recede, nature rebounds. When images of dolphins and swans supposedly appearing in newly clear Venice canals popped up on social media, it was easy to believe (though it was not entirely true) that the virus had forced people indoors and "nature" had recovered in our absence. This is the wrong climate lesson to take from the pandemic.

Humans are part of nature, not separate from it, and human activity that hurts the environment also hurts us. In China, just two months of reduced pollution is likely to have saved the lives of 4,000 children under the age of five and 73,000 adults over the age of seventy, writes Marshall Burke, an assistant professor in Stanford's earth system science department. Perhaps the real question is not whether the virus is "good" or "bad" for climate, or whether rich people will take fewer airplane flights, but whether we can create a functioning economy that supports people without threatening life on Earth, including our own.

NAMWALI SERPELL

River of Time

FROM *The New York Times Magazine*

THE KARIBA DAM is failing. Since the late 1950s, it has sat on the Zambezi River, on the border between Zambia and Zimbabwe, in one of the zigzagging gorges that ripple the land there. It provides 1,830 megawatts of hydroelectric power to both countries and holds back the world's largest reservoir. For the last decade, scientists and reporters have issued warnings about the dam's potential to cause ecological disasters—of opposite kinds. On one hand, low rainfall has yielded water levels that barely reach the minimum necessary to generate electricity. On the other hand, heavy rainfall has threatened to flood the surrounding areas. When the floodgates were opened in 2010, 6,000 people had to be evacuated.

Climate change catastrophizes the weather—and when it comes to such extremes, dams are, well, inflexible. They cannot be narrowed enough to eke more force from less water during droughts, and far worse, they cannot be expanded enough to accommodate floods. The only other ways to handle floods are to let the water flow over the top of the dam or to open up a spillway for controlled release. Neither of these measures is foolproof at the Kariba Dam because of the passage of time. The dam was built on gneiss and quartzite and is made of concrete—80 feet at its thickest point. But over six decades of the waters' rushing through spillways and crashing down on its other side have carved a pit at its base, threatening to erode its foundations. Its plunge pool is now a 266-foot-deep crater. To prevent what scientists call "unac-

ceptable scour evolution" under the dam, the plunge pool is now
being reshaped at a cost of $294 million.

Without that measure, the likelihood rises that the Kariba Dam
will not just fail but *fall*. If the dam collapses, the BBC reported
in 2014, a tsunami would tear through the Zambezi River Valley, a torrent so powerful that it would knock down another dam
100 miles away, the Cahora Bassa in Mozambique—twin disasters
that would take out 40 percent of the hydroelectric capacity in
all of southern Africa. At the same time, longer hot seasons have
drained the reservoir to record lows, and drought-induced power
cuts have become a daily reality for homes and businesses. The
World Bank is supporting efforts to secure the Kariba Dam, but
any attempts to fix or expand it risk weakening it further, which
would be disastrous in the event of a flood.

Whether the water is too high or too low, the lives of millions
of people are at stake, to say nothing of the natural ecosystem.
It's a familiar, seemingly inevitable tale of human folly: one of our
most ambitious efforts to harness the power of nature has left us
exposed to nature's vagaries.

Is this just a failure of our power of prophecy? When we talk
about climate change, we talk about our inability to predict and
control what's coming, to step into the same river twice. We're out
of time, in more than one sense: we've fallen out of rhythm with
the circulatory relations between sun and rain and earth. We've
damned ourselves, foreclosed some of the future's forking paths
—this is the aspect of time we call the subjunctive, the grammatical
mood for what is imagined or wished. A river's branches suggest to
us what *could, would, should* be. But the subjunctive mood—when it
comes to rivers, when it comes to time—doesn't move in only one
direction. If we look back, it's clear: it didn't have to be this way.

The history of the Kariba Dam is the story of a war over the past
and the future of a river. That war was fought in the 1950s between European colonial powers and the local people in a place
then called the Central African Federation or the Federation of
Rhodesia and Nyasaland. The federation was a short-lived colonial experiment—or fiasco, depending on your perspective—that
merged three adjacent territories with historically disparate relationships to the British Empire. Southern Rhodesia (now Zimbabwe) was a self-governing colony founded by the British South Af-

rica Company; Northern Rhodesia (now Zambia) and Nyasaland (now Malawi) had been demarcated as British protectorates. The decision to conglomerate the three territories into one came from the colonialists, whose motivations were exploitatively economic and crudely economical.

Colonial officers had brought some of the tribal chiefs in line by appointing them to largely nominal positions in the native authorities. But the younger, educated, radical Africans—some of whom fought for the British in World War II—wanted more say in their fate. They resisted federation fiercely. They spoke up from their positions on local councils. They staged protests and boycotts: "Down with federation! To hell with federation!" They were worried by the fact that federation would move the center of power to Southern Rhodesia, whose more deeply entrenched system of segregation, the Jim Crow–like "color bar"—Africans couldn't go to bars, hotels, or movie theaters at the same time as Europeans—seemed destined to seep into the neighboring territories if they were merged.

The choice of where on the Zambezi River to build a dam was dictated by the same gravitational shift. The river's source was in the northwest of the nascent federation, near the border with Angola and what was then the Belgian Congo. It curled down through Northern Rhodesia before heading east, following—in fact constituting—its border with Southern Rhodesia, then slanting across Mozambique to its mouth in the Indian Ocean. The largest tributary of the Zambezi was the Kafue, which flowed into it from the north at the center of the segment of the river between the two Rhodesias. Just south of that confluence of currents was a gorge known as Kariba.

From the mid-1940s on, there was debate about whether to build a dam on the Kafue or at Kariba. Northern Rhodesia had decided to begin construction on the Kafue, which was closer to the Copperbelt, a valuable mining hub and urban center. The Kafue runs through natural floodplains. A dam there—which was eventually completed in the 1970s—would be smaller and more complicated to build but cause far less trouble for the people and the environment because the area was relatively uninhabited. After the federation was formed in 1953, however, Southern Rhodesia fought for the Kariba Dam to be built first. At that crucial juncture, why did the federation's government follow the Kariba fork?

It was a question of power. A French engineer, André Coyne, advocated the Kariba site because it would supply more power, at greater value for the cost. The Southern Rhodesians also wanted the dam to be closer to the new seat of political power in the federation's capital, Salisbury. The larger Kariba Dam would be a technological triumph and a grand imperial project, raising the reputation of the backwater colonies. *Newsweek* later described it as a monument to "the know-how of Western capital": "When the Zambezi River was harnessed, the queen mother cheered."

Coyne's French company designed the double curvature dam; an Italian company, Impresit, was hired to build it; the World Bank granted a loan to pay for it. The Kariba Lake Development Company—largely made up of British personnel—was established in 1957 to conduct research and piece together some ad hoc environmental and social regulations. There was barely any assessment of the potential ecological impact of the dam, much less the human costs.

So it was only in the middle of construction that the federation's government began to take seriously the question of what to do with the 57,000 people who lived in the Gwembe Valley that was to be flooded to build the dam—a place where, for centuries, they'd fished in the Zambezi and farmed on soil made rich by seasonal floods, a place they called home.

The word *kariba* was a corruption of *kariva* or *kaliba*, a local term meaning "trap." It already named a place on the river, a massive stone slab that jutted out of the water at the opening of the gorge. One legend among the local people claimed that this rock was one of three that had once formed a kind of bridge across the river—a lintel that resembled the animal traps they used—until a flood washed the other two away. It was the sole remnant of a geological event, and from another point of view, a warning. Other legends said that this rock was the home of a river god named Nyaminyami—possibly a corruption of a local word referring to meat—with the head of a fish and the twisting whirlpool-like body of a snake. The British took one look at that big rock and decided it was the best place to build a dam, and the best word—mispronounced because they couldn't wrap their lips around the soft "b" and "l" common in Bantu languages—to explain to the Tonga exactly what a dam was.

Trap a river? The notion was so outlandish that the Tonga began to ignore the district commissioners, who despaired of convincing the villagers—only a few of whom had ever even witnessed electricity—that the dam was really going to be built, that their ancestral homes would soon be underwater. As David Howarth puts it in his blinkered but engaging 1961 history of the Kariba, *The Shadow of the Dam*, "the whole idea of stopping the river was absurd" for the Tonga: "Most of them admitted that the Europeans would probably try, but the Europeans did not know the river as the Tonga knew it; and the old men argued that if anyone thought he could stop the river by building a wall across it, it only showed he had no idea how strong the river was. Let them try . . . the river will push the wall over, or run round the ends of it."

This is exactly what happened. Seasonal rains can swell the Zambezi up to twenty times its dry-season size. In late 1956, news came from upriver that an "exceptional flood"—so exceptional it would come to be called the Hundred Years' Flood—was on its way. The water rose 66 feet and drowned the cofferdam that was in place for construction. When the waters finally subsided, only a crane had been lost, but the engineers were shaken by the unexpected and awesome sight of the torrential deluge.

They built a second cofferdam higher—but not high enough. The very next rainy season, the tributaries joined forces once more. This time the chances were deemed one in a thousand. The Thousand Years' Flood of 1958 swept away a suspension bridge, which "writhed like a snake when the water touched it." The river rose 116 feet to the top of the second cofferdam and poured over it, creating a waterfall 28 feet high. The Tonga had been roundly mocked for superstitious predictions that the "huge serpent" living in the Zambezi would "be angry with the white man's wall and knock it down." Now, the journalist Frank Clements declared: "Nyaminyami had made good his threat. He had recaptured the gorge."

The dam seemed cursed. Late in the construction, some scaffolding gave way. Seventeen workers fell into a hole and were buried in wet concrete. Some say their remains were picked out, others that they remain entombed in the dam. When the floods receded, the engineers rushed to make sure the dam was complete before the following rainy season.

This meant that the wildlife now urgently needed to be rescued

before the Gwembe Valley became the largest man-made lake in
the world. "Operation Noah," as it was messianically named by
white conservationists, managed to capture and remove 6,000
animals, though thousands more died in the floods. (This focus
on the wildlife as the principal victims has persisted as the central story of Kariba; a recent BBC article about the dam revolves
around a lone baboon "marooned" on an island in the Zambezi.)

The people proved to be more intransigent than the animals
when it came to forced resettlement. The government determined
that the Tonga were to move to Lusitu, an area to the north, and
began resettling 193 villages one at a time, carting the people and
their property there in trucks. These new lands had poor, stony
soil. There was an almost immediate outbreak of dysentery. The
Tonga way of farming, which relied on seasonal floods and leaving
land fallow, wasn't possible here. The ratio of population to land
was radically unbalanced. Traditional laws regarding the distribution of property were upended.

Those who had not yet left the Gwembe Valley, already concerned about the disruption of ancestral shrines and the lack of
adequate compensation for the loss of their homeland, now had
even less reason to leave. Some had been radicalized by the African
National Congress—a nascent, nonviolent political party whose
members agitated for the breakup of the federation and later led
the movements that decolonized its three nations. The congress
encouraged civil disobedience in the face of the relocation.

As is often the colonial way, over time the federation's persua-
sion campaign gave way to insistence, then violence. The laws
of Northern Rhodesia in fact prohibited forced removal, so the
Tonga Native Authority was cajoled into approving a legal order,
which was translated and broadcast to the people: "The Government is quite satisfied that the Lusitu plan is in your best interests
and now intends to carry out this move without delay. Those who
resist will be moved by force, using the police you see here today . . . Anybody who obstructs the move will be prosecuted. When
people have moved from a village, the huts will be destroyed."

The people rebelled. The villagers of Chisamu, who were gov-
erned by a chief named Chipepo, made a series of charges at the
police, shouting and gesturing with their spears, playing drums
and singing war songs. The standoff lasted for days, the police con-
ducting drills, Chipepo's people imitating them. "They marched

and countermarched in single file," Howarth writes, "carrying their spears like rifles on their shoulders, and instructors marched at the sides of the columns like sergeants or platoon commanders. Sometimes it looked like a parody, but perhaps they did it to convince themselves." The governor of Northern Rhodesia was brought in for an *indaba* with the leaders, but to no avail. When the constables moved in on the villagers, violence broke out. Eight Tonga were shot and killed. The people relented.

The dam was completed. The valley was flooded. Nowadays, fishing boats and "sunset cruises" slip up and down the dwindling lake above the dam. The eeriest, most beautiful thing about Lake Kariba—its main attraction for tourists—is that the submerged trees of the Gwembe Valley still stand. You can see them reaching up from the depths, branching up out of the water, forking against the sky.

"The whole might of modern technology was nearly caught by the primeval, savage forces of Africa," Clements wrote of the Kariba in 1959. With this Manichaean hyperbole, he tidily conflated the power of nature, the myth of Nyaminyami, and the resistance of the Tonga, even as he diminished all three. In the end, the might of modern technology won, escaped the trap—or perhaps became one. Many historians cast the story of the Kariba Dam as a paternalistic tale about how a zealous belief in "progress" overwhelmed a hapless tribe of what David Livingstone once called a "degraded" people. Another way to see it is that the building of the Kariba Dam redirected enormous wealth to colonial parties at the expense of the rightful dwellers of the Gwembe Valley, who are considered "development refugees" and still lack adequate access to water and electricity. As late as 2000, three of the nearby districts where the Tonga live were not connected to the national grid lines.

This dam business now directs wealth to neocolonial parties. The China National Complete Engineering Corporation is building another $449 million megadam on a tributary of the Zambezi. Within its own borders, the Chinese government is turning away from hydroelectricity and toward solar and wind energy. They know that, in the midst of a global climate-change crisis, finding alternatives to dams is better than trying to fix them.

Africans know it too. In 2014, Partson Mbiriri, then the chairman of the Zambezi River Authority, told the BBC, "It's equally important to think about solar—on the assumption, of course, that we'll continue to have sunshine." While various figures of authority—colonial, governmental, environmentalist, journalistic; then and now, well-meaning and mercenary—have all been deeply concerned to explain to Africans what will happen to us if we do not move out of the path of progress, many of them never really bothered to listen to us.

Some anthropologists, like the late Elizabeth Colson, who lived among the Tonga for decades to study the effects of resettlement, thought that they didn't even believe in Nyaminyami, that the river god was a belated import or an external projection. Regardless of its origins, the idea of a primitive deity certainly allowed colonial powers and observers to ignore the political concerns of the Tonga.

The Africans of the federation had in fact articulated a set of prescient questions and demands about the building of the dam —subjunctive possibilities.

In 1955, the Northern Rhodesian African National Congress leader, Harry Nkumbula, wrote to the queen of England, asking her to appoint a commission including Africans "to determine whether it is just that the people should be dispossessed of their land"; whether the power generated by the dam "could not be better generated by nuclear energy"; whether the compensation the people received was sufficient; and whether "the lands to which the people are being moved are equal in value" and fertility to those that would be flooded. Perhaps human folly is culturally relative.

When they were first informed about the dam, the Gwembe Native Authority also made a set of twenty-four demands respecting their rights—to land, property, reparations, protection, information. The eleventh was: "That in moving people, their choices shall be seriously considered before they shall be ignored." And when Chipepo's people staged their ultimately futile uprising, they wrote messages in English, which they sent to the district officers and the native authorities or nailed to trees on the battlefield: "We shall die in our land . . . We don't want to be removed to Lusitu or to any place. We will not go home until you dismiss your army of policemen. We will not fight with weapons but with words." What would paying attention and respect to their words have made possible?

The Tonga knew the Zambezi. They knew that a river keeps time, not like a clock but like a chronicle. They knew its sediments and grooves, the patterns of the beings dwelling within it and nearby, its might and its tendencies. Kariva rock itself was testament to the river that had knocked away its stony triplets, the river so powerful that it seemed that a god must live inside it. Whether they were syncretic or exaggerated, these local myths were still forms of geographic knowledge.

A river can channel water into an immense power. A river can also flood, spread into the spaces open to it. A river is both a singular, driving force and a distributive, branching one. The Tonga had long lived peacefully on both sides of the Zambezi, crossing back and forth to court brides, borrow food, visit relatives. They knew that you don't stop a river; you move over, through, and with it. You follow its paths. You may step into it as often as you wish, but you do not stay.

MAYA L. KAPOOR

Fish Out of Water

FROM *High Country News*

IN THE SPRING of 2016, biologists at the U.S. Fish and Wildlife Service came to a terrible realization: the Yaqui catfish, the
only catfish species native to the western United States, was on the
cusp of disappearing. After a week of searching, they could catch
only two wild fish. They estimated that, at most, just thirty fish
remained.

For approximately two decades, the last known Yaqui catfish in
the United States had been kept in artificial ponds built in and
around San Bernardino National Wildlife Refuge, on the Arizona-
Sonora border, and at a local zoo. Creatures of rivers and wetlands, they had not reproduced. Still, federal and state biologists
felt they had to try one more time. In a last-ditch breeding effort,
the agency gathered eleven fish and shipped them to a hatchery
in Kansas. Within weeks, all of them died. Eventually, even the
one geriatric catfish left on display at the Arizona-Sonora Desert
Museum had to be put down.

Today, the Yaqui catfish, a whiskery-looking creature that
evolved at least 2 million years ago and was once common enough
for people to catch for food, is functionally extinct in the United
States. There may be a few still hidden in Arizona's ponds, but
not enough to keep a population alive. According to the Fish
and Wildlife Service's 2019 five-year review of the species, it's on
the brink of global extinction; even as the catfish faces ongoing
threats in Mexico, scientists don't know enough about its basic biology to save it.

To people for whom "Sonoran Desert" conjures up images of

steadfast saguaros or sun-struck lizards, the fact that a native catfish species existed in such a dry place can be surprising. In reality, prior to European colonization, the region supported rich
waterways and aquatic communities. The current extinction crisis
speaks to an uncomfortable truth: in a land of finite resources, every choice, big or small—irrigating an alfalfa field, taking a swing
on a golf course, burning fossil fuels—means choosing what kinds
of habitat exist, even far away from town. And that means choosing
which species survive.

Now, the hunt is on to find more Yaqui catfish in Sonora, Mexico. But as the election season ramps up, the Yaqui catfish faces a
new threat: the Trump administration is racing to complete the
border wall before the 2020 presidential election, blasting desert
mountains, tearing up old-growth saguaros, and destroying the
ancestral homelands and cultural resources of tribal nations such
as the Pascua Yaqui, Tohono O'odham, and Hopi. According to
Laiken Jordahl, a staff member with the Center for Biological Diversity, wall construction will require 700,000 gallons of water each
day.

To be clear, the Yaqui catfish is no jaguar. It's no beauty; it's
not terrifying; its babies aren't even all that cute. A reclusive fish,
it never swam in more than a relatively small portion of U.S. waters. In almost a year of researching them, I still haven't gotten a
glimpse of one. When I lived in Arizona, my favorite catfishes were
different species entirely, probably channels or blues; they arrived
before me breaded and lightly fried at a small restaurant just south
of the University of Arizona campus, accompanied by iced tea,
somehow fitting into my convoluted rationalizations about being
vegetarian. And yet the Yaqui catfish's looming extinction bothers
me for the simple truth it represents: the Borderlands can't have
its rivers and destroy them too.

Biologists know surprisingly little about Yaqui catfish, dusky animals that live at the bottom of ciénegas and streams, growing up to
about two feet long. Only about 2 percent of their historic range
lies within the United States; the rest is in Mexico. Living mysterious lives in gloamy places, Yaqui catfish inhabit a world rich in
ways we humans can never imagine. Like many catfish, they are
covered with tastebuds instead of scales. Catfish are named for the
long, flexible barbels that sprout from their faces like a cat's whis-

kers, helping them feel and taste their world. Yaqui catfish may communicate with each other through drummings and stridulations, and they may hunt by tracking the electric discharges from other animals' nervous systems.

The Sonoran Desert's fishes have evolved fascinating adaptations: some give birth to live young; others snuggle down and wait out dry spells in the mud. But the past few centuries have been especially rough for them. As the Borderlands' human communities keep growing, and climate change makes the region hotter and drier, streams stop flowing and wetlands vanish. Meanwhile, introduced species, including channel catfish originally from central and eastern North America, push out, or hybridize with, native species. Like most of the Southwest's aquatic species, Yaqui catfish have struggled to survive since European colonization.

Today, a *river* in the Southwest often means a dried-out, sandy wash where trash and the skeletal remains of cottonwoods bleach in the sun. But before colonization, networks of riparian areas, wetlands, and slow-moving rivers flowed through the region, where Indigenous peoples have lived and farmed for millennia.

A combination of colonialism and human-caused climate change turned rivers and wetlands to dust. Cattle, introduced in the 1500s by the Spanish, overgrazed the land, congregating around and trampling sensitive desert river systems. Farms, mining, and the extirpation of beavers all disrupted the Southwest's rivers, which abruptly channelized in the 1800s, changing from meandering ciénegas to deeply etched arroyos. In the 1900s, enormous dam projects began sending the Southwest's water far away, irrigating California's agriculture, even as Sunbelt cities kept growing. By 1973, when the Endangered Species Act passed, such intensive pumping meant that the region's ciénegas were almost all gone, including the tiny fragment of Yaqui catfish habitat in southeastern Arizona.

Generally, the Endangered Species Act protects critical habitat for rare species, the logic being that a plant or animal can't survive if there's nowhere left for it to live. San Bernardino National Wildlife Refuge, designated in 1982, was a different approach to saving species. An archipelago of man-made pools in a sea of desert shrubland, the refuge was meant to be a place where native fish species could survive, even as their natural ecosystems drained to sand. The Yaqui catfish lived there, some for decades; what they didn't do was reproduce. These fish, creatures of deep pools and flowing rivers, seemed to need something that the artificial ponds didn't provide to breed.

Meanwhile, residents of the Southwest—including me, for approximately half my life—have washed our dishes, cleaned laundry, swum in pools, watered plants, and generally gone about our daily lives by tapping into what water persists, including the Colorado River and underground aquifers. Today in the Sonoran Desert's dry creekbeds, one can sometimes see a black band in the soil wall—all that remains of miles of rivers and wetlands.

Because only a small part of the Yaqui catfish's range lies in the United States, American researchers hope that they can collaborate with Mexico to save the species from global extinction, and maybe even donate a few more fish to the United States. But Mexico's Yaqui catfish, which historically inhabited thousands of miles of river systems throughout northwest Mexico, are disappearing too, and the race to protect them is hobbled by a basic lack of information.

On the Sonoran Desert's version of a fall day, the afternoon high hovering around 90 degrees Fahrenheit, a binational group of researchers gathered in Cajón Bonito, a canyon that cuts through former ranchland purchased by a wealthy American, Valer Clark, about a mile south of the U.S.-Mexico border. They were there to collect that missing information.

Emerged from their vehicles, the researchers gathered an assortment of buckets, field gear, and notebooks and began walking in the creek, stopping periodically to document its condition. The river clattered through the small canyon, carrying dried leaves above its sandy bed. It also carried evidence of living things that had stepped, swam, or sprouted in the water, by way of environmental DNA, or eDNA, fragments of genetic material left like microscopic calling cards. It was these cards that Thomas Hafen, a graduate student at Oklahoma State University, hoped to pick up. As they waded in the river, Hafen and his research assistant, Alex Gutiérrez-Barragán, periodically sampled the river. Wearing medical gloves, they gently scooped water into a filter designed to strain out eDNA. After a battery-powered pump had sucked through enough water, they used forceps to peel the filter out of its cup, carefully fold it up, and place it in a container for shipment to a

lab in Montana. In several months' time, they would know whether
any Yaqui catfish had passed by recently.

A friendly, quiet worker in his mid-twenties, with brown hair
and a stubbly beard, Hafen spent two years before college on a
Latter-day Saints mission trip in Mexico, becoming fluent in Spanish. Now, he was surveying as much of the remaining Yaqui catfish
habitat in Mexico as he could for traces of the elusive animals,
hoping, eventually, to be able to identify the best habitat for the
species. If Yaqui catfish breed in captivity, Hafen's research will
help identify where to release their young, and which rivers to protect.

As Hafen strained water samples, Sonoran Desert fish expert
Alejandro Varela-Romero held a bucket and binoculars and peered
into a deep, slow pool of green water swirling gently in the lee of
a boulder. A thoughtful man in his fifties, with dark hair peppered
gray and a mustache, Valera-Romero wore plastic-rimmed glasses
and a T-shirt bearing a Spanish translation of the famous quote by
evolutionary biologist Theodosius Dobzhansky: NOTHING IN BI-
OLOGY MAKES SENSE EXCEPT IN THE LIGHT OF EVOLUTION. He
called out the scientific names of tiny darting fish I could barely
see. He had a feeling that a Yaqui catfish might appear.

It certainly was hard to imagine a location more hospitable to native fish than Cajón Bonito, where smoky-black catfish could float
gently above the river bottom, sheltered by sloping banks and submerged branches. Hawks circled overhead, while songbirds called
from the bushes. Earlier that day, a skunk had excavated a small
dig below a tree stump no more than 10 feet away from where
the researchers were working, and then apparently curled up for a
nap. But even here, channel catfish—one of the biggest threats to
Yaqui catfish—had infiltrated. When Hafen got his eDNA results
months later, they would show that almost everywhere in Mexico
where he found Yaqui catfish, he'd found channel catfish too.

"In the past, there were no exotics in the [river] basins,"
Varela-Romero explained. "When the [Mexican] government
started building reservoirs, federal officers had a brilliant idea of
buying exotic fishes like channel catfish and putting them in the
reservoirs, to give to the people the opportunity to eat fish," even
though, he said, everyone already ate the native ones.

As these introduced species have pushed the Yaqui catfish to
extinction at lower elevations, the species survives higher up, in

hard-to-reach, isolated mountain headwaters. There, though, illegal drug grows threaten the fish. Varela-Romero said that in the past, while sampling Sonoran rivers for native fish, he came upon poppies growing wild on the banks. Their seeds had washed down from opium grows.

"The problem is that money from the U.S. pays for those activities," he observed. The economics can get personal: Varela-Romero's father's car was once stolen in Hermosillo, Sonora. It turned up months later in Yuma, Arizona, after being used to run drugs across the border.

Because the Yaqui catfish is dying out, Varela-Romero said, "you develop some kind of love . . . We call it *cariño* en español." He always got excited when he saw a Yaqui catfish in the wild, wondering each time how many more he might see. Today, though, he had no luck. The catfish, if any were nearby, stayed hidden.

If Hafen and his team represent cutting-edge fisheries research, with fussy portable filters and slick sampling methods, Varela-Romero embodies an older approach to natural history, his expertise earned through hours spent studying different species, counting bent spines, describing the exact color of scales, trying to think like a fish. Another biologist on the trip, Chuck Minckley, was an original member of the Desert Fishes Council, a nonprofit research organization for desert fish biologists in the U.S.-Mexico Borderlands. Minckley missed the first meeting because he was drafted for military service in the Vietnam War, but he hasn't missed one since. Now in his seventies, he waded slowly down the canyon as Hafen's crew sampled the water, pausing to rest at times on an overturned bucket. His ancient black Lab, Shadow, raced happily through the shallows, huffing like a freight train.

Varela-Romero and Minckley are determined to catch and breed Yaqui catfish in Mexico as quickly as possible, even though no one really knows how. It's easy to read their efforts as a comedy of errors, with fish found, lost, misidentified. In reality, with low budgets and improvised tools, the researchers are learning how to work with the species. The problem is that their learning curve has crashed into an extinction curve. After years of neglect, the Yaqui catfish is rare enough that every fish matters, making it hard to experiment with ways to breed them, or even to keep them alive.

One afternoon in Mexico, the researchers attempted to recover eight Yaqui catfish caught in Cajón Bonito, which had escaped from some netting in a holding pond. This involved using a broken kayak paddle to maneuver a rowboat, which Minckley had purchased as a teenager to take duck hunting, out to the middle of the pond, and pulling up yards of soggy netting.

The men excitedly hauled a catfish to shore in a bucket and gently transferred it to a shallow rectangular container. The small, soot-gray animal burrowed into the corners, seeking an escape with its mouth and barbels. We crowded over it, and I found myself getting unexpectedly emotional, looking at what might be one of the last of an understudied, barely known species, trying to escape from a plastic box. I felt foolish a few minutes later, when Varela-Romero checked its pit tag and announced that it was actually a Yaqui-channel hybrid. I wondered whether the loss of a species that looked just like one of the most common fish species on the planet really mattered.

Only later did it occur to me that perhaps, if I couldn't tell the difference between a Yaqui catfish and a channel catfish, that was because they communicate in the language of fish, not primates—that their seeming interchangeability said more about my limited understanding than it did about their limited distinctions.

The same economic pressures that push Yaqui catfish toward extinction today have wreaked havoc on local human communities for centuries. Even as American and Mexican researchers try to save the species, tribal nations such as the Pascua Yaqui are working to reestablish control over their natural resources, including the Yaqui catfish.

"If we had a tapestry of our history on this side of the border, it would probably be missing a bunch of big chunks," Robert Valencia, then-chairman (now vice chairman) of the Pascua Yaqui Tribe, told me. Valencia and I met in a conference room at the Pascua Yaqui Tribe's administrative offices, about twenty minutes away from downtown Tucson. Valencia, who is in his sixties, has dark hair and a thick salt-and-pepper mustache. He wore what he called his "end of summer" shirt—a red Hawaiian-print button-down. To Valencia, the tribe's history is too important to lose. "We can't let things go," he said. "As a matter of fact, we're doing the opposite. We're looking at—on both sides of the border—what are important parts of history we don't know, or we need to know more?"

At one point, Valencia wrote down a Yaqui word for me notebook, after I struggled to sound it out myself. "What it is, 'In the beginning,'" he explained. "We have to always ber what we had in the beginning. To me, that's number one."

To Valencia, the catfish ties Yaqui peoples to the Río Yaqui region, in part by embodying the importance of water to the tribes. As the flow of Borderlands water and species has been curtailed, so, too, has the movement of Yaqui peoples across their ancestral lands.

Considered Mexican by the United States, the Pascua Yaqui Tribe only gained federal recognition in 1978, despite its presence on both sides of the border since long before the United States or Mexico existed. In 1964, the tribe secured the land in southern Arizona that would become its reservation from the Bureau of Land Management, after surviving the Yaqui Wars, a persistent attempt —first by the Spanish government and more recently by Mexico —to kill off Yaqui tribes and use their land along the Río Yaqui for mining and large-scale agriculture. "If you read about our history, there was an unwritten extermination policy against our people in Río Yaqui," Valencia said. "They killed us, shipped us off as slaves in Yucatán, did whatever they could." People were sent as far away as the Caribbean and Morocco, never to return.

In the late 1970s, Valencia's uncle, who was tribal chairman, successfully fought to gain tribal recognition from the U.S. government. Today, the Pascua Yaqui Reservation includes approximately 2,000 acres southwest of Tucson, bordered partly by Tohono O'odham land and partly by the city's growing sprawl. Yaqui people regularly move back and forth across the U.S.-Mexico border, although that's been harder since the 9/11 terror attacks, Valencia said. Valencia's mother, who was born in the United States, grew up in the Yaqui pueblo of Tórim in Mexico. Valencia remembers her reminiscing about rivers "teeming with fish," before Mexico built three dams along the Río Yaqui.

On the one hand, the delayed recognition of its sovereignty means the tribe is only now addressing quality-of-life issues that go back decades, Valencia said. But he sees advantages to its independence from the federal government. "Because we weren't entrenched, we don't have federal programs that are based here; we didn't want any," Valencia said. "We always insist on directing whatever effort it is, whether research, whether programs, because what we found over the years, if we let other people do things for us—it fails, every single instance." Yaqui catfish conservation seems to fit this pattern: in 2015, after three years of applications, Valencia won a grant from the U.S. Fish and Wildlife Service to create monitoring and educational programs about Río Yaqui fish on both sides of the U.S-Mexico border. Not long after, he found out about the 2016 die-off of Yaqui catfish at the Kansas fish hatchery.

Working with the eight Yaqui pueblos in Mexico, Valencia wants to create family-based "microhatcheries." Partly, this is for cultural reasons—the Yaqui catfish is a traditional food—but it would also be part of the Yaqui pueblos' long battle for the Mexican government to honor their treaty rights, which include extensive control of water and other natural resources in the Río Yaqui basin. Currently, Mexico's dams force Yaqui communities in Mexico to rely on water contaminated by agricultural runoff laden with pesticides and heavy metals. According to the *Latin American Herald Tribune*, Mexico defied its own Supreme Court to build an aqueduct to pipe Río Yaqui water to Sonora's growing capital city of Hermosillo, further ignoring Indigenous claims to the watershed and its natural resources, including its fish.

The Pascua Yaqui Tribe has already built microhatchery prototypes. One sunny morning in September, I met James Hopkins (Algonquin and Métis), a law professor at the University of Arizona who directs the Yaqui Human Rights Project Clinic and has legally represented Yaqui pueblos in Mexico, in a parking lot near the Pascua Yaqui Tribe's Casino del Sol. On roads made slick and puddled by recent rain, we drove together to Tortuga Ranch, a former cattle ranch that the tribe now owns, to examine its prototype aquaponics operations. A hoop house held burbling tanks where luminous koi fish swirled under bright green vegetation. Further efforts to raise Yaqui catfish were on hold for now, until more Yaqui catfish become available. But even without the catfish, microhatcheries of fish such as tilapia or channel catfish could be important routes to nutritional independence for Yaqui families in Mexico.

"[Yaqui pueblos] want a clean, sustainable protein source," Hopkins said. "There's a huge infant mortality rate [in Mexico], primarily because all their protein from the local area—fish, dairy from local cows, goats, pigs—is going to be carrying persistent chemicals, mainly DDT."

On the U.S. side of the border, Valencia would like the Pascua Yaqui Tribe to raise Yaqui catfish in captivity, in part for a commercial market. "It's one of those things I think can be successful," he said. "The people have some knowledge of the species itself." Hopkins, though, expressed a more cynical reason for growing Yaqui catfish in microhatcheries. "Anything you're trying to return back to nature, your plan has to fit with the larger state plan," Hopkins said. For Arizona to take interest in the Yaqui catfish, there had to be a commercial value, such as a restaurant market for the fish, or an interest in sport fishing.

"I've been transparent with people like Chuck Minckley and others, saying, 'Look, if you bring the fish back, where's it going to go?'" Hopkins told me. "It's going to have to be a commodity in its own way."

One warm fall day last September, I visited San Bernardino National Wildlife Refuge, a remote landscape of rolling hills with a backdrop of sharply angled mountains at the Río Yaqui headwaters in southeastern Arizona, where the last Yaqui catfish in the United States. were caught for the failed breeding effort. Traces of catfish eDNA still turn up in ponds here, although it's uncertain whether that DNA is from living fish, or the remains of dead ones. This is where the U.S. Fish and Wildlife Service would like to reintroduce Yaqui catfish, if Mexico agrees. Up close, the ponds seemed peaceful: small, cattail-ringed pools rippling in a rising breeze.

But now, the ponds themselves are endangered, collateral damage in the Trump administration's determination to construct a border wall before the November election. Despite the Covid-19 pandemic, wall construction continues at a breakneck pace, using local water to spray down dusty roads and mix concrete. As of this writing, Customs and Border Protection has not provided *High Country News* with requested groundwater use estimates.

According to Myles Traphagen, a field biologist and GIS specialist who previously worked at San Bernardino, no environmental analysis has been conducted on the impacts of wall construction on the region's water, or on its fish. "Since there was no NEPA required, no prior studies, we are essentially navigating in uncharted territory, with no baseline for what the final effects might be," Traphagen said. Refuge staff declined to comment on the impacts of the border wall.

On my drive back to Tucson, the clouds gathered into a gray comforter, trailing rain far to the south. It was my favorite kind of dappled desert weather, the filtered light deepening the green on hillsides and softening the craggy mountains. In the distance I spotted a bright white post—a historic border marker, rising from the shrubs. Past that, the roof of a ranch house in Mexico. Trucks passed me slowly on the dirt road, spraying water to keep down the dust. It took me a few minutes to realize why: they were preparing for the construction crews to arrive, to build the border wall. Perhaps the next time I drove through, those mountain views would be gone. So, too, might the springs that the refuge's fishes relied on. Without water, there would be no future here for the Yaqui catfish.

I later called Bill Radke, who has managed San Bernardino for about two decades, and asked him whether falling groundwater levels in the refuge might be just one more threat to the Yaqui catfish, or the final nail in its fishy coffin. Radke would not comment on the effects of the border wall construction, but he acknowledged that groundwater levels have always been a big concern for the survival of fish at the parched refuge.

"You almost have to end your story that way, because I don't know that any of us can say that for sure," Radke said. "What's that adage about how far you can lean off a cliff before you fall? You don't know until you lean too far."

JULIA ROSEN

Cancel Earthworms

FROM *The Atlantic*

ON A SWELTERING July day, I follow Annise Dobson down an overgrown path into the heart of Seton Falls Park. It's a splotch of unruly forest, surrounded by the clamoring streets and cramped rowhouses of the Bronx. Broken glass, food wrappers, and condoms litter the ground. But Dobson, bounding ahead in khaki hiking pants with her blond ponytail swinging, appears unfazed. As I quickly learn, neither trash nor oppressive humidity nor ecological catastrophe can dampen her ample enthusiasm.

At the bottom of the hill, Dobson veers off the trail and stops in a shady clearing. This is as good a spot as any, she figures. She kicks away the dead oak leaves and tosses a square frame made of PVC pipe onto the damp earth. Then she unscrews a milk jug. It holds a pale yellow slurry of mustard powder and water that's completely benign—unless you're a worm.

Seconds after Dobson empties the contents inside the frame, the soil wriggles to life.

"Holy smokes!" she says, as a dozen worms come squirming out of the soil—their brown, wet skin burning with irritation. "Disgusting." I have to agree. There is something unnerving about their slithering, serpentine style; instead of inching along like garden worms, they snap their bodies like angry rattlesnakes. But the problem with these worms isn't their mode of locomotion. It's the fact that they're here at all.

Until about 10,000 years ago, a vast ice sheet covered the northern third of the North American continent. Its belly rose over what is now Hudson Bay, and its toes dangled down into Iowa and

Ohio. Scientists think it killed off the earthworms that may have inhabited the area before the last glaciation. And worms—with their limited powers of dispersal—weren't able to recolonize on their own.

For someone like me, who grew up in the Midwest seeing earthworms stranded on the sidewalk after every rain, this was a shocking revelation. With the exception of a few native species that live in rotting logs and around wetlands, there are not supposed to be any earthworms east of the Great Plains and north of the Mason-Dixon Line.

But there are, thanks to humans. We've been moving worms for centuries, in dirt used for ship ballast, in horticultural plants, in mulch. Worms from South America now tunnel through the global tropics. And European earthworms live on every continent except Antarctica. Dobson, a forest ecologist at Yale University, calls it "global worming."

But of all the earthworms people have shuttled around the world, the ones Dobson shows me at Seton Falls have scientists most concerned. Originally from Korea and Japan, they are known as *jumping worms, snake worms,* or *crazy worms.* And they have the potential to remake the once wormless forests of North America.

The perils of an earthworm invasion are hard to grasp if you've been raised to believe that earthworms are good. "They seem so symbolic of a healthy ecosystem," Dobson says. For their stellar reputation, they can thank none other than Charles Darwin. In addition to developing the theory of evolution, Darwin studied earthworms for forty years at his home in England.

With characteristic curiosity and rigor, the naturalist conducted all manner of earthworm experiments: He observed their reaction to the sound of the bassoon (none) and to the vibrations of a C note played on the piano (panic). He watched how they pulled leaves into their burrows, and tested their problem-solving skills by offering them small triangles of paper instead (most figured out how to drag them by a corner). Darwin also measured how quickly worms covered up a large paving stone in his garden with their castings. He estimated that they could move at least 10 tons of soil per acre per year.

Dirty, slimy earthworms weren't especially popular in Victorian

England. But in 1881, shortly before his death, Darwin compiled his worm studies into a book called *The Formation of Vegetable Mould through the Action of Worms with Observations on their Habits,* in which he praised the humble critters. "It may be doubted whether there are many other animals which have played so important a part in the history of the world, as have these lowly organized creatures," he rhapsodized. The book became a best-seller, giving worms' dingy public image a makeover in the process.

Gardeners now rejoice to find earthworms in their soil, and you can purchase a 1,000-pack of "Nature's Wonder Workers" on Amazon for $45. There's even an entire canon of worm-centric children's literature, including *Wiggling Worms at Work* and Richard Scarry's *Best Lowly Worm Book Ever!* But Peter Groffman, a soil ecologist at the City University of New York, says that while worms may do some good in your compost bin, they don't deserve all the credit for your bumper crops and lush ornamentals. "The earthworms are in the soil because the soil is healthy," he says. "They are not necessarily doing anything for it."

And though they can be helpful for breaking up compacted soils and breaking down organic matter, worms can also cause trouble in agricultural fields. Their burrows create channels that allow nutrients and pesticides to leak from fields into nearby waterways, and carbon dioxide and nitrous oxide to escape into the atmosphere. In fact, a 2013 review of recent research found that worms likely increase greenhouse-gas emissions.

But Darwin wasn't thinking about these things—and he certainly wasn't thinking about the consequences of introducing worms into ecosystems that had evolved without them.

The mustard pour, which Dobson had done partly for my benefit and partly just to check in on the worm population, is over a few minutes after it begins. The worms—bothered but otherwise unscathed—have disappeared back into the forest floor. So Dobson and I head back to where we left her assistant, Mark, toiling among a jungle of knee-high poison ivy and Johnny jumpseed.

He's searching for the pink flags that Dobson left here last year to mark a few dozen specimens of native plants. Her goal is to track them as they grow and reproduce to see if they show any potential of adapting to jumping worms. Mark, an undergraduate at the University of Connecticut, is doing his best. But it's his first day on the job, and also, someone has deposited a filthy, yellowing mattress over much of Dobson's research plot.

"That's the beauty of working in urban systems," Dobson jokes.

By now, it's after 11 a.m. and the heat has grown unbearable, so Dobson suggests we decamp to a nearby McDonald's. We pile into her silver Subaru Impreza, where the dashboard thermometer reads 38 degrees Celsius—100 degrees Fahrenheit. (Dobson and her car are both Canadian.) There is no gauge for the humidity, but it's stifling.

Inside the restaurant, air-conditioning and cold beverages revive our spirits. We settle into a booth, and I ask Dobson about North America's first wave of earthworm invaders. Common species like *Lumbricus terrestris,* better known as the night crawler, arrived hundreds of years ago with European settlers, and have long been welcomed in gardens and farmland. In the 1980s, however, researchers began to find European worms in the forests of Minnesota and other northern states. One hypothesis is that people spread them when they throw away extra fishing bait next to lakes and streams.

The discovery alarmed scientists. In the absence of worms, North American hardwood forests develop a thick blanket of duff —a mille-feuille of slowly decomposing leaves deposited over the course of years, if not decades. That layer creates a home for insects, amphibians, birds, and native flowers. But when worms show up, they devour the litter within the space of a few years. All the nutrients that have been stored up over time are released in one giant burst, too quickly for most plants to capture. And without cover, the invertebrate population in the soil collapses.

Where millipedes and mites once proliferated, now there are only worms. "If you were to think about the soil food web as the African savanna, it's like taking out all the animals and just putting in elephants—a ton of elephants," Dobson says.

With their food and shelter gone, salamanders suffer and nesting birds find themselves dangerously exposed. Plants like trillium, lady's slipper, and Canada mayflower vanish too. This may be because the worms disrupt the networks of symbiotic fungus that many native plants depend on, or because worms directly consume the plants' seeds. Or that native species, accustomed to spongy duff, are ill-prepared to root into the hard soil left behind when the worms have finished eating. It could be all of the above.

Perhaps most worryingly, early studies suggest that worms can sometimes halt the regeneration of trees. Josef Görres, a soil scientist at the University of Vermont, says he often struggles to find a single seedling in invaded portions of New England's famous maple forests. His theory is that the worms take out all the understory plants, leaving nothing for deer to chew on but the young trees. And that could spell trouble for the region's prized maple syrup industry. "In 100 years' time, maybe it's going to be Aunt Jemima," he says. "That's a real bad horror story for people in Vermont."

These sweeping powers are why earthworms are often called ecosystem engineers. And Dobson and her colleagues fear that jumping worms pose an even greater threat than their European predecessors. Jumping worms appear to have many of the same effects, except that they grow larger and exist in dense colonies, sometimes numbering more than 100 individuals per square meter of ground. And while European worms range throughout the upper four to six feet of soil, jumping worms stick to the top six inches or so, churning it relentlessly into a loose sediment that Dobson likens to ground beef. (Others I talked to compared it to coffee grounds.)

The disturbed soil erodes easily, dries out quickly, and generally makes poor habitat for many plants. At McDonald's, Mark pipes up to say that he noticed patches of it during his basic training at Fort Knox last year. While there, he cursed the lack of understory in which to take cover during tactical exercises.

Dobson explains that the worms act like a funnel, winnowing away the diversity of the forest. First, they take out the most sensitive native plants, leaving only hardy species like poison ivy and Virginia creeper. Then they prime the ground for invasives. Even more than their European relatives, jumping worms seem to reshape the forest from the ground up.

"Every single thing that they do is transformative," Dobson says.

Until it moves, a jumping worm looks a lot like any other earthworm: long and thin, with rosy brown skin divided into bellows-like segments. Experts will tell you to look at the clitellum—the band that holds the reproductive organs of worms, which are hermaphrodites. In European worms, the smooth, pink clitellum is found closer to the middle of the body. On jumping worms, it's milky white and sits near the head.

There are many species of jumping worms. The first arrived in the United States in California in the 1860s. Others have been in the Southeast for more than a century—long enough to earn colloquial names like the Alabama Jumper. (You can buy these online too, but worm experts advise against spreading them.)

The three species that Dobson and others worry most about are newer arrivals, and likely hitchhiked on imported plants, where they caught the attention of groundskeepers. "The gardeners were out there saying, 'I know these are earthworms, they are supposed to be good. But I swear, they are killing my plants,'" Dobson says.

An oft-repeated anecdote holds that jumping worms first appeared in Washington, D.C., among the cherry trees in 1912. By the 1940s, they were spotted at the Bronx Zoo, where one species was later raised to feed the resident platypuses. They have been in a few other New York parks for more than fifty years, which is why Dobson chose these forests to study their long-term ecological effects. For some reason, however, jumping worm populations have exploded in the last decade or so. And the worms are spreading even faster than Dobson imagined—including within city limits.

While in New York, I meet with Clara Pregitzer, a forester at the nonprofit Natural Areas Conservancy, who takes me on a tour of Forest Park in Queens. When I find her next to the statue of a somber soldier at the Richmond Hill War Memorial, she's wearing tortoiseshell glasses, a loose white blouse, and dirty jeans cuffed above leather boots. We stroll up the tree-lined avenue, then turn off onto a shady path. "This is one of my favorite parks for showing off high-quality forest," Pregitzer says.

Indeed, it's a beautiful spot. Hundred-year-old oaks stretch their boughs across the wide trails and the understory is neither bare nor overgrown. (Pregitzer says the park has good "sight lines.") The gently undulating landscape is a textbook example of the "knob and kettle" terrain left behind by retreating glaciers.

Earlier, over email, Pregitzer had warned that we might not find any jumping worms here. In 2013 and 2014, she led a survey of New York's green spaces, tallying up the trees and understory plants and noting the distribution of jumping worms. She found heavy infestations in 12 percent of the plots she studied, though they may have been present in close to a third. Less than 5 percent of the plots at Forest Park showed signs of jumping worm activity.

But within ten minutes of walking, I spot the coffee-ground soil on the edge of the trail. Pregitzer decides to do her own version of the mustard pour: a dilute solution of dish soap. Within a few minutes, a parade of jumping worms emerges from the ground. "Oh, nasty," says Pregitzer, scrunching up her nose.

We follow the worm soil away from the trail, 10 feet, 15 feet, 50 feet, into a shallow depression. There, we don't need the dish soap. Brushing away the leaf litter reveals a dozen worms in an area the size of a dinner plate. Pregitzer picks up a six-inch-long monster with a stick. "Nasty," she says again.

She is visibly unsettled by the discovery. "This is one of the best parks, and now that we are digging in a little bit and really looking for them, they're everywhere," she says, gazing around. "There have got to be, like, millions." We keep walking for another hour or so, and, much to her dismay, find only one worm-free site — on a steep hill where the topsoil has washed away.

That's a big change from a few years ago, and when I report the news to Annise Dobson, who is using Pregitzer's survey data in her own work, she immediately decides to revisit the original study sites. A week after I get home, Dobson emails me to say that she's finding worms everywhere. "I'm floored and honestly reeling from the extent of it," she says. They are now in 64 percent of the plots across the city.

It's not entirely clear how the jumping worms have spread. Conventional knowledge holds that they can't cover much ground on their own — perhaps 30-odd feet in a year, although one researcher I talk to swears he's seen a single worm move that far in an afternoon. Their cocoons, which are about the size of peppercorns, can be carried much farther in water, and scientists have noticed that invasions often march down hillsides and along waterways.

But experts suspect that the blame lies primarily with humans. It's all too easy to unwittingly transport worms and cocoons in plants, mulch, and soil, in the treads of shoes and tires, or caked onto landscaping equipment.

Perhaps we've been moving more earth in recent years, and carrying worms with it. A few major nurseries could be spreading them by accident. Or the worms may be benefiting from favorable conditions or some clever adaptation to their new environment.

Whatever the reason, jumping worms have fanned out across the northern United States over the past few decades and continue to expand their territory. They were first identified in Rhode Island in 2015 and in Oregon in 2016, though they may have arrived earlier in both places.

As of late 2017, there had only been one sighting of jumping worms in Canada, but the country's vast tracts of carbon-rich, worm-free boreal forest are already under siege by their European cousins. And scientists there know it's only a matter of time before the jumping worms follow.

Bernie Williams remembers when she discovered jumping worms in Wisconsin. October 3, 2013, was "the day that ruined many of our lives," says Williams, a worm expert at the state Department of Natural Resources.

She was leading a group of researchers and managers on a tour of the University of Wisconsin arboretum. Scientists already knew European worms had taken up residence there, and Williams led the visitors to a heavily invaded spot. But as soon as she saw the soil, she knew something was wrong. "These worms were everywhere," she says.

Over the next three years, the jumping worms stormed across twenty-five acres of forest in the arboretum, effectively eradicating their European rivals. They have now been reported in fifty-two of Wisconsin's seventy-two counties, Williams tells me, disproving predictions that the harsh winters would keep them at bay.

Many Madison residents know and loathe the worms, which can decimate flowers and vegetables. At a community garden near campus, I meet a man sifting compost through a screen as a precaution. "That usually catches 'em," he says.

Sometimes, under the right conditions, the worms can reach "infestation" levels, and come pouring out of a house's foundations "like Medusa's head," Williams says. "If you like invertebrates, you squeal with delight." Homeowners, however, tend to be less thrilled, and in bad years, Williams spends her days fielding calls from disgruntled residents.

Dobson hears similar frustrations in heavily impacted parts of the East Coast. People are seeing changes in their gardens, in their local woods, even on their kids' soccer fields (the worms can damage the roots of turf grasses). She's watched people break down in tears and pick fights with their neighbors over who's to blame for introducing the worms. "A lot of my time is taken up trying to comfort very upset people," Dobson says.

Unfortunately, there isn't much anyone can do once invasive earthworms get established. This becomes clear one day as I watch Brad Herrick and Marie Johnston crawl around for hours in Wisconsin's mosquito-infested woods, plucking jumping worms from inside a series of two-foot-wide metal enclosures. Herrick and Johnston, both researchers at the UW arboretum, want to test one of the few promising weapons against jumping worms: a low-nitrogen fertilizer called Early Bird, commonly used on golf courses. To assess its effectiveness, they've been manually removing all the worms from each of twenty-four high-walled rings before adding back a known number of victims. (When I ask Herrick what they do with the evicted worms, he says, "We gently chuck them.")

The problem is that the cocoons act like a seed bank. Mature worms lay them in the fall before they die, and the cocoons hatch throughout the spring and summer, providing a seemingly endless source of young worms. "It's like a two-headed monster," Herrick says.

Herrick and Johnston each claim an enclosure and start sweeping through the leaf litter. In his first, Herrick finds thirty-seven jumping worms; Johnston counts fifty-two in hers. Then they move on to the next. This is their sixth removal attempt, and Herrick is baffled by the high number of holdouts. "Maybe they are climbing in," Johnston muses, only to look over and see a worm crawling up the outside of a nearby enclosure. "Crap!"

Even if Early Bird is effective, Herrick says, it will only be useful for small infestations in gardens or urban landscapes. Scientists are wary of applying it to forests. Prescribed burning shows some promise, but everyone agrees that by far the best solution to the worm problem is to stop spreading them in the first place.

Johnston and Herrick recently published a study showing that heating compost and soil to 104 degrees Fahrenheit effectively kills both worms and cocoons. That's something that mulch and compost companies could do. The bigger challenge is educating the public, which Herrick has made a personal mission.

The day I visit, he and Johnston give a talk to a few dozen teachers who are visiting the arboretum for a training workshop. That same night, he drives three hours across the Illinois border to speak with a group of master gardeners. He tells them to buy only mulch and compost that have been treated to kill stowaways, and to avoid city compost made of leaves collected from sites all over town. He urges them to inspect potted plants for jumping worms and to buy bare-root varieties whenever possible. Some scientists go even further, advising local garden clubs not to hold plant swaps at all.

Under its invasive-species rules, Wisconsin has officially banned the transportation and sale of jumping worms, as has New York. But Williams says the state focuses more on helping limit worm movement than on punishing people who spread them, since it's usually unintentional.

The day after our meeting, she heads up north to talk to a group of loggers about the risks of invasive worms and what they can do to stop them. She recommends using a broom to knock soil off trucks and tires. She's under no illusion that this will solve the problem, but "you'll slow them down," she says.

In general, it's hard to change people's opinions about earth-worms. "We have this Eurocentric mindset that whatever is good in Europe has got to be good here," Dobson says. "People don't take it really seriously unless they're actually seeing the impacts on the ground."

In a perverse way, the jumping worm has been something of a blessing: unlike European worms, they elicit an intense feeling of repulsion in most people. "They only get more gross over time," Dobson says.

"They are the stuff of nightmares," says Justin Richardson, a soil biogeochemist at the University of Massachusetts at Amherst, who studies heavy-metal accumulation in worms. Remembering his first encounter with them, he says, "It was like *Night of the Living Dead.*"

Bernie Williams is the only person I talk to who has anything nice to say about jumping worms. "They're much faster. They're aerodynamic, almost. They're smooth. They're handsome," she says. "I really think they are an amazing animal."

Williams's enthusiasm for worms knows few bounds. She tells me that the annelid phylum, to which earthworms belong, has been around for at least 500 million years. It contains more than 7,000 species of worms, which come in all different shapes and sizes, including one native to eastern Washington State that reportedly grows up to three feet long. And, as if that weren't enough, earthworms have five hearts—or the worm equivalent. (However, it's a myth that cutting a worm in half makes two; it usually just makes one dead worm.)

Williams understands the concerns about invasive worms' effects on forests, and she spends a lot of her time teaching people how to prevent their spread. But she also takes a more pragmatic view. "We are going to have to live with them," she says.

It's a sad truth about most biological invasions. If humans catch the problem early on, or if the intruders are confined to an island, we can sometimes eradicate an unwanted visitor. But more often than not, the offending organism infiltrates the landscape long before we fully grasp the threat. Think of chestnut blight, which toppled 4 billion trees in the early twentieth century, or the Burmese python, now a top predator in Florida's Everglades.

Given that earthworms won't likely fell giant trees or swallow your cat, I wonder if they will fit into the category of lesser-known offenders, like house sparrows and starlings. Both were introduced in the late 1800s from Europe to New York. The sparrow was brought over for pest control and old-world nostalgia, the starling by a group of Shakespeare enthusiasts intent on importing every species mentioned in his writings.

The birds outcompete native songbirds for nest sites and hurt agricultural yields by eating seeds, pooping on grains stored in silos, and transmitting diseases. By one estimate, starlings cause $800 million in damage every year. But the birds have not set off alarms in most ordinary citizens. They are just part of American life.

Will worms be the same? There is no question that forests are changing in fundamental ways as a result of the invasion. "Do they store less carbon? Probably. Are they more susceptible to drought? Probably," Peter Groffman says. "Are they supporting a different suite of biodiversity? Yeah, they are." But the impacts are invisible to most of us, and represent metamorphosis as much as destruction.

Perhaps it's not a matter of whether worms are good or bad. Maybe such projections of human values—from the Victorians to Darwin to today—are what got us into this mess in the first place, Dobson says. "What I've come to realize is that that's not the right way to think about anything in an ecosystem."

*

On my last day in New York, Dobson picks me up and we battle our way along the Gowanus Expressway toward Staten Island
and more city parks. With brick warehouses and shiny skyscrapers
rising around us like mutant trees, it's tempting to dismiss New
York's parks as urban anomalies with little bearing on the fate of
the rest of the continent's forests. But Dobson pushes back against
that idea.

For one thing, she says, the ecosystems aren't that different;
native hardwoods dominate the overstories of both, and the un-
derstories of New York's parks often hold surprising troves of rare
plants like true Solomon's seal. More importantly, the stresses urban forests experience—including biological invasions, pollution,
and human disturbance—presage those that other North American forests will encounter in the future. "What happens here is
going to predict potentially what happens beyond," Dobson says.

Eventually, the traffic thins and we cross the Verrazzano-
Narrows Bridge, passing from a world of glass and concrete into
one overrun by lush vegetation. We head south and soon arrive at
a quiet sweep of forest called Wolfe's Pond.

Dobson parks on an empty street and we take a few minutes
to ready ourselves. I follow her lead, tucking my pants into my
socks to protect against ticks, as she prepares three jugs of mustard slurry. Then we set off into the woods, using the GPS on her
phone to locate one of Pregitzer's plots.

We find the site next to a muddy wash, surrounded by dented
beer cans and faded plastic debris. The ground is solid poison ivy
and Virginia creeper, both mowed down to ankle height by deer.
The soil looks like the dregs in a French press. "Wow, this is so
wormy," Dobson says. She throws down the PVC frame and pours
out half of the mustard mixture. Then she sets a ten-minute timer
and starts counting. Within five minutes, she's at twenty-four. She
dumps out the rest of the bottle, and another wave appears. "Oh,
there's thirty," she says. "Thirty-one."

Dobson suspects that this is a relatively new invasion, like the
one I saw at Forest Park. Worm infestations tend to peak and then
decline. Their numbers climb quickly when resources are abundant. Then, after the worms eat themselves out of house and
home, the population density drops. "This is the top of the pa-
rabola," Dobson guesses.

Next, we head to another plot, across the road. Upon setting foot in the woods, we can tell it's different. We retreat to the sidewalk and clean the treads of our shoes with sticks to avoid transporting cocoons into what appears to be a worm-free forest.

Dobson points out dainty Canada mayflower growing along the trail. She pulls out a handful of duff and lights up when she sees the thin white tendrils of hyphae—symbiotic fungi—weaving through the humus. "There are so many beautiful things here!" she says.

We continue a few hundred yards up the trail to another of Pregitzer's plots. Dobson lays down the frame, carefully fitting it over a few fragile plants. Then she pours. We hold our breath, waiting to see if anything moves. After a minute, a jumping worm emerges, then another. She counts four in all. For the first time, Dobson appears crestfallen. This forest seems so healthy. She looks over her shoulder, back toward the worm-infested spot across the road.

"This might very well look like that in a few years," she says. While it may not be good or bad, it will certainly be different.

NORA CAPLAN-BRICKER

Long May They Reign

FROM *The Atavist*

One

THE PHENOMENON THAT some people in Brookings, Oregon,
would later call a miracle began in early July 2019, when the same
monarch butterfly appeared in Holly Beyer's yard almost every day
for two weeks. Beyer recognized it by a scratch on one wing. She
and a friend named it Ovaltine, inspired by *ovum*, for the way it en-
crusted the milkweed in Beyer's garden with eggs. Each off-white
bump was no larger than the tip of a sharpened pencil. Clustered
together on the green leaves, they looked like blemishes, as if the
milkweed had sprouted a case of adolescent acne.

Brookings sits on Oregon's rugged coast, and squarely within
the monarch's habitat. Every spring and summer, several generations of butterflies breed, lay eggs, and die, each in the span of
about a month. The last generation of the year is different. Come
fall, rather than produce offspring, it migrates south. Beyer, a pe-
tite retiree with a trace of red in her gray hair, is part of a local
group who promote butterfly-friendly gardening practices—planting native flowers, for instance, and forgoing pesticides.

Most female monarchs disperse their eggs as widely as possible,
but for unknowable reasons, Ovaltine laid almost 600 in Beyer's
yard. Under normal conditions, fewer than 5 percent of monarch
eggs survive to adulthood. Beyer wanted the marvel she had witnessed from her deck to have a happier ending. She snipped the
laden leaves and brought them inside, to shield the eggs from
wind, rain, and predators. Before long she had hundreds of cat-

erpillars, then hundreds of butterflies. She released them into the wild, and Brookings, with a human population of just 6,500, was suddenly ablaze with orange wings. A person could be taking the trash out or crossing a parking lot and see a flash, like a struck match, from the corner of their eye.

Soon the monarchs had blanketed Brookings in "the second eggsplosion," as Beyer put it. "Every milkweed plant"—the only flora that monarch caterpillars eat upon hatching—"got egg-bombed." Over the next several weeks, Beyer counted nearly 2,700 eggs in her yard. Based on other people's reports from their own gardens, she estimates that there were 5,000 more across Brookings.

Beyer dutifully gathered the eggs laid on her property and put them in a maze of mesh crates she'd set up on the small deck of her 400-square-foot apartment. As the caterpillars hatched, she gave her life over to their care. The tiny creatures do nothing but eat and evacuate, and Beyer spent every day harvesting milkweed to go in one end and sweeping away the droppings that came out the other. "I would start at 10 a.m. and wouldn't finish feeding them until six at night," she told me. "I lost fifteen pounds because I forgot to feed myself."

A friend of Beyer's sent out a grassroots SOS, begging anyone in Oregon with experience hand-rearing monarchs to come and take some of the remaining eggs off her hands. That's how Amanda Egertson heard about the eggsplosion. A trained ecologist, Egertson is the stewardship director of a land trust in central Oregon. She lives in the city of Bend, almost 300 miles away from Brookings. She called her husband and asked if he could skip work for a day or two. They packed their kids into the car and started driving.

Waiting for them in Brookings was a makeshift incubator for 110 monarch eggs: a foil lasagna pan lined with damp paper towels to keep the milkweed leaves placed inside it from wilting. It would be up to Egertson to usher the butterflies into life. If she succeeded, the monarchs hatched under her care would be the first she'd seen that year. In the weeks leading up to her trip to Brookings, she'd scouted for flickers of orange as she traversed the land trust. For the first time she could remember, she hadn't seen a single one.

*

While Egertson was retrieving the lasagna pan, I was a continent away, sitting for hours every day on my in-laws' porch in Massachusetts. From there I could see thick patches of butterfly weed, a variety of milkweed that grows wild on their hilltop property. It was waist-high in places, and covered in starburst bunches of brilliant orange flowers. Each tiny blossom had one row of tangerine petals stretching down and another row stretching up, like little dancers with raised arms. The plant is sometimes called orange glory, which suits it. When my husband's father cut the field, he mowed carefully around the flowers. Once, he motioned me off the porch swing where I was reading to show me a monarch feeding on the blossoms.

The monarch *(Danaus plexippus)* is a large butterfly, and among the slowest-moving in North America. It takes no special skill to spot one or to identify what you've seen. If you went to an American elementary school, you probably learned in science class how a monarch egg becomes a caterpillar becomes a chrysalis becomes a butterfly. Some lepidopterists (butterfly experts) disdain monarchs in the way that anyone with esoteric tastes looks down on what's popular. "People call them the cockroaches of butterflies," one scientist told me.

But the world has dulled your capacity for wonder if you cannot be awed by monarchs, which undertake one of the longest annual migrations of any insect on earth. A paper clip weighs a gram; a monarch weighs about half that. The thickness of a piece of paper is one-tenth of one-thousandth of a meter; the thickness of the monarch's distinctive orange wings—veined in black and dappled with white spots at the edges—is measured in microns, or millionths of a meter. Monarchs can travel even on wings that appear too torn for flight, and scientists estimate that in good weather they can cover 30 miles or more in a single day. East of the Rockies, millions of butterflies migrate thousands of miles, from southern Canada to central Mexico, where oral traditions suggest that the incandescent insects have blanketed forests every winter for centuries. West of the Rockies, a smaller number of monarchs fly south as the weather cools and spend the winter huddled in trees on the central coast of California. No one knows how they find the groves to which they return every year.

The monarch was once as common as it is beautiful—the most ordinary of extraordinary things. As a child, I saw them all the time in the warm months, drinking from weedy flowers at the edges of cornfields. Now, though, the population is in precipitous decline. This is true all over North America, but especially out west, in places like Oregon and California. In 2018, a count of western monarchs turned up only 27,218, fewer than 1 percent of the number recorded in the mid-1980s. Worse still, that figure failed to clear an existential threshold of sorts: ecologists had recently warned that the western monarch's risk of extinction could intensify if the population fell below 30,000.

The causes of the decline are many and man-made: loss of habitat, increased use of pesticides, the acceleration of climate change. On the broadest scale, these forces overlap with the reasons that the island where my in-laws live floods more severely with every passing year. Visiting them last summer, I spent more time outside than I had since my childhood, but the pleasure of sunny days was darkened by dread. The monarchs in particular brought me back to a time before I knew about climate change or lived with the awareness that I might someday witness a mass extinction. The idea of a future without them started to represent everything I was frightened to live through.

It's not news that our impending environmental cataclysm requires urgent action, especially by world leaders and fossil-fuel companies—people and entities with the power to fundamentally change the way we use our planet's resources. The steps we can take as individuals won't be sufficient; they won't even be significant unless millions of people follow suit. For me at least, this made it hard to commit to even small forms of environmental action. I would attempt something—composting, or taking the bus more, or cutting meat out of my diet—only to find that it didn't allay my sense that I was doing nothing. It was like trying to ride a bike when the gears wouldn't catch. I wanted to push the pedal down and feel myself move.

I envied my father-in-law's steady sense of purpose as he mowed around butterfly weed so that monarchs could feed on the flowers. It was a modest act of stewardship that brought him great satisfaction when butterflies landed on the patches of growth he'd conserved. I, too, wanted to do something that mattered in ways I could see and feel.

Near summer's end, I read a news article in which a biologist made a case for monarch conservation that went beyond butterflies. Karen Oberhauser of the University of Wisconsin–Madison described monarchs as a flagship species, an animal that captures imaginations and induces people to care about its fate. "We've surveyed people and asked, 'How much would you pay to save monarchs?'" Oberhauser said when I called her. "It's up there with whooping cranes, polar bears, and wolves—all these charismatic vertebrates." When people like my in-laws protect monarch habitat, they assist other species they may never have heard of. By extension, they support entire ecosystems. "On some days, I feel like maybe we won't save monarchs, but if we *try* to save them, we're going to do good for the world," Oberhauser said.

Was this what I had been looking for—an animal to lend its shape to my formless sense of environmental grief? In the book *What I Don't Know About Animals*, novelist and essayist Jenny Diski observes that humans have been turning animals into symbols ever since the beginning of language, employing them as tools "to think about anything and everything." Maybe I could make monarchs my personal shorthand for something otherwise too large to grasp. I didn't yet know about Beyer and Egertson, who were upending their lives with a feverish energy that comes from believing you can make a difference, but I had arrived at a similar idea: I hoped that, if I trained my attention on a single creature, I would figure out what it meant to do my part. Maybe fear of a loss specific enough to imagine would impel me to act. And once I started, maybe I wouldn't stop.

Two

Back home with the eggs, Egertson cut air holes in giant Tupperware bins, which covered the floor of her son's bedroom. "He gets the best morning sunlight," she explained. After the eggs hatched, the caterpillars grew quickly, sometimes doubling in length in the span of forty-eight hours. They molted their striped skin five times in two weeks. Egertson returned to the store for more bins, then more again. When she and her family ran out of milkweed to feed the caterpillars, they called on neighbors to harvest it from their yards.

Eventually, Egertson carried the Tupperware condominiums upstairs to the master bedroom and opened the doors to the deck to expose her charges to fresh air and light. At night, when the temperature dropped into the forties, she and her husband piled their bed high with blankets. During the day, when the sun heated the room to 90 degrees, they stripped down to tank tops.

After two weeks, the caterpillars started to crawl to the roofs of their enclosures and hang upside down in a J shape, like a collection of fishhooks. One by one, each creature began to pulse, and its skin went translucent. Slowly, its striped outer layer peeled back from its head, revealing a sticky, twisting green mass whose gyrations gradually stilled. Over the next day, it inflated and hardened into a perfect jade pendant adorned with gold flecks. This was the chrysalis, a shell that looks like a sarcophagus but is really a womb. Inside, a butterfly was taking shape.

Egertson tried to stay close to home so she wouldn't miss the moment when a monarch emerged. She swims at 4 a.m. every morning, and she perched her cell phone on the edge of the pool so she could check it after each lap. One morning she got the text she'd been waiting for and "flew out of the pool," she told me. She raced home, "wet suit and all," and made it in time.

In the hours before a monarch "ecloses," or emerges from its chrysalis, its wings become visible through the shell. The lacquer splits, and the butterfly pushes itself out on long black legs. It pumps fluid from its abdomen through the veins of its wrinkled, misshapen wings, which slowly unfurl into four fiery orange and black fans. Within a few hours, they harden and dry. If the process is interrupted and the wings remain crumpled, the monarch won't be able to fly.

Handling the new butterflies with great care, Egertson and her family affixed tags to their wings—lightweight but durable stickers bearing serial numbers that scientists use to track the insects' journey if they are spotted again. Researchers have been tagging butterflies since the 1940s, when the zoologist Fred Urquhart decided to trace the until-then-unknown route of the eastern migration. The organization he founded, now called Monarch Watch, remains the largest tagging project east of the Rockies. Egertson had requested stickers from a lab at Washington State University that hosts one of several tracking operations out west. When she applied the tags to the butterflies, her fingers came away dusted with glittering scales.

Egertson's family gave each insect a name along with a serial number: Michael Phelps, for the Olympic swimmer's butterfly stroke; Chopin, for one of Egertson's favorite composers. Egertson's eleven-year-old son, who had shared his room with the caterpillars, named one monarch Flamingo, after the species for which he had dyed his own hair neon pink.

Egertson and her kids released Flamingo in a park in Bend on a sunny day in September. Afterward, Egertson went to a tattoo artist to have butterflies inked onto her foot and ankle. They were a fiftieth birthday gift to herself, a reward for making it through what she told me was a difficult decade. "I've been smitten with butterflies my whole life," she said. "At first glance they seem so fragile, like the wind can just blow them any which way. Like they don't have a lot of say about where they end up. But in fact they do. They're incredibly resilient, powerful creatures."

As Egertson gritted her teeth against the sharp scratch of the tattooist's needle, she hoped that Flamingo was drifting southward —borne, at least that day, on a light breeze.

It's not strictly rational to devote one's excess energy to protecting a species loved mostly for being beautiful, especially when so much of the world is dying. Insect populations in particular are in free fall, and monarchs are far from the hardest-hit species among those that scientists are able to monitor. Entomologists estimate that humans have identified as few as 20 percent of all insect species, meaning that millions of unique creatures could be swept off the planet without our knowing that they existed in the first place. No one can make a case for ensuring the butterflies' survival based on particular usefulness. Though they get lumped in with pollinators, they are bad at the job. They don't even come close to rivaling bees, without which farmers would need to hire human workers to pollinate fruit trees.

Karen Oberhauser's argument about flagship species helps explain a recent burst of interest in monarchs. Many conservationists choose to focus their efforts on keystone species, which anchor ecosystems (starfish, for example, keep tide pools from being overrun by mussels), or indicator species, which reflect the health of a landscape (dragonflies can't live on polluted streams). But prioritizing flagship species like monarchs is a practical choice for advocates competing for the public's attention. It helps that, unlike some endangered species, monarchs don't require governments to set aside large tracts of land for them. They don't need pristine conditions or continuous wilderness. What they need are options: milkweed on which to lay eggs, and nectar plants on which to feed all along their migratory path. An ecologist I interviewed estimated that monarchs might be able to pass through a landscape that is just 1 percent habitat—that is, it might be enough for an urban neighborhood to offer up a pot of milkweed or a flowering window box for every few hundred feet of concrete.

This fact—call it the allure of tangibility—has helped spur tens of thousands of people across North America to get involved in conservation efforts on the monarch's behalf: people like Beyer, Egertson, and my in-laws. "Pretty much anyone can help," said Emma Pelton of the Xerces Society, the leading advocacy organization for insect conservation. "This is an area where individuals can have an impact." One activist I spoke to, a former fourth-grade teacher in Illinois, had converted an old school bus into a traveling classroom that she drove around the corn belt proselytizing about planting milkweed for monarchs. "It's like Ms. Frizzle's Magic School Bus," she told me.

Hand-rearing, however, is a part of the crusade to save monarchs that most scientists reject. For starters, it can sow confusion among researchers. A bonanza of butterflies in a place like Brookings, where monarchs might have been scarce but for human interference, foil attempts to track—and thus support—the shrinking western population. "We can't trust the sightings we have," Pelton said. "In such a critical year, after the population collapses, it's really frustrating. It's a huge loss to our ability to understand where the population is so that we can help it." Some experts also fear that butterflies hatched in captivity may be inferior navigators, making them less likely to survive their long migration. Other studies suggest they may be at higher risk of spreading disease. Most concerning, monarchs have evolved to produce hundreds of progeny that don't make it, employing a different biological strategy than large mammals that lavish energy on each individual offspring. By protecting eggs and caterpillars that might have been picked off in the wild because of an inherent weakness—and by encouraging inbreeding, which reduces genetic diversity—the people hand-rearing monarchs almost certainly introduce inferior genes into a struggling population. "You need to be a really fit monarch to make it," Pelton explained. "The last thing we want to do is make them weaker."

Experts consider mass rearing a misdirection of energy: to save a species you must protect its habitat, not keep a few creatures alive on your porch or in your bedroom. Egertson, who oversees the planting of native flowers on her land trust, understands these concerns—to the point that she hesitated before taking the trip to see Beyer. Though she had raised a few butterflies in the past, just for the joy of it, she had qualms about hatching more than 100 in captivity. "Any time you tamper with nature, you have to wonder if you're doing the right thing," she told me. "But if you were to ask me, do I regret participating in captive rearing, the answer is absolutely not." The severity of the crisis made her want to do something that felt more immediate. "The numbers are *so* low that, if we can boost them at all, maybe that will make a difference," she said.

Pelton told me that she worries about quashing the energy of volunteers who are just trying to help. It makes sense that adding individual lives to a plummeting total seems like the best, most direct thing a person can do. "It's hard for people to grasp whole ecosystems—that's hard for me even in my backyard," Pelton said. "We all pivot toward something we can grasp, like an animal. But the question is if we can look a little wider. People spend hours every day rearing caterpillars. If you spend hours every day doing anything, you could be a fantastic community organizer working to reduce pesticides, or have an amazing garden that helps lots of animals, not just monarchs."

To avoid becoming paralyzed by the number and scope of the environmental problems we face, it's often necessary to narrow our focus. How do we also keep a view broad enough to see how our actions fit into, and sometimes work against, a larger effort? It's a difficult balance. To avoid losing hope, people need to experience their individual power to change things. But with the fight to save monarchs, as with so many crises, little of the work that needs doing can be done alone.

Three

Monarchs migrate solo. The first challenge facing the butterfly that Egertson's son named Flamingo was likely crossing the Cascade Mountains running from southern Canada to northern California.

Clearing a 10,000-foot peak is well within a monarch's capability; although the eastern and western migrations rarely mix, butterflies have even been known to cross the Rockies. On cool nights, finding shelter from the wind and other elements would have been imperative. Once the sun set and the temperature dropped below 55 degrees, Flamingo's powerful wing muscles would become paralyzed; another 15 degrees and he would no longer be able to crawl. If he were knocked to the ground in the night, he could become prey for mice or voles. They would eat his narrow body and leave his wings in the dirt like a discarded costume.

There were also human dangers to contend with, starting with busy highways. Along the eastern migratory route, millions of Flamingo's kind become roadkill every year; according to one study, collisions with cars in Oklahoma, Texas, and northern Mexico deplete the monarch's numbers by as much as 4 percent. Out west, researchers see less evidence of significant losses along highways, but the smaller population can little afford any at all. It complicates the picture that one promising initiative to restore monarch habitat is to plant flowers on roadsides, since the land there has few other uses.

After a few weeks, Flamingo probably reached the wide, flat floor of California's Central Valley. Most western monarchs are funneled through this corridor, which some forty years ago was an inviting place: a rich patchwork of grasslands, dotted with bright blooms in all but deepest winter, threaded with streams and rivers, and soaked with sun almost 300 days of the year. John Muir famously described the Central Valley as "the floweriest piece of world I ever walked, one vast level, even flower bed." But in recent decades, industrial farming has ironed out all but the last inches of wild land, replacing ungoverned prairies with perfect rows of produce. Roughly a quarter of America's food comes from the Central Valley, including 40 percent of our fruit and nuts. The region is also California's fastest-growing in terms of population. The World Wildlife Fund (WWF) estimates that more than 99 percent of the valley's standing grass consists of green lawns and cereal crops. Cultivation has crowded out native grasslands, with their goldenrod, milkweed, and thistle.

Water is also a scarce commodity. Vast wetlands once fed by the Sacramento and San Joaquin Rivers shrank by more than 90 percent in the past century as the water was diverted to irrigate fields.

A seven-year drought, which ended in 2019, dried up even more marshlands, endangering birds as well as butterflies. The northern end of the valley is famous for vernal pools unlike any on earth: rain fills them in winter, and they evaporate slowly in summer, leaving rings of wildflowers each time the water level drops. The WWF estimates that more than two-thirds of these unique ecosystems have disappeared, either drained for agricultural use or leveled to make way for pastures and fields.

Where Flamingo once would have found water and nectar there were instead gleaming cattle and tidy crops treated with pesticides. Crops like strawberries and almonds are rendered unsellable by even the slightest damage, so farmers make liberal use of chemicals to keep them pristine. When the Xerces Society tested milkweed across the valley, it found pesticide in every sample, including plants grown in private gardens by people who claimed they had never sprayed. Even if Flamingo managed to find food, he risked consuming poison along with his meal.

Art Shapiro knows all about poison. He started counting butterflies in 1972. He was new to the West Coast back then—he'd moved for a job in the zoology department at the University of California, Davis, not long after finishing his doctorate at Cornell—but he was accustomed to spending long days searching for insects. Growing up in an unhappy home on the outskirts of Philadelphia, he would slip out the door with a field guide in his pocket and lose himself looking for flashes of color in an undeveloped expanse of land across the street from his house. As a college student in the 1960s, he studied phenology, the scientific term for biological seasonality: how subtle cues such as temperature and sunlight tell fruit trees when to bloom, insects when to hatch, and birds when to migrate. Shapiro began to dream of creating an enormous data set. If he could track many butterfly species over many years—wet years and dry ones, hot years and cold ones—he would be able to see which aspects of a climate exerted the most control over the life cycle.

In California, he selected five sites at various elevations, each with its own diverse ecosystem, and made the rounds every two weeks, weather permitting. He compiled a list of 160 species of native butterflies to monitor, monarchs among them. Unlike scientists who tag monarchs to track their migration, Shapiro's goal was to compare the size of butterfly populations from one year to the next. The methodology could hardly have been simpler: he visited the same sites on the same schedule and noted the number of butterflies he saw. Since Shapiro didn't drive, each location had to be accessible by public transportation; he sometimes hiked several miles from a bus stop to get where he needed to go. In those early days, he rarely failed to find his quarry. At a single stop, he would frequently see as many as thirty species. Along with monarchs, there were skippers and sulfurs, swallowtails and painted ladies, henna-colored lustrous coppers and periwinkle Melissa blues.

He planned to do the project for five years, since that was all the time he'd have if he didn't get tenure. When he was offered a permanent place at the university, he decided to keep going. Gradually, Shapiro added five more sites, until his study covered a large swath of the Central Valley. As local bus systems grew less reliable, he asked graduate students for rides. He became a fixture at gas stations and dive bars all over his route, his annual arrival a welcome sign of spring. We spoke over the phone for this story, but in pictures Shapiro looks like a hermit in a Georgian-era painting —weathered face, wild hair, enormous white beard—except that he's often wearing a southwestern-patterned shirt.

After a few decades, he realized that he had inadvertently conducted what might have become the world's longest continuous butterfly study, rivaled only by one of similar vintage in the United Kingdom. He soon noticed something else: the number of butterflies at his locations was declining. At first, Shapiro wasn't too worried. Insect populations are naturally "bouncy," meaning that numbers can dip in years with unfavorable weather and rebound quickly in good years, since each female lays hundreds of eggs. But in 1999, the populations of multiple butterfly species crashed simultaneously, their totals plummeting well below any natural ebb that Shapiro had witnessed before.

Shapiro strongly suspected that the butterflies were suffering the effects of neonicotinoids, a class of insecticides that were introduced in the early 1990s and entered wide usage toward the end of the decade. Neonics, as they're called, are systemic, meaning that plants absorb them into every cell from bud to stem, and they can be long-lasting, building up year after year and persisting in water and soil even if a farmer stops using them. A growing body of research would ultimately support Shapiro's hypothesis, showing that even in doses too small to kill outright, neonics shorten the life spans of insects and make them weaker fliers and foragers. Because the pesticides attack a creature's nervous system, they interfere with navigation, which matters for species like bees, which must find their way back to their hives. The major varieties of neonics are now banned for outdoor use in the European Union but remain popular in the United States, where they have been connected to the widespread collapse of bee colonies.

They also have especially pernicious effects on migratory species like monarchs. Karen Oberhauser of the University of Wisconsin remembers the world before butterflies began disappearing. In 1997, a boom year, she flew to Mexico to see eastern monarchs in their overwintering grounds. She witnessed millions of roosting insects, clustered on every inch of oyamel fir trees, their combined weight bending the boughs. Seeing them gathered together, Oberhauser felt a shiver of fear: they seemed so vulnerable, as if a single blow could erase them from the earth. In the years that followed, she noticed how much habitat monarchs had lost in the agricultural fields that form a large part of their summer breeding grounds. She realized that crops genetically engineered to be resistant to herbicides allowed farmers to blanket their land with chemicals, which eradicated millions of acres of milkweed that once flourished between the soybean plants and cornstalks. Monarch numbers had fallen in tandem. Oberhauser began raising money and advocating for conservation efforts. "A lot of what's driving monarch loss is changing agricultural practices. Addressing that is going to require policy changes," she told me.

In California, Shapiro continued to see worrying signs. Monarchs' seasonal behavior seemed to be shifting. The last generation of butterflies that hatches each fall is called a super generation. During their migration, they enter a state known as reproductive diapause, conserving the energy that other generations expend breeding so they can live six months or more—enough time to make it through winter and breed first thing the following spring. But Shapiro started to hear stories of winter roosts breaking up earlier and earlier. Whereas they used to stay put until March, now they were disbanding in late January or early February. Were they taking their cue from the ever milder weather? Where would they find nectar and milkweed, when most native plants break ground in March at the earliest? Would a freak storm batter them to death or a cold snap freeze them? No one could guess how many migrators might be surviving the winter only to die without finding somewhere to lay their eggs in the spring.

As Shapiro described it, 2018 was the year when "everything went in the toilet." He witnessed the worst butterfly season he'd ever seen. He counted only twelve monarchs in all of his stops, and for the first time ever he didn't see a single monarch caterpillar. The once-common species had become so rare so abruptly that every individual insect seemed to matter. Would the next year be the one when Shapiro finally saw no monarchs at all?

Four

By October 2019, if he survived that long, Flamingo might have faced fire. In Sonoma County, along the Central Valley's western edge, the season's worst wildfire consumed almost 80,000 acres. Acrid smoke blanketed much of the Bay Area. What does a monarch make of a forest fire? Does it waste precious energy flying around it, or risk getting caught in the conflagration? Still heading south, Flamingo might have ridden the same winds that carried embers across the landscape, wisps of fire that shone even more brightly than his vivid wings.

A few weeks after the Sonoma fire died down, I was preparing to fly to California too. I had arranged to join an annual effort, organized by the Xerces Society, to monitor western monarchs in their winter habitat—groves where butterflies, in an unsolved mystery of migration, return to the same roosts year after year. Some coastal cities have built around these stands of trees, even if they're in the middle of town. The first scheduled stop on my itinerary was Ellwood Mesa, a sandy bluff of eucalyptus groves just west of Santa Barbara. But by Thanksgiving morning, it too was in the path of a wildfire. Thousands of people in the surrounding county had evacuated. "There's no way anyone can take you there," a municipal employee told me when I called from my home in Boston. Ellwood Mesa was vulnerable to stray embers and to mudslides. When I asked if I could go alone, there was silence on the other end. "There are signs telling people to enter at their own risk," the employee said finally.

I hung up feeling thwarted. My husband tried to comfort me: Even if the fire interfered with my plan to see a species imperiled by climate change, didn't that only prove my story's point? I packed a bag and boarded my flight. I landed in Los Angeles and turned on my phone to find a text from another city employee: a snowstorm had dampened the blaze. I could go to Ellwood Mesa after all.

Xerces calls its annual monitoring effort the Thanksgiving count because it takes place over three weeks in November and early December. More than 100 volunteers visit upwards of 240 sites where monarchs are known to roost. The count is not unlike Art Shapiro's work—volunteers note every butterfly they can find, then tally the numbers into a single snapshot of the total population that made it to the coast for the winter. If they see any with tags—a rare, exciting event—they can compare the serial number with an online database and determine where the butterfly came from.

In 1997, its first year, the count recorded more than 1.2 million monarchs. Two years later, that figure fell to fewer than 250,000, despite an increase in the number of sites being monitored. Though the population still fluctuates, it hasn't broken 300,000 since 2000. It plunged to the historic low of 27,218 in 2018. Volunteers visit most count sites only once; if a site is subject to more frequent monitoring, Xerces uses the highest number observed in a single day. Between this and the fact that some butterflies might be spotted twice if they move between neighboring groves, the final tally is more likely to overestimate the monarch population than to underrepresent it.

Once upon a time, Ellwood Mesa attracted more than 100,000 butterflies each year. When I pulled into the site's parking lot on a Tuesday morning, the sky was overcast. The clean, sweet smell of eucalyptus washed over me. Ellwood Cooper, who once owned this land and for whom the site is named, was a rancher and horticulturalist who helped introduce eucalyptus to the United States. He raised his first trees here in the 1870s. Cooper envisioned the quick-growing eucalyptus as an invaluable source of lumber. It turned out to be brittle and prone to decay, but it did provide an ideal winter home for monarchs, which were observed on the West Coast in growing numbers as eucalyptus spread in the late nineteenth century. Today, the tree is widely considered a scourge on the landscape. With its shaggy bark and fragrant oil, it is quick to catch fire. But it is also monarchs' preferred home for the winter; though the insects roost in other trees, such as cypress, they choose eucalyptus groves over forests with exclusively native species. This has put monarch advocates in the odd position of trying to protect a beloved native butterfly by fighting to plant a despised invasive tree.

The volunteer coordinator for the monarch count in Santa Barbara County was a woman named Charis van der Heide, a monarch biologist and environmental consultant for the city of Goleta. She wore a straw hat and hiking boots, and the rest of her attire was dotted with images and emblems of butterflies: a patterned scarf at the neck of her purple parka, a crocheted keychain dangling from her backpack. I followed her down a sandy path into a grove of trees, where the light grew dimmer and the smell heavier and loamier as we followed a muddy streambed. The ground was blanketed with strips of gray bark—"eucalyptus are messy," Van der Heide said—but many of the branches above us were bare. According to Goleta officials, one in five trees here died during California's long drought. Of those that remained, many were ailing. Giants 180 feet tall leaned against their neighbors or bent into archways over the path.

Changes in the grove have profound consequences for monarchs. Butterflies choose where to roost with extreme sensitivity. New generations not only go to the same groves and trees as the previous year's butterflies—they alight on the same branches. They seek a precise microclimate, a perfect alchemy of humidity, temperature, wind speed, wind direction, and light. Every time a tree falls, the delicate balance shifts.

We entered a clearing. Van der Heide, who is in her late thirties, with wavy chestnut hair and a broad, friendly face, pointed out the features that once drew monarchs here. Perhaps because they are meandering fliers—they flap-flap and glide, flap-flap and glide—they choose groves with high, vaulted ceilings that are "cathedral-like," Van der Heide explained, gesturing upward. The natural architecture gives them space to flit and float. But a thinning canopy may not provide sufficient protection from winter storms. The grove we stood in was once enclosed on three sides by thick walls of trees; now trunks crisscrossed the forest floor, leaving openings everywhere.

Van der Heide told me that she wanted to plant more eucalyptus where we stood. The controversy surrounding that approach didn't bother her. "We conserve what we love," she said—even if a favorite natural phenomenon might not exist as we know it without human influence. She echoed Oberhauser's point about flagship species, arguing that fighting for monarchs could help a wide array of pollinators that share their habitat. "Having a little pragmatism about what people can get behind can serve you in a larger way," she said.

But even with new eucalyptus to draw them, would the butterflies return in large numbers? As the climate changes, many species are expected to shift their habitat ranges northward and upward, to higher elevations, chasing the conditions for which they've evolved. Monarchs might soon abandon the known groves altogether. Maybe they already have. Some count volunteers told me that, in their most optimistic moments, they imagine the butterflies aren't declining—they're hiding, and we just have to find them. But it's not clear where exactly they could have gone. "If you look in the hills, we don't have trees up there," Van der Heide told me. "They've all burned."

Van der Heide scanned the trees around us, looking for monarchs. "Okay, so we have a few," she said, handing me her binoculars. I looked and looked, but I couldn't see them. Van der Heide set up a scope and showed me a cluster of twenty-seven. The outsides of their folded wings were tawnier than I expected, muted enough to blend in with dead leaves. Somehow they all knew to hang at the same angle, so that their identical wings formed an intricate pattern. Occasionally, a pair of wings opened, looking almost red in the deep gloom of the grove, then closed back into the tessellation.

We followed a path deeper into the woods. Two weeks earlier, Van der Heide had counted 250 butterflies on the mesa, most of them concentrated in the single grove we were now headed to. That wasn't many for a site where she, along with other volunteers, had counted 47,500 butterflies in 2011. But it beat finding only 27. Van der Heide seemed optimistic that she could show me more.

We reached a small overlook. In past years, volunteer docents brought tours here to gaze into the grove below, as if it were an amphitheater. Van der Heide opened her backpack and pulled out a binder, opening it to a picture taken on this spot in 1975. The photographer had aimed a lens up the trunk of a tree entirely concealed beneath thousands of butterflies, a carpet of orange wings that led straight to the sky. People who rode horses here decades ago have described similar scenes—of sitting, frozen in awe, as butterflies descended on their mounts, drawn by the smell of sweat. Imagine a horse that looked for an instant as if it were made of butterflies, at risk of dissolving into a flurry of wings.

Van der Heide raked the branches with her binoculars. We stood there for a long time. She didn't move to take out her scope or fill the silence with effusive talk. She just looked, swung around, and looked again. "Wow, I'm not seeing any right now," she said. "That's really hard." The hundreds she'd seen here before the snowstorm that put out the fire were gone, possibly washed away. She turned from the empty grove. "That's really hard," I heard her repeat under her breath.

As we emerged from the woods, we ran into a group of teenagers on a field trip, led by members of the state's Department of Fish and Wildlife, one of whom conducts the Xerces monarch count in nearby Ventura County. "Did you see any?" the group asked as we drew level.

"I saw 27," Van der Heide told them.

"How have your numbers been?"

"So low."

"So low," the woman from Ventura echoed sadly.

"Two weeks ago, I saw 250 across Ellwood Mesa," Van der Heide said.

One of the Fish and Wildlife employees whistled. "That's crazy," he said. "Two years ago, there were literally thousands."

As the kids started down the path, Van der Heide's Ventura counterpart lingered with us to talk about her cadre of volunteers, who were finding so few butterflies that their work took almost no time. To keep them engaged, she kept sending them to reconfirm what they'd seen, or rather what they hadn't. "I feel like I should do something," she said as she took off after her group. "But I don't know what to do."

Five

I was struck by the strangeness of the assignment I'd made for myself: I'd come to observe an absence, to look at the lack of what used to be. In Pismo Beach, another count site, people described the nadir of 2018 in anguished terms. "It was like, *uh-oh, where are they?*" count coordinator Jessica Griffiths recalled. Sites that had hosted thousands of monarchs the year before had only a few hundred, or a few dozen. So far in 2019, Griffiths had seen more than she did the year prior, but nothing approaching what, until recently, she'd considered normal. "I'm surfing the line between relieved and bummed," she told me.

As we spoke, we walked a serpentine path through a eucalyptus grove where she had seen a cluster of monarchs two weeks earlier. Now all we found were pairs of wings, left behind by whatever had eaten their owners. None of the wings had tags. None of them had flown here from a makeshift nursery in Oregon.

I did see a tag in another part of Pismo Beach, on a tree blanketed with more than 1,000 butterflies, a big cluster by California's new standards of scarcity. I was visiting the city's monarch sanctuary, a tiny park in the dunes where tourists stroll around a viewing area smaller than a city block, sturdy fences separating them from the trees. My guide at the site, a California State Parks employee, beckoned me under a barrier to see the monarchs huddled on the leeward side of a cypress tree.

I willed myself to feel wonder and failed. I couldn't help but compare what I saw before me with the images I'd seen in photographs, of the same site shrouded in more than 100,000 butterflies. It's the abundance—the mysterious gathering of monarchs that flew hundreds of miles alone—that makes the migration astounding. One butterfly can make your breath catch; a roost of 100,000 can transport you into a dream. For me the diminished cluster did neither. The monarchs looked drab and windblown. They reminded me of a tattered piece of cloth torn from a quilt, a sign of the undamaged thing that should have been.

Then I saw a flash of orange too bright to be natural. I squinted and found it again, a neon spot on a wing at the heart of the cluster. I shouted for the State Parks employee. "Look, I found a tag," I said, too excited to care that I was being pushy. My heart was suddenly beating very fast. "I'm pretty sure it's a tag." We took turns pressing an eye to the lens of a scope, struggling to find the right butterfly. When we did, neither of us could make out the minuscule numbers that would tell us where the monarch had come from—who had tagged it and sent it on its way.

It started to rain, first a few drops, then harder. The wind picked up, ruffling the branches so that the monarch swayed in and out of the scope's view. My guide warned me that we didn't have much time; the equipment could be damaged if it got too wet. I wanted to stay. I told myself I would stand there as long as it took—never mind the equipment, or interviews, or getting soaked. I felt a wild excitement at the chance to be part of something, to finish what whoever had tagged the butterfly started, to add a data point to the store of common knowledge. I wanted this creature, which had worked so hard to sustain a dying migration, to accomplish something more than its own survival.

Eventually, we made out most of the number, through the zoom lens of a camera. We climbed into a State Parks truck, wet and shivering. Back at her office, my guide checked the tag's digits and discovered that the butterfly had already been sighted a few weeks before. So much for my contribution.

Sitting in my rental car with the heat on blast, I asked myself why I had crossed the country, trailing my invisible cloud of carbon emissions. For a moment, I'd been sure that my presence mattered—that I'd landed in the right place at the right time. As my sense of significance ebbed, I thought of Emma Pelton, urging me to think about western monarchs as a whole population, not to fixate on particular butterflies. Was there a corollary, one having to do with accepting that my individual impact might be beyond my reckoning or take place outside my view?

I had assumed that monarch advocates found motivation in seeing the imprint of their efforts, but that turned out not to be entirely true. During our time together, Van der Heide brought up her anxieties about climate change, her uncertainty about the world her young children would live in as adults. I asked if devoting her days to butterfly conservation made her feel that, in some way, she was helping stave off that dark future. Not really, she replied. "I feel like I'm at the tail end of something amazing," she said of the monarchs' migration to California. "I feel like I'm recording the end of something, and in twenty years people won't even know that there used to be monarchs here."

When the Xerces Society tallied the final numbers, it found that the 2019 monarch count had barely improved from the year before. Volunteers had seen 29,418 butterflies, still below the estimated extinction threshold of 30,000. When I spoke with Pelton, she told me that she and her colleagues were focused on figuring out where monarchs go if they are indeed leaving their winter roosts early, and how to get milkweed and nectar plants into the ground in those places. The size of each year's first generation of monarchs matters exponentially for the number of butterflies that will set off on the migration come fall.

A familiar monarch might have been among those that emerged from the roosts. In late fall, a researcher in Santa Cruz spotted a monarch with a white sticker—a Washington State University tag —on its wing. It was perched on a twig of a Monterey cypress tree. The serial number indicated that the butterfly had flown roughly 500 miles, all the way from Bend, Oregon. It was Flamingo—he'd survived the migration.

Egertson indulged in a moment of pure joy when she heard the news, jumping up and down in her office. "For me, that was one of the most profound experiences," she said. She compared the feeling to giving birth to her children, or to the rush that comes from doing things that scare her. "If this tiny creature that weighs no more than a paper clip can fly from Bend to Santa Cruz, then I most certainly can do whatever it is that I'm facing," she told me. Recently, she'd been trying to overcome a lifelong fear of public speaking. She agreed to address a room of 500 people about monarch conservation. To get through it, she made hundreds of pairs of antennae and asked the audience members to wear them. She told me it was the hardest thing she'd ever done.

By February, twenty-two of Ovaltine's descendants had been sighted in California. Holly Beyer, in whose yard all this began, hoped that the number of survivors was evidence that hand-rearing doesn't necessarily produce monarchs with diminished navigation instincts. For Pelton, the news didn't allay her concerns. If the monarchs raised in Brookings were genetically inferior, and enough of them survived the winter to pass their weaker traits to a new generation, that could make matters worse for the species in the long run. What looked on its face like a small success could pose a danger to the entire monarch population out west.

For the time being, Egertson was avoiding the controversy, devoting her energy instead to interventions she felt sure about. When we spoke, she'd recently placed an order for thousands of native flowers and plants, including milkweed, to brighten her land trust. She described with relish the exhaustion that descends during a day of hard work outside. "I love that feeling," she told me. "When your back is sore, and you're looking at an old road-bed that has now come to life as a meadow. It feels really good."

Six

By the time I sat down to write this story, the world was facing a faster-moving disaster than the monarchs' decline. People every-where were sheltering in their homes to avoid catching the novel coronavirus or spreading it to anyone else. Some environmental-ists saw cause for hope in the speed with which ordinary people took action and in the vivid illustrations of our interdependence across the planet. The question seemed to be whether we would succeed in maintaining a sense of urgency once the present dan-ger had passed.

The response to the Covid-19 crisis suggests that we are capable of the kind of collective action that could slow the advance of cli-mate change and repair other forms of ecological devastation. But it also illustrates the limits of what we can do as long as our leaders keep denying reality—as do most politicians, on both sides of the aisle, when it comes to the enormity of the environmental catastro-phe before us. Without political action and economic reforms, the world will keep growing warmer, until monarch butterflies, like so much else, disappear for good.

Karen Oberhauser has provided input for the U.S. Fish and Wildlife Service as it considers whether to list the monarch as a threatened species, which would expand the regulatory protec-tions and funding sources available for its conservation under the Endangered Species Act. A decision, now several years in the mak-ing, is expected by the end of 2020. "We can try to get as many people as possible doing things that will support monarch conser-vation, but I believe, in the long run, it's going to take regulation," Oberhauser told me. She wouldn't share her opinion on the spe-cific matter before the government, but she talked about reversing the fortunes of monarchs by limiting herbicide and insecticide use and by paying farmers to devote parcels of land to conservation instead of agriculture. Meanwhile, the Xerces Society has helped at least thirty local governments pass policies protecting insects from neonics and other pesticides—a modest start, but perhaps a model for the future.

As long as the coronavirus traps us at home, the only environmental actions within reach for most people may be the smallest ones. In March, I got a message from an ecologist via an email list of monarch enthusiasts. "We appear to have entered a new era of uncertain duration," Chip Taylor wrote of the lockdowns beginning in many U.S. states. "Yet, we must carry on with monarch conservation—somehow." He urged everyone to order flats of milkweed and flowers.

"Gardening gets you out of the house," he wrote. "Social distancing and quarantines will ground many of us, confining us to our properties yet giving us time to garden for monarchs." I forwarded the email to my in-laws, who were hunkered down on their hilltop, and called local hardware stores to find flower seeds that I could plant in pots on my fire escape.

I knew that I could faithfully water my flowers and still the summer could pass without a single monarch finding them. Maybe, if that happened, I would once again feel like I had done nothing. But it felt like doing something to cover my face with the cloth mask a friend had made me and walk to the store in a light April drizzle. Maybe, like Egertson, I had looked to the monarchs for courage—a sense of steadiness that would permit me to act without knowing whether what I did would matter, or how.

I bought seeds to grow dusky purple lavender, mauve coneflowers with glaring red eyes, and pom-pom-shaped Lilliput zinnias in orange, yellow, and pink. And even though the packet said they would take at least a year to bloom, I bought the seeds for orange glory. At home, in the closet, I had a stack of clay pots. I would fill them with soil as soon as the weather turned.

How long did Flamingo live? There were no sightings to tell us that he made it through the winter. If he survived long enough to find a mate, the female butterfly would have had to locate milkweed on which to lay her eggs. As I write this in early May, it's possible that Flamingo's descendants are retracing the path of his migration, fanning north and east in successive generations. If they are sufficiently numerous, maybe a few will survive to reach Oregon.

I'm still chasing a sense of satisfaction in the small things I can do. It feels like the only way to face the possibility of their futility.

In California, I tried to find pleasure in wending through forests, scanning patches of sky for flying monarchs, even as I braced myself to see empty blue. I noticed myself getting better at looking for butterflies. At Ellwood Mesa, I'd struggled to pick them out of the gloom, but in the groves of Santa Cruz—my last stop, like Flamingo's—my eyes went right to them. At the very least, I was learning how to bear better witness. And what I saw was not entirely absence.

I didn't see Flamingo, but I did see a cluster of perhaps 2,500 monarchs on a towering cypress tree. I was in a field near Lighthouse State Beach, and the air smelled of salt and rang with the cries of seagulls. At first the tree was in shadow, and every monarch sat folded. But then a cloud moved. When sunlight slanted across the upper branches, the monarchs opened their dazzling wings one by one. They blazed like little lanterns. I watched one drift up into the sky, weightless, and I felt it: the joy of living on this damaged planet, and a will to witness whatever comes next.

ROSANNA XIA

A Toxic Secret Lurks in Deep Sea

FROM *The Los Angeles Times*

NOT FAR FROM Santa Catalina Island, in an ocean shared by divers and fishermen, kelp forests and whales, David Valentine decoded unusual signals underwater that gave him chills.

The UC Santa Barbara scientist was supposed to be studying methane seeps that day, but with a deep-sea robot on loan and a few hours to spare, now was the chance to confirm an environmental abuse that others in the past could not. He was chasing a hunch, and sure enough, initial sonar scans pinged back a pattern of dots that popped up on the map like a trail of breadcrumbs.

The robot made its way 3,000 feet down to the bottom, beaming bright lights and a camera as it slowly skimmed the seafloor. At this depth and darkness, the uncharted topography felt as eerie as driving through a vast desert at night.

And that's when the barrels came into view.

Barrels filled with toxic chemicals banned decades ago.

Leaking.

And littered across the ocean floor.

"Holy crap. This is real," Valentine said. "This stuff really is down there."

"It has been sitting here this whole time, right off our shore."

Tales of this buried secret bubbling under the sea had haunted Valentine for years: a largely unknown chapter in the most infamous case of environmental destruction off the coast of Los Angeles—one lasting decades, costing tens of millions of dollars, frustrating generations of scientists. The fouling of the ocean was so reckless, some said, it seemed unimaginable.

As many as half a million of these barrels could still be under-water right now, according to interviews and a *Times* review of his-torical records, manifests, and undigitized research. From 1947 to 1982, the nation's largest manufacturer of DDT—a pesticide so powerful that it poisoned birds and fish—was based in Los Ange-les.

An epic Superfund battle later exposed the company's disposal of toxic waste through sewage pipes that poured into the ocean —but all the DDT that was barged out to sea drew comparatively little attention.

Shipping logs show that every month in the years after World War II, thousands of barrels of acid sludge laced with this synthetic chemical were boated out to a site near Catalina and dumped into the deep ocean—so vast that, according to common wisdom at the time, it would dilute even the most dangerous poisons.

Regulators reported in the 1980s that the men in charge of getting rid of the DDT waste sometimes took shortcuts and just dumped it closer to shore. And when the barrels were too buoyant to sink on their own, one report said, the crews simply punctured them.

The ocean buried the evidence for generations, but modern technology can take scientists to new depths. In 2011 and 2013, Valentine and his research team were able to identify about sixty barrels and collect a few samples during brief forays at the end of other research missions.

One sediment sample showed DDT concentrations forty times greater than the highest contamination recorded at the Super-fund site—a federally designated area of hazardous waste that of-ficials had contained to shallower waters near Palos Verdes.

The world today wrestles with microplastics, bisphenol A (BPA), per- and polyfluoroalkyl substances (PFAS), and other toxics so unnatural they don't seem to ever go away. But DDT—the all-but-indestructible compound dichlorodiphenyltrichloroethane, which first stunned and jolted the public into environmental action—persists as an unsolved and largely forgotten problem.

Signs warning of tainted fish to this day still cover local piers. Recent studies show our immune systems may be compromised. A new generation of women—exposed to DDT from their moth-ers, who were exposed by their mothers—grapples with the still-mysterious risks of breast cancer.

The contamination in sea lions and dolphins continues to stump scientists, and the near-extinction of falcons and bald eagles shows how poisoning one corner of the world can ripple across the whole ecosystem.

Decades of bureaucracy and competing environmental issues have diverted the public's attention. Valentine hoped digging up physical evidence from the seafloor would get more people to care, but calls and emails to numerous officials since his discovery have gone nowhere.

Rallying for the deep ocean is not easy, Valentine acknowledged, even though we rely on the health of these waters far more than we know: "The fact that there could be half a million barrels down there . . . we owe it to ourselves to figure out what happened, what's actually down there, and how much it's all spreading."

Once hailed as a major scientific achievement, DDT combated both malaria and typhus during World War II. It was so potent that a single application could protect a soldier for months. The U.S. Army's chief of preventive medicine, Brigadier General James Simmons, famously praised the chemical as "the war's greatest contribution to the future health of the world."

Manufacturers rushed to supply the postwar demand—including Montrose Chemical Corporation of California, which opened its plant near Torrance in 1947. The chemical industry was celebrated at the time for boosting the nation into greater prosperity and preventing crop failures across the globe. The United States used as much as 80 million pounds of DDT in one year.

But there were two edges to this sword. A top U.S. Department of Agriculture scientist had urged the military not to allow DDT insecticides for commercial use without further research, worried about "the effect they may have on soils and on the whole balance of nature."

Even Swiss chemist Paul Hermann Müller, who won a Nobel Prize in 1948 for discovering DDT as a pesticide, cautioned that he himself did not fully understand how the chemical interacted with the living world. Decades of painstaking study still lay ahead for biologists, he said.

Rachel Carson, a marine biologist, heeded these words in 1962 and ignited a movement against what she called "the reckless and irresponsible poisoning of the world that man shares with all other creatures."

Her revolutionary book *Silent Spring* evoked the sudden silence of songbirds missing in the skies—alerting unknowing people to the dangers of long-term exposure, even in tiny doses, to a chemical that they could not physically avoid.

DDT is so stable it can take generations to break down. It doesn't really dissolve in water but stores easily in fat. Compounding these problems is what scientists today call "biomagnification": the toxin accumulating in the tissues of animals in greater and greater concentrations as it moves up the food chain.

Consider phytoplankton, the microscopic algae that are the base for almost all food webs in the ocean.

DDT-contaminated phytoplankton gets eaten by zooplankton, which fish and whales consume by the thousands.

In 1969, shipments of jack mackerel from Southern California were recalled because DDT levels were as high as 10 parts per million, or ppm—double what the U.S. Food and Drug Administration considered safe for consumption at that time.

Tumors started appearing on bottom-feeding fish like white croaker.

In that same year, California brown pelicans, which eat the fish, laid eggs on Anacapa Island with chemicals broken down from DDT averaging 1,200 ppm.

Scientists discovered that the chemicals led to eggshells so thin that the chicks would die. Bald eagles had also vanished from the Channel Islands, along with peregrine falcons and the brown pelicans. Similarly, sea lions with more than 1,000 ppm in their blubber were giving birth to pups prematurely. Bottlenose dolphins had concentrations as high as 2,000 ppm.

Montrose executives aggressively defended DDT through the 1960s as the public reckoned with these alarming new concerns about food chains and poisoned ecosystems.

They said in letters and editorials that DDT played a vital role in society when properly used and was not a serious threat to human health. They accused environmentalists of scare tactics and misleading information and touted the company's reputation of making the best DDT in the world—a technical grade sold to other firms that would then dilute it into specific insecticides.

The company was supplying governments from Brazil to India, they said, and even the World Health Organization. International malaria eradication programs turned to Montrose for supplies.

But after years of intense inquiries, government officials said they were convinced that the chemical posed unacceptable risks to the environment and potential harm to human health. In 1972, the United States finally banned the use of DDT.

Demand was still strong in other countries, however, so the chemical plant in Los Angeles kept churning out more. Montrose managed to operate for another ten years before the factory, looming over Normandie Avenue near Del Amo Boulevard, finally went dark.

In the early 1980s, a young scientist at the California Regional Water Quality Control Board in Los Angeles heard whispers that Montrose once dumped barrels of toxic waste directly into the ocean. People at the time were hyperfocused on the contamination problems posed by poorly treated sewage, but Allan Chartrand was curious about the deep-sea dumping and started poking around.

He called Montrose, and to his surprise, the staff pulled out all their files. He and a team of regulatory scientists combed through volumes of shipping logs, which showed that more than 2,000 barrels of DDT-laced sludge were dumped each month. They did the math: between 1947 and 1961, as much as 767 tons of DDT could have gone into the ocean.

"We found actual photos of the workers at two in the morning dumping—not only dumping barrels off of the barges in the middle of the Santa Monica Basin," he said, "but before they would dump the barrels, they would take a big ax or hatchet to them, and cut them open on purpose so they would sink."

On a recent morning, Chartrand rummaged through stacks of yellowing papers and reports detailing everything he had discovered so many decades ago. Now a seasoned eco-toxicologist in Seattle, he never understood why all this information wound up gathering dust—undigitized and largely forgotten.

He pulled out faded reports that his team had published from 1985 to 1989, summarizing what they had found at Montrose and in the water quality control board's own records. "This makes my heart sing," he said, as he reread conclusions that still resonate today.

Chartrand said he was astonished to learn this kind of activity was allowed. Federal ocean dumping laws dated back to 1886, but the rules were focused on clearing the way for ship navigation. It wasn't until the Marine Protection, Research, and Sanctuaries Act of 1972, also known as the Ocean Dumping Act, that environmental impacts were considered.

Dumping industrial chemicals near Catalina was an accepted practice for decades.

Landfills could hold only so much, and people were concerned about burning toxics into the air—but the Pacific Ocean seemed a good alternative. Explosives, oil refinery waste, trash and rotting meats all went into the ocean, along with beryllium, various acid sludges, even cyanide.

Dilution is the solution to pollution, the saying used to go, but at what cost? The ocean covers more than 70 percent of the planet, but it can absorb only so much. What we eat, what we breathe is ultimately dictated by what we do to the sea.

"It's just sad, sad, sad," Chartrand said. "When stuff's being dumped offshore like that, it's in the dead of night, nobody's seeing it. It's out of sight, out of mind."

For years, a company called California Salvage docked at the Port of Los Angeles, loaded up Montrose's DDT waste and hauled everything out to sea. Workers were instructed to dump in a designated spot, dubbed Dumpsite No. 1, that was about 10 nautical miles northwest of Catalina.

But compliance inspections were infrequent, and crews sometimes took shortcuts. Chartrand discovered notes from California Salvage indicating they had decided to dump elsewhere because Dumpsite No. 1 was in line of a naval weapons firing range.

The report concluded that these companies likely dumped in closer, much shallower waters.

"Our report caught them red-handed," Chartrand said. "Here I was this young guy—newly married, just had my first kid, got my new job at the water quality control board—heard about this dumping, went down to Montrose . . . and it very quickly got so much bigger than me."

In 1990, a few years after Chartrand compiled his reports, the Environmental Protection Agency teamed up with the state and launched a court battle against Montrose and a number of other companies under the Superfund law. Environmental groups ex-

pected the lawsuit—the largest in U.S. history alleging natural re-
source damages from chemical dumping—to be a landmark case
in resolving coastal pollution issues.

Chartrand and dozens of others were pulled in to testify. Science
was disputed in court, evidence debated, expertise challenged. In
numerous depositions, former factory workers were grilled on how
they operated.

Bernard Bratter, a Montrose plant superintendent, described
how they would call California Salvage to dump its acid waste in
bulk: "The trucks would come in, we'd load the trucks, they would
then haul them down to the harbor where they had their barges,
and the truck would unload into the barge, and when there was
enough liquid in the barge, they'd haul the barge out to the specified area in the ocean and release the acid."

Montrose officials, who had filed counterclaims, asked the court
to exclude the evidence presented on ocean dumping—arguing
that such dumping wasn't relevant.

They said the government's natural resources damage claim
was based solely on the release of DDT through the sewer system
to the Palos Verdes shelf, and that attorneys could not prove that
Montrose's disposal of DDT-contaminated waste into the deep
ocean actually hurt various bird species.

They also questioned Chartrand's calculations of how much
DDT went into the ocean and made the point that there was nothing secret or illegal about the dumping at the time. The government, they said, allowed this to happen.

In an interoffice correspondence in 1985, Samuel Rotrosen,
Montrose's president at the time, wrote that "it is true that from
1947, when the plant started up, until sometime in the 1950s we
disposed of our waste sulfuric acid at sea through California Salvage Company who barged it out to state-approved dumping areas.

"We stopped this disposal after we installed our acid-recovery
plant, at which time we sold the acid to fertilizer makers," he said.
"Because our acid contained traces of DDT (50–250 ppm) . . . the
fertilizer producers would no longer take it, and so we disposed of
it at landfills."

As the court battle waged on, a handful of curious scientists
kept trying to solve the DDT questions at the bottom of the ocean.

Chartrand did not have a deep-sea robot, but he figured out

a way to collect sediment samples and clumps of tar by dragging a large otter trawl net along the seafloor. He also took samples of rattails, kelp bass, and other fish from different depths of the ocean.

He called Robert Risebrough, a legend among DDT scientists whose testimonies in the 1960s and early 1970s helped Congress understand why the chemical should be banned. Risebrough, a UC Santa Cruz research ecologist at the time, ran the samples and authored a sweeping study. He confirmed the existence of considerable concentrations of DDT chemicals in both the sediments and the "tar cakes" by the dumpsites.

It was unclear how much the DDT could move through the water at such depths, where there is little oxygen, he said, but the dumping was close enough to the Channel Islands that the upwelling of deeper water common in this area could stir up what enters the food chain.

And if the barrels were indeed punctured, he added, some of the sludge could have leaked out on its way down to the seafloor.

He had a strong suspicion that the disappearance of bald eagles from Catalina was connected to the dumping operations, but he didn't have the data to confirm it. DDT contamination was also significantly higher in birds that fed on fish, compared with birds that ate mostly rodents and prey on land—another clue that the DDT from the ocean dumping was harming wildlife.

He called for more studies to connect the dots, but Chartrand had run out of funding. Chartrand held on to what he could—even the remaining samples that neither he nor Risebrough could bear to throw away. Some of that deep-sea sediment has yet to be tested.

"They're in a deep freeze now, but because it's DDT, even though it's been thirty, forty years, they're still valid," Chartrand said. "If we could get the funding, those are still worth running."

M. Indira Venkatesan, a geochemist at UCLA who studied how chemicals moved through the sea, had taken one of these samples in the early 1990s and run her own analyses. She, too, concluded there must be a DDT source in the ocean much larger than just what had come out of the sewage closer to shore.

She collected additional sediment cores from the seafloor by a manual pulley that her technicians and graduate students spent hours pulling up. Her team distinguished the DDT "fingerprint" for Montrose's ocean-dumped waste and discussed the upward and downward diffusion of DDT in the sediments.

"It gets resuspended and remobilized. That's why you see it all over the basin," she said. "I knew, I just knew, this DDT source was significant, just from the chemical analysis, but we couldn't show the extent of the dumping, nor the number of barrels."

Back in court, the arguments were focusing on the more tangible: the hundreds of tons of DDT and PCBs, another toxic chemical, that had been released two miles off the coast of Palos Verdes where the sewage emptied into the ocean. Many saw the need to make this public health problem—much closer to shore, with visible harm to humans and the ecosystem—a top priority.

The site—spread across more than 17 square miles—was declared a Superfund cleanup in 1996. About 200 feet deep, it was considered one of the most complicated hazard sites in the United States—at least three times deeper than similar Superfund sites in Boston and New York harbors.

By late 2000, the parties decided to settle. They negotiated a consent decree midway through trial—no sides admitting fault, with an agreement that more than $140 million would be paid by Montrose, several other companies that owned or operated a share of the plant, and local governments led by the Los Angeles County Sanitation Districts.

The settlement—one of the largest in the nation for an environmental damage claim—would pay for cleanup, habitat restoration, and education programs for people at risk of eating contaminated fish.

"This Decree was negotiated . . . in good faith at arm's length to avoid the continuation of expensive and protracted litigation and is a fair and equitable settlement of claims which were vigorously contested," according to the decree, which mentioned that the damage claim includes "any ocean dumpsites used for disposing of wastes from the Montrose Plant Property."

Attorneys representing Montrose, when contacted by the *Times*, declined to comment on the new underwater data and noted that the ocean claims related to the DDT operation were resolved twenty years ago. Litigation continues to this day over other impacts from the former plant. In August, a $56.6 million settlement was finally reached over groundwater contamination.

Back at UCLA, on a recent morning in the geology building, Venkatesan thought ruefully back to those DDT years. KCBS had run a local news series on the barrels, and the *Times* followed the story for a brief period.

The information caught the attention of Assemblyman Tom Hayden (D–Santa Monica), the 1960s activist turned lawmaker who married Jane Fonda and was remembered as "the radical inside the system." For a few years, he pushed for more information about the barrels and an action plan, but so many unchecked environmental problems demanded attention back then.

Even Venkatesan got pulled away. As public concerns shifted from water to air pollution, her research focus changed to aerosols.

She had tried for a while longer to get the word out—giving public lectures in Santa Monica bookstores and telling whoever would listen that the deep ocean also needed healing.

"I didn't know what to do with this data; I felt bad," she said. "As scientists, we thought we could leave it to the politicians and the government to do their job . . . But if the government is not proactive, then people don't care. If people don't care, then the government doesn't do anything."

Now that she's retired, her filing cabinets—filled with her work since she started in 1975—have been moved into a basement at UCLA. She recently reviewed the data that the UC Santa Barbara researchers had uncovered with deep-sea robots, which validated Chartrand's estimates, as well as her own.

She held out her hands and said she was trembling with excitement, knowing that people might care about this issue again.

"Disposing any waste, where you don't see and forget about it, does not solve the problem," she said. "The problem eventually comes back to haunt us."

One afternoon in Santa Barbara, hunched over a computer humming with data, Valentine and Veronika Kivenson, a PhD student in marine science, scrolled through the eerie images they had gathered underwater.

They leaned in to examine an icicle-like anomaly growing off one of the barrels—a "toxicle," they called it—and wondered about the gas that bubbled out when the robot snapped one off. To have gas supersaturated in and around these barrels so deep underwater, where the pressure was ninety times greater than above ground, was unsettling. They couldn't help but feel like they were poking at a giant Coke can ready to explode.

One thing was clear, Kivenson said: this stuff is spreading. She had tried to collect sediment many meters from the barrels as a baseline to compare the samples collected right next to the source. But the baseline turned out to also have similarly high concentrations of DDT—most of them higher than the permissible threshold established by the National Oceanic and Atmospheric Administration.

"These barrels do seem to be leaking over time," she said. "This toxic waste is just kind of bubbling down there, seeping, oozing, I don't know what word I want to use . . . It's not a contained environment."

So much of this data, collected in 2011 and then again in 2013, came down to timing and good luck: the underwater robots had been on loan for a different project, but that research cruise was ahead of schedule, so they had a window of extra time to explore.

A scientist involved in the discovery of the *Titanic* happened to be on board, so he helped them program the robots on where to go and how to search for the barrels. A marine geochemistry lab at Woods Hole Oceanographic Institution ran the samples, and Kivenson, whose graduate fellowship and tuition were the only funding for this research, analyzed them for her PhD.

She tracked down the patent for the DDT acid waste that supposedly went into the barrels. She combed through eBay for out-of-print research books on ocean dumping and flipped through rolls of microfilm in the archive rooms of court buildings and government agencies.

She validated Venkatesan's conclusion that the DDT near the barrels did not have the same characteristics as the Superfund site—ruling out the possibility that this was just DDT from Palos Verdes that somehow traveled farther into the ocean and settled onto the deep seafloor. One key difference was that the barrel samples contained no PCBs, which are abundant in the contamination near the sewage outfall.

Each barrel seemed to contain acid waste with about 0.5 percent to 2 percent technical-grade DDT—which, at half a million barrels, would amount to a total of 384 to 1,535 tons of DDT on the seafloor. The distribution was patchy; one hot spot had a concentration of DDT that was forty times higher than the highest level of surface sediment contamination recorded at the Superfund site.

All told, she concluded that the total amount of DDT from the dumping seemed comparable to the estimated 870 to 1,450 tons that had been released through the sewer.

But in the end, these are still extrapolations—we don't know how much is actually down there, said Kivenson, who published these findings last year in the journal *Environmental Science and Technology* and is now a postdoctoral fellow at Oregon State University. Logical next steps would be to somehow map and identify just how many barrels there are, determine any hot spots, and study how much the chemical is leaking and spreading and accumulating.

Valentine tried calling those with the power to do something about these barrels: the EPA, which has been in charge of cleaning up the Superfund site. But the EPA, it turns out, hasn't even figured out what to do with the DDT problem that got all the attention and millions of settlement dollars. After more than twenty years of meetings and high-level studies, the site off the Palos Verdes shore has become its own controversial saga.

A pilot experiment more than a decade ago to bury the DDT under a thick cap of clean sand showed mixed results. Then sampling in 2009 suggested that most of the DDT had mysteriously vanished—prompting a burst of headlines and more internal paralysis. The longtime project manager unexpectedly retired, and many of the scientists who had dedicated decades of their careers to the chemical have also either retired or moved on.

Many, when reached, said they had not been involved with the site for a number of years.

"I feel like something's happened at the site; it just sort of died. It's been very weird," said Robert Eganhouse, a research chemist at the U.S. Geological Survey who had been studying the Superfund site and the breakdown rates of DDT since the 1970s.

His last meaningful exchange with the EPA was in late 2016, when he submitted an immense amount of data and a final synthesis report for the site—a research endeavor that took more than eight years and cost millions of dollars. To this day, Eganhouse, who recently retired, is not quite sure what the EPA did with this information.

Judy Huang, the Superfund's project manager for the past decade, when reached by the *Times,* directed questions to regional
headquarters.

In an email, an EPA spokeswoman said the agency had suspended capping efforts and collected new data that showed twice
as much DDT as the 2009 results. The EPA is now reassessing its
approach: "We are updating our evaluation of the mechanisms of
how the DDTs and PCBs in the sediment impact human health
and the environment in this complex system."

In the meantime, projects to restore local kelp forests, wetlands,
seabirds, and underwater habitats have been supported over the
years with the settlement money, as well as education outreach
that helped prevent anglers and vulnerable communities from eating poisoned fish.

Fish remain contaminated, but the concentrations seem to be
slowly going down, according to findings from the EPA's most recent five-year review of the site, released last fall. The bald eagles
and peregrine falcons are coming back after years of human assistance, and nature seems to be healing itself over time.

After all these years of costly stops and stalls, some think a so-
called monitored natural recovery approach might just be the best
solution. The EPA plans to start a new feasibility study that aims to
lead to a final cleanup strategy. That study is not expected to be
published for another four years.

Mark Gold, who had championed the DDT problem as a marine
scientist since the 1990s, could barely find the words to describe
how he felt about the attempted cleanup of the Palos Verdes shelf.

"To have the EPA say, twenty-five years later, that maybe the best
thing to do is to just let nature take its course is, frankly, nothing
short of nauseating," he said.

When asked about the barrels, he was so shocked he had to
pause and grab a calculator to process the amount of DDT that
could be in the deep ocean. At an absolute minimum, he said,
there needs to be further investigation into how much is actually
down there and how much this dumping has harmed the ecosystem.

Gold, who is now Governor Gavin Newsom's deputy secretary
for coast and ocean policy, said he had heard stories of illegal
dumping back when he was helping state and federal officials
build the case against Montrose. But there was no firsthand evi-

dence in the 1990s, he said, nor a sense of whether it was five barrels, ten, or twenty.

"Nobody in their worst nightmares," he said, "ever thought there would be half a million barrels of DDT waste dumped into the ocean off of L.A. County's coast."

For scientists today, DDT poses a new generation of complications. Dilution, it seems, just means the problem reaccumulates elsewhere. In the environmental health laboratory at San Diego State's School of Public Health, Eunha Hoh recently discovered the chemical had wound its way into dolphins in unexpected ways.

Marine mammals, like humans, nurse their young and live long lives. Slow to evolve, their long-term health is a window into the lasting impacts of chronic exposure and accumulation—and how these chemicals get passed on to babies. As some of the largest predators of the sea, they're also an important indicator of the ocean's overall health.

So when Hoh sampled the blubber of eight adult dolphins that had lived deeper off the coast of Southern California, she was surprised to find significant amounts of forty-five DDT-related compounds. Every dolphin she tested had washed up dead—and had accumulated much more of these chemicals than dolphins tested in Brazil and elsewhere around the world.

"DDT contamination—is it really going down in Southern California? Can we really say that, or are we missing something," said Hoh, who also serves on the California Ocean Protection Council's science advisory team. "Sure it was banned decades ago, it might be manageable globally, but Southern California? We're different. Our ocean is so much more polluted with DDT. We cannot just say, 'That's done; we can move on to other things.'"

Hoh's expertise is in discovering new chemicals, but she remains mystified by how DDT keeps reappearing in new and unexpected ways. Where, she often wonders, is all this DDT coming from?

When she first heard about the barrels scattered across the seafloor, it was as if someone finally handed her missing pieces to a puzzle that had never quite added up.

The questions came tumbling out. If that much more DDT is out there but forgotten, and no one knows to study it, she said, how will we ever understand the true legacy of this chemical?

Current monitoring shows that the local ecosystem, on the

whole, is stable. But what's unclear are these long-term unknowns, said Keith Maruya, who co-authored the dolphin study and retired last year as the Southern California Coastal Water Research Project's head of chemistry.

"It's not like something's going off the cliff. But what we don't know is whether these things are going to have a longer-term, more subtle effect—are some populations really, really slowly going to be declining?" he said. "We don't know the answer. Moreover, we don't really have the tools yet to answer that question fully."

He jolted up in his chair when the discovery of the barrels came up in a recent conversation.

"Wow. Wait, how many did they find? I need to write this down."

He jotted a few numbers, then silently compared this with the known quantity of DDT dumped at the Superfund site.

"If nobody accounted for this second source . . . if you've got twice the amount," he said, thinking aloud. "It's such a staggering number, but what does this mean? . . . The bottom line is always going to be: So what? We have a chemical out there, so what?"

At the Scripps Institution of Oceanography, in a developmental biology and environmental toxicology lab overlooking the sea, Amro Hamdoun has been pondering this question for much of his life.

He's found through molecular studies that "persistent organic pollutants," like flame retardants and DDT, can block a key protein from eliminating toxins from the human body—a clue, perhaps, into why they bioaccumulate. Even in small amounts, these contaminants could interfere with the human body's natural ability to defend itself.

Hamdoun teaches *Silent Spring* and DDT to his students as an example of how the world used to be—but can't help but wonder how much the jobs and science of the future will be dealing with these messes of the past.

"There's a broader problem of thinking of the ocean as this unlimited garbage dump that's going to take up our CO_2, take up our mercury, deal with the plastic that we don't throw away properly, be a dumping ground for pesticides, deal with whatever is in runoff—and that our health is going to be separable from that," he said. "But what we're learning more and more is that our health and the ocean's health are pretty inseparable."

At what point, he asked, does it become our prerogative, as people who live in a shared society, to decide what it is that we want to put in our environment—and our bodies?

He leaned forward in his chair, hands clasped, head bowed, like Valentine and Chartrand and so many who came before.

"These chemicals are still out there, and we haven't figured out what to do," he said. "They are an issue, and we still don't have a plan."

MARINA KOREN

SpaceX Is Taking over a Tiny Texas Neighborhood

FROM *The Atlantic*

BOCA CHICA, TEXAS —Mary McConnaughey was watching from her car when the rocket exploded on the beach. The steel-crunching burst sent the top of the spacecraft flying, and a cloud of vapor billowed into the sky and drifted toward the water.

McConnaughey and her husband had planned to drive into town that day in late November, but when they pulled out onto the street, they noticed a roadblock, a clear sign that SpaceX technicians were preparing to test hardware. She didn't want to miss anything, so she turned toward the launchpad, parked her car at the end of a nearby street, and got her camera ready.

The dramatic test was a crucial step in one of Elon Musk's most cherished and ambitious projects, the very reason, in fact, he founded SpaceX in 2002. Weeks earlier, Musk had stood in front of the prototype—164 feet of gleaming stainless steel, so archetypically spaceship-like that it could have been a borrowed prop from a science-fiction movie—and beamed. He envisions that the completed transportation system, a spaceship-and-rocket combo named Starship, will carry passengers as far away as Mars. A few months before the explosion, hundreds of people came to the facility in South Texas, on the edge of the Gulf Coast, to see the spaceship, and thousands more watched online. "It's really gonna be pretty epic to see that thing take off and come back," Musk gushed at the event, as if he were seeing the finished Starship in front of him.

McConnaughey was there, and even posed for a picture with Musk. At the end of the night, she made the short trip home to her house on a small road lined with stout palm trees. McConnaughey lives in Boca Chica Village, a tiny neighborhood located in startling proximity to SpaceX's facilities. Many of the village's residents have lived there for years, long before SpaceX arrived, some before the company even existed.

Friction between next-door neighbors is quite different when one of them is a rocket company. Instead of an ugly fence, there might be an ugly fence with massive tanks of cryogenic liquid behind it. When residents find papers stuck in their front door, the notes don't ask them to keep the noise down or clean up after their dogs; they warn them that their windows could shatter.

Boca Chica's residents have learned to live with a rocket company, or at least tolerate it, over more than five years. But SpaceX's work is about to become even more disruptive. (The explosion certainly made that clear.) So the company has offered to buy their homes. Some have taken the offer. Others, such as McConnaughey, have rejected it, even as Musk prepares to launch a giant rocketship just a short hop from their houses. SpaceX is already hard at work on the next Starship prototype, and Musk says the company might launch it into orbit as soon as this year. "We love Texas," James Gleeson, a SpaceX spokesperson, said in a statement, "and believe we are entering a new and exciting era in space exploration."

Few people in this part of South Texas could have predicted the recent trajectory of their life when SpaceX moved in. They have become space fanatics and legal experts, Musk supporters and thorns in his side, trying to make sense of their place in a strange story that could someday end millions of miles away from Earth. All because they got new neighbors.

"They're here to stay," McConnaughey told me, "and they want us to leave."

Boca Chica is an unincorporated community of about forty houses, mostly one-story homes with soft-orange brick exteriors, on the southernmost tip of Texas. There are no shops or restaurants or amenities of any kind around, including municipal water pipes; Cameron County regularly trucks in gallons of water, which is stored in outdoor tanks. Many residents are retired; they spend summers in northern states and flock south for the winter like migratory birds, eager for the peaceful stillness of the coast-line.

The only way to reach the village is via State Highway 4, a two-lane road that runs through mostly empty land. It originates to the west, in the city of Brownsville, and disappears into the shores of Boca Chica Beach, an eight-mile stretch of unspoiled sand, free of boardwalks and souvenir shops. About three miles south, through thick desert brush, is the Rio Grande, winding like a curled ribbon along the border. On a clear day in the village, you can see straight to Mexico.

The residents of Boca Chica first learned of SpaceX's plans at a public meeting in the spring of 2012. SpaceX was preparing to fly cargo for NASA to the International Space Station for the first time, and in anticipation of increased demand for the company's services, Musk wanted to build "a commercial Cape Canaveral"—a launch site all SpaceX's own, where Falcon 9 rockets could fly as many as twelve times a year. South Texas was one of several areas under consideration, in part because of its proximity to the plan-et's equator, which spins faster than the poles, providing depart-ing rockets with an extra boost. SpaceX also has a long history in Texas; it has tested rocket engines at a facility in McGregor, north of Austin, for nearly two decades.

Hundreds of people went to the meeting in Brownsville, accord-ing to the *Brownsville Herald*. Some had concerns about the local fauna—Boca Chica sits in a national wildlife refuge, where each year more than 500 species of migratory birds funnel through and sea turtles come ashore to lay their eggs. But most people spoke in support of the project, which SpaceX promised would bring hundreds of jobs to the area. To residents of Boca Chica Village, the whole thing felt like a pep rally. For Brownsville, one of the poorest cities in the country, SpaceX seemed to offer an unlikely dream: the opportunity to turn a border town into a twenty-first-century space city.

"Most of the kids that are fortunate enough to get a college education usually leave the area and they don't come back," Eddie Treviño, the county judge for Cameron County, told me. Treviño grew up in Brownsville, left for college, and then returned for good. "SpaceX may draw kids to either come back or maybe to stay," he said.

A few days earlier, SpaceX had bought its first piece of land from the county. Texas had heavily courted SpaceX since 2011 with millions of dollars in incentives and legislation that would limit public access to beaches along the Gulf. SpaceX, the thinking went, could commandeer the coastline as needed. A review by the Federal Aviation Administration eventually found that rocket operations wouldn't cause any "significant" environmental impacts, clearing the way for SpaceX to get started. In the spring of 2013, hundreds of people showed up to another meeting, some in LAUNCH BROWNSVILLE T-shirts, and a state official read aloud a letter of support from then-governor Rick Perry.

Company reps did try to reassure the few villagers who attended the meeting about being so close to a launch site. "They said that we would be okay, that we wouldn't even have to wear hearing protection," McConnaughey said. "They wanted to be good neighbors."

SpaceX broke ground at the beach in the fall of 2014, and soon trucks made daily trips into Boca Chica, packed with soil that would provide a sturdy foundation for a launchpad on a bedrockless shore, less than two miles from the village. State Highway 4, unaccustomed to so much traffic, stretched and cracked, so crews from the Texas Department of Transportation followed, patching the holes. A pair of massive antennae, shaped like mushrooms and larger than buildings, were shipped in from Cape Canaveral to track SpaceX missions.

McConnaughey found herself spending hours outside nearly every day, a camera dangling from her shoulder. She had never considered herself a photographer, and usually got behind the lens only on family vacations. Now she was snapping pictures of hardware and sweaty technicians, like a wildlife photographer angling to capture an elusive creature.

She posted the photos to a forum on nasaspaceflight.com, a community for space fans, with a watermark of her username, BocaChicaGal, in pink font. She learned a new language, writing on the forum—and eventually to her thousands of new Twitter followers—about leg mounts and bulkheads and stainless-steel coils. She learned to look for signs of activity, such as a raised construction crane, and stayed when she saw them, sending her husband into town to run errands without her. She usually goes to Michigan during scorching Texas summers, but she stayed put last year, intent on capturing the activity.

"A lot of it is just right place, right time, being observant," McConnaughey said. "You have to pay attention to what's going on. You just have to be patient and ready and wait."

Elsewhere, SpaceX was making progress—punctuated by a fiery blaze or two. A Falcon 9 rocket exploded during takeoff. The company recovered a rocket booster after launching it into orbit, gently guiding it back to the ground, an industry first. Then another rocket blew up. The company got better at reusing boosters and returning them in one piece. In Florida, the inaugural launch of the Falcon Heavy, the most powerful rocket in operation today, was a milestone for SpaceX, and, it turned out, for Boca Chica Village. "I think it gives me a lot of faith for our next architecture, our interplanetary spaceship," Musk said at a press conference after the launch in 2018. Forget about routine Falcon 9 launches; he wanted to test this new vehicle, the Starship, at the company's location in South Texas.

"We've got a lot of land with nobody around, so if it blows up, it's cool," Musk said.

That December, a mysterious tower rose at SpaceX's construction site in Boca Chica. Users on nasaspaceflight.com exchanged guesses about what the cylinder of patched-together metal could be. A water tower, probably. But then workers sliced holes into the structure and gave it legs.

Gene Gore drove over early one morning. The sun hadn't yet risen when he saw a gleam in the dark, next to the mystery structure. "I'd never seen a rocket, but it looked like the nose cone of a rocket," Gore told me. "That was the first time anybody had seen it. And then it was apparent. *Oh, they really are building a rocket here.*"

Gore, a South Texas native, has run a surfing company on nearby South Padre Island, a touristy resort town, since 1995, and operates a webcam for surfers checking up on the waves. "When they started building and you could see it, I'm like, 'Whoa, we gotta point the camera over there,' which kind of pissed off the surfers," said Gore, who often surfs at Boca Chica Beach. "So I pointed it back, and then that pissed off the space people, and then I'm like, 'I gotta buy another camera.'" On a clear day, you could see the earliest Starship prototype from the southernmost tip of the island, a tiny smudge of silver jutting into the blue sky.

Brownsville residents started stopping by too. Austin Barnard, a student at Texas Southmost College, discovered SpaceX by accident, when a binge of Carl Sagan videos led the YouTube algorithm to suggest something from Musk. For Barnard, State Highway 4 is "the highway to Mars," the first leg of a trip that humanity is destined to take. He likes to walk around the fenced-in SpaceX facilities, soaking in the sounds of the construction. At home, before he starts his homework, Barnard sometimes sits in his room in silence, practicing for the quiet isolation of a months-long mission to the planet.

Around the time the "water tower" showed up, Rosemarie Workman and her husband were returning from Brownsville with groceries when their car was stopped near the village.

Residents are used to checkpoints; there's a U.S. Border Patrol stop on State Highway 4 when you're heading toward Brownsville and away from the Mexican border. This checkpoint was different. Instead of a border agent in a dark-green military uniform or a car-sniffing German shepherd, a SpaceX employee approached their car and asked the couple whether they were "on his list." Workman was stunned. She and her husband have owned their home in Boca Chica for twenty years. "I should not have to be on a list," Workman told me. "Besides that, why would a SpaceX employee have authorization to stop me on a state highway and tell me I can't go through?"

To meet safety requirements from federal and county regulators, SpaceX sets up two checkpoints during launch operations, including testing. Only SpaceX personnel and village residents can pass through the first checkpoint, about 15 miles out from Boca Chica on State Highway 4; homeowners can add names to a list, but visitors must stay on their host's property during road closures or they risk arrest for trespassing. No one is allowed past the second checkpoint near the village, beyond any homes and closer to the launchpad, an area the FAA says isn't safe during testing.

The road closures became a fact of life in Boca Chica. So did the intrusions of SpaceX operations, which reached a new intensity. Nearly every day, beneath the sound of wind blowing through dry grass and the staccato chirps of blackbirds, there was the clanging, whirring, and buzzing of construction equipment, the high-pitched beeping of trucks in reverse and cranes climbing high, the music the workers put on to entertain themselves. The work lasted through the night, beneath the glow of industrial lights. On windy days, pieces of plastic wrap drifted away from the construction site and stuck to the yucca trees. Some residents started calling and emailing county offices, state officials, federal agencies—anybody who could tell them whether any of this was sanctioned.

All the residents I spoke with—close to a dozen—told me a version of the same story. Before SpaceX, the village felt like a coastal paradise, contentedly dislodged from civilization. At night, the only light came from the distant hotel towers on South Padre Island to the north and the Milky Way overhead. Now the place felt almost claustrophobic.

In August, the Cameron County sheriff dropped off notices in the village. When they hear a police siren, the memo said, residents should step outside their home. Better yet, they should consider leaving for the day. SpaceX was scheduled to conduct an important test of *Starhopper,* an early Starship prototype, the one mistaken for a water tower. "There is a risk that a malfunction of the SpaceX vehicle during flight will create an overpressure event that can break windows," the notice read. "It is recommended that you consider temporarily vacating yourself, other occupants, and pets from the area." Before the flight, which was, at the time, SpaceX's biggest Starship test, the FAA required SpaceX to increase its liability insurance, from $3 million to $100 million.

Starhopper rose into the sky, climbing to a height of nearly 500 feet before levitating back down. McConnaughey watched from a safe distance, along with some SpaceX workers, who whooped and cheered as they watched. "It was just amazing," she said.

Musk was in awe too. "Congrats SpaceX team!!" he tweeted. "One day Starship will land on the rusty sands of Mars."

The letters, printed on SpaceX letterhead, arrived a few weeks later. "When SpaceX first identified Cameron County as a potential spaceport location, we did not anticipate that local residents would experience significant disruption from our presence," the note read. "However, it has become clear that expansion of spaceflight activities as well as compliance with Federal Aviation

Administration and other public safety regulations will make it increasingly more challenging to minimize disruption to residents of the Village." (The FAA is in charge of approving SpaceX's launch activities in Boca Chica.) The letters came with contracts, offering homeowners a deal to buy their home. Those who sold, SpaceX promised, were welcome to return for private, "VIP" launch-viewing events of Starship.

The village residents had suspected that SpaceX would want them out someday. But some were still shocked by the letters, especially after they read what was inside.

SpaceX had commissioned an appraisal of their properties, without their knowledge, and was now offering them three times the resulting market value. The process would be handled by JLL, a commercial real-estate company. SpaceX gave the residents two weeks to respond.

Some signed the contracts, but many didn't want to leave. A sale at triple a property's worth certainly sounds generous. But the appraisers SpaceX hired hadn't even stepped inside their homes, the residents said, to see the new plumbing and air-conditioning systems they had installed, or the fresh tiling and gleaming backsplashes. JLL conducted a second assessment, and this time appraisers took pictures of the interiors. But when residents saw the new evaluations, some were insulted—the comparable homes the appraisers had listed included foreclosed houses and properties with foundation issues. (SpaceX declined to comment on property negotiations.)

SpaceX still tried to be neighborly, in a way. The company invited residents to Musk's Starship presentation days after the letters came. Those who RSVP'd rode in a sleek black van to the control center, where Musk showed off the latest Starship prototype. Afterward, the residents were herded into a room stocked with soft drinks and snacks, while elsewhere Musk took questions from reporters—including about buying out the village. When he was done, he joined them for about half an hour and, with the help of a SpaceX lawyer, tried to reassure them about the buyout process. Musk told the residents that they could stay in Boca Chica as long as they were willing to put up with the "inconveniences" of living next door to a spaceship construction site, McConnaughey said. He warned them there would be more.

A week before Thanksgiving, Celia Garcia-Johnson got on the phone with Elizabeth Clampitt, a senior vice president at JLL and the residents' point of contact for the buyouts. Garcia-Johnson has loved Boca Chica Beach since she was a little girl growing up in nearby Brownsville, and bought two homes in the village, one in 1991 and the other in 2004, with the hope that someday her two sons would inherit them. Like some of her neighbors, she found and paid an appraiser to conduct another review, which came up with higher values. JLL rejected the appraisal. So Garcia-Johnson told Clampitt that she wasn't taking SpaceX's offer. Clampitt said SpaceX wouldn't come back with another.

Garcia-Johnson described the conversation: Clampitt told her, "Well, I'm going to talk to you like I would talk to my mother or my aunt, someone that's close to me. If you don't sell at our prices, we're going to close the books on you, and the houses are going to the county." (SpaceX declined to comment on this conversation, and Clampitt did not respond to a request for comment.)

SpaceX can't force the residents to leave, but the county can. In 2013, county commissioners established a corporation "to assist in the promotion and development of a spaceport project" in Cameron County. Under Texas law, the corporation has the authority to exercise the same right that lets governments take over private property and compensate its owners. When we met, Treviño, the county judge, told me that while he sympathizes with the residents, the use of eminent domain in Boca Chica Village is "probably a distinct possibility." The law is on SpaceX's side. A 2005 Supreme Court ruling expanded the definition of *public use,* the legal justification for eminent domain, to include economic development, and since then, states have taken advantage of that leeway: Texas, for instance, claimed dozens of homes to make room for a new stadium for the Dallas Cowboys. "That is something, unfortunately, that happens way more than it should," Renée Flaherty, an attorney at the Institute for Justice, a nonprofit law firm in Washington, D.C., told me.

Flaherty first heard about Boca Chica in the fall. "Those people were already there and [SpaceX] brought a nuisance to them, and now it's escalated to the point where the nuisance is so severe that they're telling them that they have to leave their own property," said Flaherty, a Texas native herself. If the county moves ahead with legal proceedings, she believes that the remaining residents would have a case for eminent-domain abuse.

"We're watching the situation very closely," Flaherty said. (She has been in touch with several residents. She doesn't represent them, but is considering making a trip to Boca Chica.) "I don't like to make threats, and I never do make threats, but we are watching. I think it's probably a very good thing if they know that someone has an eye on them."

Today, Boca Chica is busier than ever: as Musk put it recently, SpaceX is "going max hardcore" on production of the new Starship prototype. Fresh cracks have appeared in the asphalt of State Highway 4, and this month, SpaceX asked federal regulators for permission to launch the newest Starship vehicle almost halfway to the edge of space sometime in 2020. Musk wants to put people on board this year too.

In the village, residents see SpaceX workers moving around the homes the company now owns, lugging in furniture and paint and carrying out old carpeting. And judging by the cars that stay parked in the driveways overnight (including Teslas, another product in the Elon Musk firmament) residents suspect that the employees have begun sleeping there. At one property, a chain-link fence has been replaced with a wooden one, and someone strung up small twinkle lights—the kind that might illuminate a cozy outdoor party—over the yard. As some residents are packing up to leave, SpaceX is settling in.

For McConnaughey, watching her neighbors trickle away has been difficult. She felt a twinge of sadness when she saw SpaceX workers trimming the grassy median on her street. "It's just something that *we've* always done," she said.

Cameron County is a far cry from Cape Canaveral, but the infrastructure to support a departure gate from Earth is slowly emerging. The county has spent millions of dollars to spruce up a park at the southern end of South Padre Island as a prime viewing spot for future launches, installing an amphitheater, event center, pavilion, and boardwalks. SpaceX has begun inviting people to Stargate, the control center in Boca Chica, to meet and interview with recruiters. Last month, the company also held a business fair in Brownsville so that the company could "learn about select products and services available from Rio Grande Valley vendors." Gore attended, hoping, he said half-jokingly, to become SpaceX's official surfing instructor.

Like others who visit the SpaceX facilities regularly, Gore and Barnard, the space-eyed college student, know some of the residents, including McConnaughey, whom they consider a friend. They know what the homeowners are going through, and they feel bad for them. But Barnard sees their situation through a more cosmic lens. A common argument for space travel, especially Musk's version of it, is that it is inevitable. Of course people will someday leave Earth and build homes on other planets, and Musk is the one who can get them there. Humankind is poised to become a spacefaring civilization—to find not only survival beyond Earth, but a happy life—and for Barnard, Boca Chica might be the cradle. "It sucks," he said. "But the needs of the many outweigh the needs of the few. *Star Trek*."

As SpaceX plants roots in Cameron County, the company seems to be growing impatient about the people—about twenty—who don't want to leave. In mid-January, David Finlay, SpaceX's senior director of finance, came to town and stopped into the homes of residents who haven't sold. "He sat in my living room and he apologized because they're making us leave," but he said SpaceX needed them out soon, Garcia-Johnson told me. When she asked about eminent domain, Finlay said he could tell Texas officials that SpaceX might leave if it can't maintain operations here, which could prompt the county to start eminent domain proceedings. Garcia-Johnson hasn't changed her mind.

SpaceX declined to discuss the specifics of its negotiations with residents in Boca Chica and the possibility of eminent domain, and emphasized its commitment to its Texas locations. "Every single SpaceX rocket and spacecraft is tested in Texas before flying to space and back again," said Gleeson, the SpaceX spokesperson. "South Texas will play an increasingly important role in our efforts to help make humanity multi-planetary."

Like Garcia-Johnson, McConnaughey never wants to give up her home in Boca Chica, but she would do it for what she feels is a fair price, enough to find a similar home with the views and the quiet she has enjoyed in the village for years. But she can't imagine being anywhere else, especially now. She loves documenting what

SpaceX is doing, even though she knows the company wants her to leave.

McConnaughey recognizes that her desires might seem difficult to reconcile, but she doesn't feel conflicted about her hobby, which she said feels like an addiction. She doesn't do it for the love of SpaceX. (She doesn't see the appeal of going to Mars either—"isn't it –80 degrees there?"—though she understands why others do.) She does it for the nasaspaceflight.com community, who encouraged her to post more pictures when she first started, and for other space fans who don't have the view that she does. Her photos, alongside pieces by Barnard and Gore, are now on display at the Brownsville Museum of Fine Art, as part of an exhibit chronicling SpaceX's transformation of Boca Chica. Last month, all three mingled alongside company reps at a black-tie gala to celebrate its opening.

The week I visited Boca Chica, about a month before the small explosion in November, SpaceX workers had removed the nose cone and the fins of the Starship prototype, giving it the appearance of an uncapped lipstick. It was a warm, cloudless day, and the ground seemed to sizzle in the heat of the sun. Workers in hard hats and reflective vests arrived in pickup trucks. A security guard drove over and asked me to step away from the fence. If something were to fall from the crane, he said, we'd both be in trouble.

The air-conditioned houses in the village were blissfully cool in comparison, and residents showed me around their homes and backyards, pointing out their seashell collections and where they drank their coffee each morning with views of the nearby bay. McConnaughey brought out a stack of notices about public hearings and launch activities she had received over the years, a paper trail documenting the residents' rocky coexistence with SpaceX, culminating with a glossy booklet, prepared by JLL, telling them how much their homes are worth. "Nice little wooden shack," McConnaughey said, referring to one of the properties used for comparison. "It looks lovely."

The next day was windy and 20 degrees cooler. Heavy rain had passed through overnight, scattering palm fronds on roadways, and the bay had swelled and risen, flooding State Highway 4 just a few miles out from the village. There was no getting out, or in, until the water subsided hours later. At a hotel in Brownsville, where

SpaceX puts up employees who come to work from out of town, a few technicians from the company's facilities in California sat in the lobby. They had driven over to Boca Chica in the morning but were told to leave as the water crept up onto the highway. One of them told me they had come up so quickly on the flooding that they almost lost control of the car.

I asked them what they were going to do for the afternoon. They shrugged. "Rest," one said. Another pointed to a brightly colored liquid in a plastic cup in front of him. "Drink."

For a few hours that day, Boca Chica Village was cut off from the rest of the world. Residents said they'd never seen the bay flood this badly before, but they didn't seem to mind. They could pretend that the village was theirs again, even as they prepared to face the reality of giving it up. McConnaughey, Garcia-Johnson, Workman, and others said they'll stick it out for as long as they can, but they know that Boca Chica has become something else, something harder to recognize. It is a strange existence, to move through the familiar routines of their days without knowing how many are left in their shifting paradise. A security guard near the big antennae told me that some of the residents wave to him during their daily walks, and he knows some of them by name. I asked him whether he knew that SpaceX was trying to buy their homes. "I thought the county made a decision for them already," he said.

JIAYANG FAN

The Friendship and Love Hospital

FROM *The New Yorker*

MU ZHIXIA DISCOVERED the lump in her left breast on an unseasonably warm night in March of 2014. At twenty-seven, she was strong and healthy, and hadn't seen a doctor since giving birth to her son, Xuan, two years before. But her mother, Sulin, told her not to take chances and marched her to their local hospital, in Pingding, a small city in the province of Shanxi. A doctor conducted a swift examination and wrote a prescription. "Doesn't she need a scan?" Sulin asked. "No need!" the doctor responded. "The medication will be enough."

Zhixia dutifully took the pills, but after a few months the lump was still there, so Sulin accompanied her to a hospital in Yangquan, a nearby industrial city of 1.5 million people. The doctors said that she needed immediate surgery. As is typical with dire diagnoses in China, they did not tell Zhixia that she had breast cancer, informing only her mother. Sulin, in turn, assured her daughter that the growth was benign.

After the operation, a biopsy revealed that the cancer had spread. The doctors put Zhixia on a course of chemotherapy, and she was hospitalized for several weeks. A year later, the cancer returned, and the doctor who had prescribed the chemo remarked casually that if they had followed it up with radiation the outcome might have been better. Sulin wanted to know why they hadn't done that, but she felt too intimidated to say anything.

In the next three years, Zhixia had four more long stays in the hospital, emerging frailer each time. The cost of her treatments, 100,000 yuan (almost $15,000), plunged the family into financial crisis. The cancer progressed to her lymph nodes, her lungs, her bones. Her body became so ravaged that she was almost unrecognizable. When her son was taken to visit, he had to be prompted to call her Mother.

Shanxi is in the heart of China's coal country, and has disproportionately high rates of esophageal and lung cancer. Zhixia was only five months old when her father, a farmer, died of esophageal cancer. (Sulin remarried, but her second husband succumbed to lung cancer.) Still, Zhixia grew up to be a sunny, optimistic woman. Moonfaced, with high cheekbones, she liked to say that she met her father whenever she looked in the mirror. She quit school after seventh grade and worked various jobs to help support the family. When she was twenty-five, she met her husband, a coal miner named Zhang Wei.

Three years into Zhixia's illness, in the spring of 2017, Wei felt a pain in his back so severe that he couldn't lift up their son. He didn't go to a doctor: caring for Zhixia left little time, and he figured that he'd hurt himself while swimming. Two weeks later, the pain was so bad that he couldn't get out of bed. When he finally went to a doctor, he was informed that he had a blood disorder. The doctor, who suspected late-stage leukemia, told him to check in to the hospital right away. Wei said that he needed to keep working, to pay for his wife's treatment.

Wei died on a brisk fall day, three months later. What pained Zhixia the most was knowing that he had been alone at the end. His mother was too distraught to enter his hospital room, his father had been at work in the coal mines, Zhixia had been receiving another round of chemo, and her own mother was busy caring for Xuan. In the days after, Zhixia told her mother, "Please don't let me die." By then, she knew that she had cancer: her father-in-law, who was illiterate, had inadvertently let her see one of her medical reports.

Zhixia's doctors told Sulin to begin thinking about funeral arrangements. In desperation, she started asking around about other medical facilities, and a neighbor told her about a man named Li Youquan, who had opened a small private hospital on the outskirts of Yangquan. Its name was Yangquan You'ai Hospital—*you'ai* means "friendship and love"—and it had a unit devoted to hospice care, a concept still unfamiliar in China.

Few cultures relish talking about death, but in China the subject remains taboo. Mentioning it is considered so unlucky that dying people are often reluctant to discuss arrangements with their families or even to make wills. (Last year, *The Farewell*, an American film about a Chinese family that uses a wedding as an excuse to gather around a terminally ill grandmother without arousing her suspicions, was a breakout hit in the West, but it was largely ignored in China, where such stories are commonplace.) As a result, fewer than 150 institutions specialize in end-of-life care, in a country where nearly 20 percent of the population—a quarter of a billion people—is sixty or older. The United States, with some 70 million people over sixty, has more than 5,500 such institutions.

In China, the family has traditionally provided care for the vulnerable: "Raise a child against old age; stockpile grain against famine," one proverb counsels. Confucian expectations of filial piety remain strong, but for most Chinese they have become increasingly difficult to fulfill. Dizzying economic expansion has made China's population ever more mobile, and the one-child policy, in force from 1979 to 2015, means that many adults have no siblings with whom to share the burden of caring for relatives. Hundreds of millions of workers who have moved to the country's booming cities cannot do much more for aging parents back in remote villages than wire whatever money they can spare.

Rural areas also lack adequate public health services. Close to half the population lives in the countryside, but about 80 percent of China's medical facilities are concentrated in cities. Health-care costs have risen sharply in recent years, and Chinese patients must navigate a byzantine system of government coverage. Most people have basic insurance, but anything beyond routine care usually requires steep out-of-pocket payments.

The early stages of the coronavirus pandemic brought to light some of the dysfunctions of China's medical system, including underinvestment in primary care clinics and overreliance on huge, rigidly bureaucratic urban hospitals. But, if the coronavirus exposed the country's health care challenges in their most acute form, the quieter crisis in end-of-life care reveals a chronic underlying condition, whose symptoms are at once brutally economic and deeply cultural. Prosperity and medical advances have transformed the way Chinese people live, but they have done little to address the question of how they should die.

*

Li Youquan named the Friendship and Love Hospital for an earlier iteration, which was founded, like most of China's first hospitals, by Western missionaries. American representatives of the Church of the Brethren arrived in the area in 1910, and their hospital trained generations of doctors and nurses. It closed not long before the Communists came to power, in 1949, and expelled foreign missionaries. But, as Li told me when I visited him last summer, almost everything in China runs in cycles: "Sooner or later, what was banned will be reborn."

Li is a sturdily built man in his early fifties, with alert eyes set in a frank, expressive face, and he comes from a family of farmers. His route to providing palliative care was a circuitous one. In the late eighties, he attended a vocational school that specialized in traditional Chinese medicine. After graduating, he left traditional medicine behind and did an internship at the largest hospital in Yangquan, where he encountered an ultrasound machine for the first time, and was amazed. "In Eastern medicine, there is so much interpretation and guesswork," he said. "But with ultrasound you could actually see inside a patient's body." After scraping together enough money to buy a machine, he started operating a clinic out of his house, near the village where he was born. Charging a couple of dollars per scan, he found that there was good money to be made detecting tumors and pregnancies.

At the time, in the early nineties, Deng Xiaoping's market reforms had been opening state-run services to private enterprise. The Communists had established near-universal health care, which, though often rudimentary, increased the average life expectancy from thirty-five, in 1949, to sixty-nine, in 1990. Under the economic reforms, however, hospitals were allowed to ramp up what they charged for their services, and government health care spending shrank. The process was anything but linear—during a crackdown on unlicensed operators, Li's clinic was closed and his machine confiscated—but the burden of cost continued to shift toward patients, and the gap between standards of care in the cities and those in the hinterlands widened.

Li spent the next decade working at state clinics, until, in 2005, he saw another chance to launch an entrepreneurial venture. In the aftermath of the SARS epidemic, the government embarked on the country's most ambitious health care reforms in more than a generation, in part to address the rural-urban divide. Li built the Friendship and Love Hospital on a plot of undeveloped land next to his home. Initially, there were only twenty beds. "I knew I wanted to offer something that other, bigger hospitals didn't," Li told me. "But I hadn't really figured out what that was." Still, the business thrived, and, five years later, in 2010, he decided to hold a celebration for the centennial of his hospital's missionary namesake. It had occurred to him that publicizing the institution's American roots might attract investment and expertise from the United States. The thought prompted him to send emails to various organizations associated with missionary work.

One of these emails reached Li Ruoxia, a recent graduate of a Lutheran seminary in Dubuque, Iowa. She was struck by the coincidences: she, too, was from the Yangquan area; she had written a master's thesis on the heritage of Christian communities in Shanxi; and her husband, an American anthropologist, had written his doctoral thesis on the rituals of aging and filial piety in rural China. Ruoxia had been volunteering at a nursing home in Iowa—she later worked at a hospice—and, as she corresponded with Li Youquan, she started to think about the lack of such services back home. "To die seems so private, but the process is embedded in a larger system," she told me. In China, death was met first with denial and then with stoicism. "But in a hospice death is accepted, so its trauma is slightly eased," she said. In 2012, she and her husband moved to Yangquan, so that she could introduce hospice care to Li's hospital.

At Friendship and Love, Ruoxia began hiring a team to act as social workers, both for the hospital's resident patients and for discharged patients facing their final days at home. She led workshops on family support and bereavement counseling, and handed out translations of American writing on the subjects. Sometimes she asked team members to write letters to themselves or their loved ones to practice communicating difficult emotions. Her recruits were women from the countryside, most of whom had first come to the hospital to care for dying relatives. Few of them had finished high school, but they could do simple health-aide tasks and instruct patients and families on the importance of hygiene. Just as important, the women understood the people they helped. They spoke the same dialect, shared the same outlook, knew the local customs, and had an intuitive sense of how best to broach the subject of dying.

One of the first team members was Liu Meiying, a cancer sur-
vivor in her early sixties. After receiving her diagnosis, more than
twenty years ago, she'd quit her job at a factory and joined a fledg-
ling patient-support group in Yangquan. The group, known as the
South Mountain Anti-Cancer Club, offered support that was rare at
the time, arranging outings and gatherings where members could
talk about treatment options and the day-to-day impact of their
illness. "Back then, no one had ever heard of the word 'hospice,' "
Meiying said. "But, with our regular visits to our dying members,
that's what we were trying to do, in our clumsy way."

Ruoxia's hospice unit was unusual enough to seem suspect to
conservatively minded people. Early on, when the hospital registered with Shanxi's Civil Affairs Bureau, the officials were particularly concerned about what the unit would be called, and Ruoxia
had to come up with a name that was suitably euphemistic. She
knew that phrases like "hospice care" and "palliative care" wouldn't
do, but even "comfort care" was judged to hint too aggressively at
a failure of treatment. Eventually, the government settled on Elder
Navigation Center. "The bureaucrats don't even know sometimes
why they are so resistant to try new things," Ruoxia told me with a
bitter laugh. "Usually, they just know it's safer."

By the time Zhixia arrived at Friendship and Love, last April, can-
cerous growths were compressing her spine to the point where
she could no longer feel the lower half of her body. She experienced severe nausea—even spoonfuls of porridge were hard
to keep down—and had lost so much weight that her emaciated
frame seemed lost in the hospital's striped pajamas. Pain from ulcers in her mouth kept her from sleeping. Sometimes, early in the
morning, she became delirious, and thought that she heard ghosts
on the other side of the wall. Still, she had no thought of giving
up. "Why aren't they treating the source of my illness?" she fumed
to her mother when her doctors focused on her bedsores. "They
should be trying to fix that first!"

After Zhixia's experiences in other hospitals, the work of Ruoxia's team seemed strange to her. She was perplexed that the social
workers talked so much about her feelings rather than about her
disease. But their unusual methods were not always unwelcome.
Instead of examining her, Ruoxia offered to wash her hair, and,
when Ruoxia mentioned her young daughter, Zhixia took out her

phone to show off pictures of her son. When Ruoxia arrived bearing nutritional shakes and cartons of yogurt, Zhixia suggested,
with a weak smile, that perhaps her mother could eat them: "Her
belly is my belly, her heart my heart."

One morning, a few days into Zhixia's stay, Meiying got a text
message from her: "Auntie. Urgent! I need to talk to you!" When
Meiying saw the time stamp on the message—4 a.m.—she feared
the worst and rushed over to the hospital. "Are you feeling okay?"
she asked breathlessly when she got there. But Zhixia seemed better than usual and greeted her with a bright smile.

Sulin was in the room too, fussing over her daughter. As soon
as Sulin left, to get some breakfast in the cafeteria, Zhixia lowered
her voice and said, "Can you help me with a task, big sister?" She
told Meiying about something that she'd just read on her phone.
Constantly thinking about how she might rescue her family from
the financial hole into which her illness had sunk them, she'd
come across a government-assistance scheme that claimed to help
those in dire need. She wanted Meiying to help her apply. "My
mom is strong, but I don't think she realizes how impossible our
money situation will be," Zhixia confided.

Meiying nodded and promised to do her best. She was happy
to help, and felt satisfaction that Zhixia was opening up more and
more to her and the other aides. On the way out, Meiying ran into
Sulin, her eyes puffy and red-rimmed. She hadn't gone to breakfast after all but was just waiting there in the corridor, unable to
leave her daughter. When Sulin moved to go back in, Meiying put
a hand gently on her shoulder and told her that it would do her
good to get some breakfast. "I'll stay with Zhixia for as long as you
need," she said.

Early one morning last June, I accompanied Meiying on a visit to
a patient everyone called Brother Zhang, who suffered from advanced cirrhosis of the liver. Brother Zhang lived in a remote part
of the countryside where his family had farmed corn for generations. As we drove north out of Yangquan, the industrial cityscape
gave way to squat, sagging houses of weathered brick. Brown-gold
cliffs composed of Shanxi's distinctive sandy loess soil punctuated
fields of millet and sorghum. An hour later, we got out of the car
and walked the last mile, up a hilly dirt path and then on broken
stone steps enmeshed in roots. Brother Zhang's house was what is

known in the region as a *yaodong*, or cave home—a courtyard arrangement of vaulted rooms carved into the loess hillside.

Zhang, a lean stalk of a man with a hawkish, wind-parched face, came out to greet us. Though his gait was sometimes unsteady, he moved about the premises quickly, showing me his well, a giant millstone, a barn where he'd once kept donkeys, and herbs and peppers hanging to dry under the lintel of a doorway. "It means something when city folk visit them," Meiying murmured. Zhang boasted that he could still haul a respectable amount of corn on his back. I believed him; he had the wiry strength of a frail person made strong by circumstance.

Zhang lived alone. His wife had left him years ago, and his two sons rarely visited: the older son was a coal miner, with a family of his own; the younger, in his late twenties, suffered from mental illness and usually stayed with relatives. Zhang took us inside the single room that served as his living quarters. The bare stone walls arched overhead, like those of a wine cellar. A bulky old TV set was pushed up against his bed, on which all his earthly possessions were piled. Meiying noticed a pack of cigarettes and playfully chided him, but Zhang said, with a grin, that, at this point, smoking was probably keeping him alive rather than killing him.

When I asked Zhang about his health, he had trouble explaining the details of his condition, something that Meiying told me is common among the people she sees. Until he was in his mid-fifties, he had never stepped inside a hospital, and rarely thought about his body as anything more than a machine from which he needed to extract as much work as possible. But one day, five years ago, he had coughed up enough blood to soak an entire handkerchief. When he visited the First People's Hospital, in Yangquan, he was told that an operation was required, to biopsy a tumor on his spleen. He could not afford it, and, besides, there would be no one to care for him while he recovered. Seeking a second opinion, he ventured to Taiyuan, the provincial capital, but the verdict was the same. In the past three years, Zhang has twice stayed at Friendship and Love to receive intravenous infusions. The second time, he checked himself out early, against his doctor's orders. It was harvest season, and Zhang had no one else to reap his corn.

I noticed a pair of faded calendars, years out of date, hanging above Zhang's bed. One featured the face of President Xi Jinping. The other showed that of Jesus. Some Christians had given him the calendar a few years ago, he said, promising that Jesus could save him. Zhang started attending their meetings. "But Jesus did not cure me," he said scornfully, adding that the Christians were no help at all with obtaining medication or IV infusions, or with harvesting the crops in his absence. "If neither Jesus nor his followers could do those basic chores, what are they good for?"

Our visit unfolded almost like a social call, but Meiying slipped in questions about the things she'd come to check up on. Zhang admitted that he still had trouble keeping food down. She asked after his younger son, who had recently been threatening suicide. Meiying had done what she could to help, getting in touch with the son on WeChat and telling him not to do anything rash. She advised Zhang to focus on taking care of himself, but his son's problems and other family disputes agitated him. Every so often, he picked up his cigarettes and then, with a glance at Meiying, put them down. The senselessness of suicide, as he desperately clung to life, baffled him. "Life might be short and brutal, but the point is to survive," he said.

At the end of our visit, he gave Meiying a bag of plums that he had picked from his own trees. After we left, Meiying observed that Brother Zhang's case was not unusual. As people moved where the work was, the only people left in the villages were the very old, the very young, and the disabled. Shi Lizhen, the head nurse of the geriatric department at the first hospital Zhang visited, told me that when she started working, more than thirty years ago, elderly patients typically had six or seven children, two or three of whom could devote themselves entirely to their parents' care. Those days were gone, and the hospital's resources were so scarce that they couldn't take in patients who didn't have around-the-clock caregivers. Some patients hired health aides, but many were too embarrassed to ask their children to help with the cost. Shi, who is fifty-seven, mentioned that her daughter lives in Beijing, 700 miles away, and is a mother herself. "When I look at my patients now, immobile in bed, I think about myself and what my future will look like," she said.

One day during my stay in Yangquan, Li Youquan and I drove two hours to Taiyuan, to visit Shanxi Tumor Hospital, the province's largest hospital and one of the first in China to specialize in cancer care. In its vast atrium, amid a sea of red banners bearing socialist slogans, there was a no-smoking sign and a poster warning of the link between smoking and cancer. Nearby, a man sat staring at his phone, a lit cigarette dangling from his mouth. In China, the messages on posters are sometimes a better guide to what is not happening than to what is. Other signs referred to anti-corruption policies and stipulated grave penalties for bribery and profiteering in health care settings. But, in a dimly lit stairwell, there were phone numbers and notes scribbled on a concrete wall, advertising the services of people who could falsify almost any kind of medical paperwork—patient records, diagnoses, prescriptions, bills.

Shi Lizhen had told me that, in addition to outright corruption, there were many gray areas. Doctors at public hospitals receive paltry remuneration for patient consultations but much more when they order major procedures. As a result, surgeries are suspiciously common. For hospitals, drug sales are an important revenue stream, so overprescription is rife. Shi sometimes encounters patients who are fully recovered but are still being prescribed medication. She discreetly suggests that they ignore their prescriptions, but they usually tell her that they are afraid of offending the doctors. "So we reach a compromise," she said. "The patient buys the medicine, so that the doctor gets paid, but they'll just take it home instead of using it."

In Taiyuan, Li and I went to the Shanxi hospital's residential compound to visit his friend Song Jianguo, the former head of the hospital's respiratory department. Song, who was in his early sixties, had just retired, after having received a diagnosis of stage IV stomach cancer. Greeting us at the door of his apartment in pajamas and slippers, he explained that he had just finished a round of chemotherapy. He had sharp cheekbones and exuded a placid, scholarly air. In his left nostril was a thin nasogastric tube.

When I apologized for bringing up end-of-life care, Song laughed dryly. "It's a subject we should talk about more openly in this country," he said, pointing out that, even in a hospital of this scale, there was no consistent palliative care. Again, distorted incentives were part of the problem: doctors earned far less for prescribing pain medication than for ordering chemotherapy or surgery. There were cultural factors too. Many patients in severe pain were wary of opioids, which they associated with addiction, and China's newfound wealth inspired unrealistic expectations. "There's this very optimistic idea that, if we spend enough, diseases will be cured at the rate that new skyscrapers are built and bullet-train tracks are laid," Song said. "But that's not how the human body works." Richer patients couldn't accept that money wouldn't guarantee survival, and those who couldn't afford treatments, he said, "sometimes jump from their window to spare their family the burden of caretaking and the expense."

Song took a shallow, labored breath. He worried that a deepening distrust of doctors was undermining end-of-life discussions: "It's impossible when the patient or the patient's family is thinking at every turn, *Oh, is the doctor saying there's nothing we can do because that's really the case or because he doesn't think he'll earn enough to be worth his effort?*" Song adjusted his nasal tube. "Everyone should know what's coming. When that day comes, we have to know the difference between giving up and letting go."

One day in May, at Friendship and Love, Zhixia was visited by one of the aides, a woman in her fifties named Cuihe, a former kitchen worker in a restaurant who had come to work in the hospital after her mother had been a patient there. After chatting for a bit, Cuihe read aloud from a book about Mother Teresa. When Cuihe mentioned that she and the others had been reading it in a training workshop, Zhixia said, "I didn't like school, so I don't know very much." Cuihe smiled and said that she hadn't had much schooling either. After reading a chapter about faith, Cuihe asked Zhixia what she thought the passage was about. "I guess it's about the relationship between a parent and a child, about love and our attitudes to love," she said, and then added, "It's about death too. I think I will die soon."

Cuihe asked if she was afraid of death. In Zhixia's five years as a cancer patient, no one had ever asked her that. "When the pain overwhelms me, I'm not afraid," Zhixia said. "I just want it to stop so badly that, if death can take away the pain, I almost prefer it." She let out a ragged breath and tried to sit up. "What do you think death is?" Cuihe asked softly. Zhixia thought for a moment and, with Cuihe's help, sipped water through a straw. "I don't know," she responded. "Lately, I've been thinking about time passing, and what time is like after death. When I'm suffering, every second feels like a year." Zhixia tried to smile, but parting her chapped lips made them bleed.

Later, Meiying, who was Zhixia's favorite aide, asked if Zhixia had told her son about her illness. "He knows nothing," Zhixia replied. She said that she'd been taking as many pictures with him as she could, and had organized her and her husband's medical records so that when her son was old enough he could read them. As Zhixia spoke, she started to cry, but muffled her sobs so as not to disturb her mother, who had nodded off in a chair next to her bed. She told Meiying that she'd thought about writing a letter to her son, but worried that she wouldn't be able to express what she felt. "It's okay to keep it simple," Meiying said, and suggested that it might be even better to talk to the boy directly. "The way you speak to him is what he will remember years from now."

The oldest hospice in China, Songtang Care Hospital, opened in Beijing in 1987. Its founder, Li Songtang, now in his seventies, likes to recount how he came to realize that the dying had needs beyond the merely medical. During the Cultural Revolution, he was sent, as a seventeen-year-old, to Inner Mongolia, to work as one of the "barefoot doctors," whom the Communists dispatched to provide basic care in underdeveloped communities. There he met an exiled professor from Beijing, who was terminally ill with stomach cancer. In the last days of his life, the man became obsessed with clearing his name of the political crimes of which he'd been accused. Li, desperate to give him peace of mind, eventually came up with a lie, saying that he had persuaded authorities to expunge the charges. Li told me that he has never forgotten the way the dying professor grabbed his arm in gratitude. The party's exoneration, Li realized, was "his only medicine and salvation."

"Dying is not the scary part," Li told me, as we sat in a conference room at the hospital. "It's the uncertainty, the anxiety, of feeling utterly out of control." His facility houses about 300 patients, and, throughout the years, hospital personnel have visited from all over the country to observe how it functions. The staff's daily check-ins with patients are focused less on medical requirements than on soothing anxieties and fostering a sense of connection. When Li showed me around, we came across a late-stage-Alzheimer's patient making her way down a corridor, clutching the railing affixed to a wall. She had to work a substitute factory shift, she explained urgently, because her daughter was out sick. "Your daughter is very lucky to have such an able mother," Li gently replied.

Li noted that, despite Songtang's success, not many places have replicated its model. "Hospice care is not economically prudent," he said. "People who derive the most benefit from it don't live long enough to advocate for it. And the sense of cultural taboo about death deters the living from promoting it." The facility has had to move a dozen times because of complaints from its neighbors; once, protesters who blamed the hospice for bringing a curse on the neighborhood smashed its windows.

An avid collector of antiques, Li proudly pointed out how the hospital building incorporated many pieces of Qing-dynasty architecture. He started salvaging these in the nineties, when Beijing's construction boom indiscriminately razed pagodas and temples. "Chinese society is caught between the old and the new paradigms," Li reflected. "It hasn't decided what it wants to discard and what it wants to import." As he spoke, he took long drags from a cigarette. Smoking was a habit he'd picked up in his teens, to cope with the stresses of being a barefoot doctor, but he told me that he wasn't worried about cancer. "We all carry cancer within us," he said. "It's relentless checkups at predatory hospitals that actually kill you."

I toured a number of other upscale facilities in Beijing and began to wonder how my mother, back in New York, would fare in such places. For almost a decade, she has suffered from ALS, the fatal neurodegenerative disease that causes progressive, and eventually total, paralysis. In Beijing, a publicist led me around a facility called Golden Heights—its interior a riot of floral chintz and gilding—and told me that its clientele included the business elite and retired TV stars. There was a gym, a calligraphy room, Ping-Pong tables, a grand piano, and a "nostalgia room" filled with old calendars, enamel basins, and a black-and-white TV from the eighties. A brochure detailed some of the costs: a deposit of 100,000 yuan ($15,000) and a room charge that started at 10,000 yuan a month. Standard health aides could be hired for 20,000 yuan a month, but employing people capable of caring for someone with late-stage ALS would cost much more.

My mother and I had had conversations, while she could still speak, about what was to come. Back in China, she'd been a doctor herself, a pulmonologist in an army hospital, and she told me that the most unbearable way to die is to be deprived of the ability to breathe. As was customary at the time, she had often told white lies to terminal patients about their true diagnoses and had seen them brought again and again to the ER to receive oxygen, only to suffer through their last days in half gasps. So now, at Golden Heights, I raised the subject of do-not-resuscitate orders, explaining to my guide that my mother would prefer comfort care to extraordinary measures. She assured me that if my mother were in crisis she would be taken to the ER. "But what if she signs a document specifying that she does not want to be resuscitated?" I asked. With an expression of serene forbearance, the woman began shaking her head before I could finish the question. "It's our job to save her life," she insisted, adding, by way of explanation, "Our goal at Golden Heights is to be humane, which means that we will do everything we can to save her."

Under Xi Jinping, the Communist Party has tried to restore the centrality of Confucian thought to national life, including the importance of filial piety. In 2013 it issued the Elderly Rights Law, which threatened legal consequences for children who did not visit their parents. The legislation was met with widespread derision. It was government policy, after all, that had encouraged migration to urban centers, generating unprecedented prosperity but also depriving the elderly of a social safety net. Not for nothing does Confucianism see the faults of a child as reflecting the failures of the parents.

Li Ruoxia, the leader of the hospice unit at Friendship and Love, believes that, these days, an obsession with filial piety usually does more harm than good. "The consideration here isn't necessarily the well-being of the parent but the reputation of the child," she told me. "It's the *performance* of filial piety." It isn't unusual for more money to be spent on elaborate funerals than on making patients comfortable in their final days, and several doctors told me that people often ask them to keep their ailing parents alive for the sake of their retirement checks. Ruoxia's recruits often found themselves in the middle of fraught situations. One man told Meiying about a safe where he kept his life savings. "That made his children and grandchildren very nervous," she recalled. "I told them that if they were gentler with the old man he might not feel like they were just waiting for him to die so they could have his money."

The unit's slogan is "Start the conversation!" because, Ruoxia explained, "in Chinese families, there's sometimes this expectation that, if you really care about someone, you can intuit what they want without them asking." Recently, though, the unit has experienced communication failures of its own. Its members conceived of themselves as a kind of family. "Once I am trusted, I become the go-to person for a family member," Meiying said. "I can't abandon them when I've worked so hard to allow them to let me in." Ruoxia, however, believed that they should be observing clearer boundaries between the professional and the personal. "If their commitment to their job is completely contingent on their feelings about me, how do we, as an organization, grow into something bigger?" she said.

Tensions escalated when Cuihe—who was the designated nutritionist, on account of her restaurant experience—suggested that, instead of preparing patients' meals in the hospital kitchen, it would be easier if she cooked them at home. Ruoxia asked Cuihe to sign a liability waiver making clear that the food's preparation had not been subject to hospital inspection. Cuihe was deeply insulted. "Does she think I'm trying to poison the patients?" she said, relating the story to me months afterward. For Ruoxia, the piece of paper was just a matter of protocol. For Cuihe, it was a sign of personal distrust.

The way that care work distorts boundaries was all too familiar to me. After my mother's diagnosis, I moved her from Connecticut, where she lived alone, to New York, where I worked and lived. I found a two-bedroom apartment in an accessible building and hired home aides to care for her when I was at work. I planned her daily schedule, her meals, her physical therapy, and vacations to places she had always wanted to visit but could not afford.

My therapist told me, "It's like you two are so enmeshed that you share a psychic world." I tried to explain that, though the situation might seem unhealthy, I was fulfilling an important duty. In Chinese, 孝, the character for *xiao*, the Confucian concept of filial piety, is made up of part of the character for "old" above the character for "child," as if the latter were carrying the former. In previous generations, to care for one's parents in their old age was not a choice but a given. My actions were predicated on a Chinese conception that individuals live in the service of the larger, unified entity: the family.

Still, my mother often strenuously protested against my efforts. It was all so much fuss and so expensive, and I think that she didn't want to believe that her decline would come so quickly—preparation made real an eventuality she was still half hoping might not arrive. When I overrode her, I thought of myself not as defying her wishes but as guiding her toward her true desires. To do what she asked rather than what I thought she wanted would have been the real dereliction of duty. It took a long time for me to see that, as her body was gradually robbed of mobility, my wholesale renovation of her life might be depriving her of something that she was just beginning to value at life's end: her autonomy.

I thought of this again when I was at the luxurious Beijing nursing home that could cater to residents' every desire except the desire to one day end their suffering. A paternalistic system of government robs people of the ability to make meaningful choices, infantilizing them to the point where they no longer feel responsible even for the choices that they do make. Sophia Zhang, a gerontologist at Peking University, told me that Chinese doctors are uniquely positioned to see how decades of volatile and violent politics have left their marks on individuals. "People who were very poor in their youth can finally eat to their hearts' content," she said. "To them, three meals of fatty pork and unlimited liquor and no need to do backbreaking labor is the good life. They feel betrayed when I say that they need to curb those appetites."

The idea of dying with dignity is premised on the idea that there is a difference between the quality of life and its length. "That's a profound shift for a generation of people entering their old age, for many of whom life has been just about collective survival," Zhang told me. The ability to take stock of one's life and make choices about its final stage requires the kind of power and responsibility that most Chinese have never had.

At the beginning of May, Zhixia asked to go home. When I visited her the next month, with Meiying and Cuihe, both of them told me that none of the doctors had expected her to survive more than a week after being discharged. "It says something about the body's desperation to live," Meiying observed quietly.

Zhixia's room was a shrine to her son, its walls plastered with gold-and-red achievement certificates; a table and a dresser were crowded with framed photographs of a boy with Zhixia's wide, high cheeks. When we came in, Zhixia was propped up on cushions, swiping on her phone. Her feet were swollen, and her hip bones protruded sharply from underneath a thick cotton blanket. Sulin, her mother, eagerly reported that, in the past two days, Zhixia had eaten a little more soup than usual. When Meiying asked about her bedsores, Sulin lifted her daughter on her side to reveal a lesion on her back the shade of a darkening banana.

I followed Sulin into the kitchen, where she was preparing scallion pancakes for us and porridge for Zhixia. While we waited for the water to boil, I asked her about her plans for the future. "Zhixia specified that she wants to be cremated rather than buried," she said. "She doesn't want to be bitten by the bugs underground." Her eyes reddened as she spoke, and I gently clarified that I meant her own future: What about her old age? Sulin paused and looked up from the stone floor. She told me that, for the past five years, she had not thought about anything but her daughter. Outside, an old man wandered into the courtyard and squatted down to drink from a faucet. It was Zhixia's father-in-law. "He's half deaf, and his wife passed away not long ago," Sulin told me. "This house is his, or Zhixia and I wouldn't have a place to live. But, if I wasn't here, he wouldn't have a cook. We are a misshapen family—two white-haired parents raising an orphan." She smiled dryly as she said this, but then gave a heaving sob. "I know I've already sewn her funeral clothes, but I don't actually believe she'll die," she said. "I just can't believe it."

Zhixia hung on for seventeen more days. She died on a July morning, with her mother and son by her side. By then, she couldn't recognize them, or speak, swallow, or close her eyes. But the June day when I saw her had been a "good day," according to Sulin. Sunlight was streaming through an open window, and it was possible to imagine a string of summer days stretching on forever. Zhixia wanted only to talk about a math test her son had recently aced, the cucumbers and the tomatoes that had ripened in the yard outside, and a romance novel that she was reading on her phone. Sulin tried to usher us into the kitchen to eat the pan-cakes while they were still hot, but Zhixia's eyes suddenly flickered wide open. "No, please, don't let me be alone," she said. Everyone stopped, but it was Meiying who climbed onto the bed, bending down close to Zhixia's face. "You are not alone," she said, stroking Zhixia's arm with fingertips so light that they seemed barely to graze skin. "Not for a second."

SARAH ZHANG

The Last Children of
Down Syndrome

FROM *The Atlantic*

EVERY FEW WEEKS or so, Grete Fält-Hansen gets a call from a stranger asking a question for the first time: What is it like to raise a child with Down syndrome?

Sometimes the caller is a pregnant woman, deciding whether to have an abortion. Sometimes a husband and wife are on the line, the two of them in agonizing disagreement. Once, Fält-Hansen remembers, it was a couple who had waited for their prenatal screening to come back normal before announcing the pregnancy to friends and family. "We wanted to wait," they'd told their loved ones, "because if it had Down syndrome, we would have had an abortion." They called Fält-Hansen after their daughter was born —with slanted eyes, a flattened nose, and, most unmistakable, the extra copy of chromosome 21 that defines Down syndrome. They were afraid their friends and family would now think they didn't love their daughter—so heavy are the moral judgments that accompany wanting or not wanting to bring a child with a disability into the world.

All of these people get in touch with Fält-Hansen, a fifty-four-year-old schoolteacher, because she heads Landsforeningen Downs Syndrom, or the National Down Syndrome Association, in Denmark, and because she herself has an eighteen-year-old son, Karl Emil, with Down syndrome. Karl Emil was diagnosed after he was born. She remembers how fragile he felt in her arms and how she worried about his health, but mostly, she remembers, "I thought he was *so* cute." Two years after he was born, in 2004, Denmark became one of the first countries in the world to offer prenatal Down syndrome screening to every pregnant woman, regardless of age or other risk factors. Nearly all expecting mothers choose to take the test; of those who get a Down syndrome diagnosis, more than 95 percent choose to abort.

Denmark is not on its surface particularly hostile to disability. People with Down syndrome are entitled to health care, education, even money for the special shoes that fit their wider, more flexible feet. If you ask Danes about the syndrome, they're likely to bring up Morten and Peter, two friends with Down syndrome who starred in popular TV programs where they cracked jokes and dissected soccer games. Yet a gulf seems to separate the publicly expressed attitudes and private decisions. Since universal screening was introduced, the number of children born with Down syndrome has fallen sharply. In 2019, only 18 were born in the entire country. (About 6,000 children with Down syndrome are born in the United States each year.)

Fält-Hansen is in the strange position of leading an organization likely to have fewer and fewer new members. The goal of her conversations with expecting parents, she says, is not to sway them against abortion; she fully supports a woman's right to choose. These conversations are meant to fill in the texture of daily life missing both from the well-meaning cliché that "people with Down syndrome are always happy" and from the litany of possible symptoms provided by doctors upon diagnosis: intellectual disability, low muscle tone, heart defects, gastrointestinal defects, immune disorders, arthritis, obesity, leukemia, dementia. She might explain that, yes, Karl Emil can read. His notebooks are full of poetry written in his careful, sturdy handwriting. He needed physical and speech therapy when he was young. He loves music—his gold-rimmed glasses are modeled after his favorite Danish pop star's. He gets cranky sometimes, like all teens do.

One phone call might stretch into several; some people even come to meet her son. In the end, some join the association with their child. Others, she never hears from again.

These parents come to Fält-Hansen because they are faced with a choice—one made possible by technology that peers at the DNA of unborn children. Down syndrome is frequently called the "canary in the coal mine" for selective reproduction. It was one of the first genetic conditions to be routinely screened for in utero, and it remains the most morally troubling because it is among the least severe. It is very much compatible with life—even a long, happy life.

The forces of scientific progress are now marching toward ever more testing to detect ever more genetic conditions. Recent advances in genetics provoke anxieties about a future where parents choose what kind of child to have, or not have. But that hypothetical future is already here. It's been here for an entire generation.

Fält-Hansen says the calls she receives are about information, helping parents make a truly informed decision. But they are also moments of seeking, of asking fundamental questions about parenthood. Do you ever wonder, I asked her, about the families who end up choosing an abortion? Do you feel like you failed to prove that your life—and your child's life—is worth choosing? She told me she doesn't think about it this way anymore. But in the beginning, she said, she did worry: "What if they don't like my son?"

In January, I took a train from Copenhagen south to the small town of Vordingborg, where Grete, Karl Emil, and his thirty-year-old sister, Ann Katrine Kristensen, met me at the station. The three of them formed a phalanx of dark coats waving hello. The weather was typical of January—cold, gray, blustery—but Karl Emil pulled me over to the ice-cream shop, where he wanted to tell me he knew the employees. His favorite ice-cream flavor, he said, was licorice. "That's very Danish!" I said. Grete and Ann Katrine translated. Then he zagged over to a men's clothing store and struck up a conversation with the clerk, who had just seen Karl Emil interviewed on a Danish children's program with his girlfriend, Chloe. "You didn't tell me you had a girlfriend," the clerk teased. Karl Emil laughed, mischievous and proud.

We sat down at a café, and Grete gave her phone to Karl Emil to busy himself with while we spoke in English. He took selfies; his mother, sister, and I began to talk about Down syndrome and the country's prenatal-screening program. At one point, Grete was reminded of a documentary that had sparked an outcry in Denmark. She reclaimed her phone to look up the title: *Død Over Downs* (Death to Down Syndrome). When Karl Emil read over her shoulder, his face crumpled. He curled into the corner and refused to look at us. He had understood, obviously, and the distress was plain on his face.

Grete looked up at me: "He reacts because he can read."

"He must be aware of the debate?" I asked, which felt perverse to even say. *So he's aware there are people who don't want people like him to be born?* Yes, she said; her family has always been open with him. As a kid, he was proud of having Down syndrome. It was one of the things that made him uniquely Karl Emil. But as a teenager, he became annoyed and embarrassed. He could tell he was different. "He actually asked me, at some point, if it was because of Down syndrome that he sometimes didn't understand things," Grete said. "I just told him honestly: Yes." As he's gotten older, she said, he's made his peace with it. This arc felt familiar. It's the arc of growing up, in which our self-assuredness as young children gets upended in the storms of adolescence, but eventually, hopefully, we come to accept who we are.

The decisions parents make after prenatal testing are private and individual ones. But when the decisions so overwhelmingly swing one way—to abort—it does seem to reflect something more: an entire society's judgment about the lives of people with Down syndrome. That's what I saw reflected in Karl Emil's face.

Denmark is unusual for the universality of its screening program and the comprehensiveness of its data, but the pattern of high abortion rates after a Down syndrome diagnosis holds true across Western Europe and, to a somewhat lesser extent, in the United States. In wealthy countries, it seems to be at once the best and the worst time for Down syndrome. Better health care has more than doubled life expectancy. Better access to education means most children with Down syndrome will learn to read and write. Few people speak publicly about wanting to "eliminate" Down syndrome. Yet individual choices are adding up to something very close to that.

In the 1980s, as prenatal screening for Down syndrome became common, the anthropologist Rayna Rapp described the parents on the frontier of reproductive technology as "moral pioneers." Suddenly, a new power was thrust into the hands of ordinary people—the power to decide what kind of life is worth bringing into the world.

The medical field has also been grappling with its ability to offer this power. "If no one with Down syndrome had ever existed or ever would exist—is that a terrible thing? I don't know," says Laura Hercher, a genetic counselor and the director of student research at Sarah Lawrence College. If you take the health complications linked to Down syndrome, such as increased likelihood of early-onset Alzheimer's, leukemia, and heart defects, she told me, "I don't think anyone would argue that those are good things."

But she went on. "If our world didn't have people with special needs and these vulnerabilities," she asked, "would we be missing a part of our humanity?"

Sixty-one years ago, the first known prenatal test for a genetic disorder in the world took place in Copenhagen. The patient was a twenty-seven-year-old woman who was a carrier for hemophilia, a rare and severe bleeding disorder that is passed from mothers to sons. She had already given birth to one infant boy, who lived for just five hours. The obstetrician who delivered the baby, Fritz Fuchs, told her to come back if she ever became pregnant again. And in 1959, according to the published case study, she did come back, saying she couldn't go through with her pregnancy if she was carrying another son.

Fuchs had been thinking about what to do. Along with a cytologist named Povl Riis, he'd been experimenting with using fetal cells floating in the yellow amniotic fluid that fills the womb to determine a baby's sex. A boy would have a 50 percent risk of inheriting hemophilia; a girl would have almost no risk. But first they needed some amniotic fluid. Fuchs eased a long needle into the woman's abdomen; Riis studied the cells under a microscope. It was a girl.

The woman gave birth to a daughter a few months later. If the baby had been a boy, though, she was prepared to have an abortion—which was legal under Danish law at the time on "eugenic grounds" for fetuses at risk for severe mental or physical illness, according to Riis and Fuchs's paper describing the case. They acknowledged the possible danger of sticking a needle in the abdomen of a pregnant woman, but wrote that it was justified "because the method seems to be useful in preventive eugenics."

That word, *eugenics,* today evokes images that are specific and heinous: forced sterilization of the "feebleminded" in early-twentieth-century America, which in turn inspired the racial hygiene of the Nazis, who gassed or otherwise killed tens of thousands of people with disabilities, many of them children. But eugenics was once a mainstream scientific pursuit, and eugenicists believed that they were bettering humanity. Denmark, too, drew inspiration from the United States, and it passed a sterilization law in 1929. Over the next twenty-one years, 5,940 people were sterilized in Denmark, the majority because they were "mentally retarded." Those who resisted sterilization were threatened with institutionalization.

Eugenics in Denmark never became as systematic and violent as it did in Germany, but the policies came out of similar underlying goals: improving the health of a nation by preventing the birth of those deemed to be burdens on society. The term *eugenics* eventually fell out of favor, but in the 1970s, when Denmark began offering prenatal testing for Down syndrome to mothers over the age of thirty-five, it was discussed in the context of saving money—as in, the testing cost was less than that of institutionalizing a child with a disability for life. The stated purpose was "to prevent birth of children with severe, lifelong disability."

That language, too, has long since changed; in 1994, the stated purpose of the testing became "to offer women a choice." Activists like Fält-Hansen have also pushed back against the subtle and not-so-subtle ways that the medical system encourages women to choose abortion. Some Danish parents told me that doctors automatically assumed they would want to schedule an abortion, as if there was really no other option. This is no longer the case, says Puk Sandager, a fetal-medicine specialist at Aarhus University Hospital. Ten years ago, doctors—especially older doctors—were more likely to expect parents to terminate, she told me. "And now we do not expect anything." The National Down Syndrome Association has also worked with doctors to alter the language they use with patients—"probability" instead of "risk," "chromosome aberration" instead of "chromosome error." And, of course, hospitals now connect expecting parents with people like Fält-Hansen to have those conversations about what it's like to raise a child with Down syndrome.

Perhaps all of this has had some effect, though it's hard to say. The number of babies born to parents who chose to continue a pregnancy after a prenatal diagnosis of Down syndrome in Denmark has ranged from zero to thirteen a year since universal screening was introduced. In 2019, there were seven. (Eleven other babies were born to parents who either declined the test or got a false negative, making the total number of babies born with Down syndrome last year eighteen.)

Why so few? "Looking at it from the outside, a country like Denmark, if you want to raise a child with Down syndrome, this is a good environment," says Stina Lou, an anthropologist who has studied how parents make decisions after a prenatal diagnosis of a fetal anomaly. Since 2011, she has embedded in the fetal-medicine unit at Aarhus University Hospital, one of the largest hospitals in Denmark, where she has shadowed Sandager and other doctors.

Under the 2004 guidelines, all pregnant women in Denmark are offered a combined screening in the first trimester, which includes blood tests and an ultrasound. These data points, along with maternal age, are used to calculate the odds of Down syndrome. The high-probability patients are offered a more invasive diagnostic test using DNA either from the fetal cells floating in the amniotic fluid (amniocentesis) or from placental tissue (chorionic villus sampling). Both require sticking a needle or catheter into the womb and come with a small risk of miscarriage. More recently, hospitals have started offering noninvasive prenatal testing, which uses fragments of fetal DNA floating in the mother's blood. That option has not become popular in Denmark, though, probably because the invasive tests can pick up a suite of genetic disorders in addition to Down syndrome. More diseases ruled out, more peace of mind.

But Lou was interested in the times when the tests did not provide peace of mind, when they in fact provided the opposite. In a study of twenty-one women who chose abortion after a prenatal diagnosis of Down syndrome, she found that they had tended to base their decisions on worst-case scenarios. An extra copy of chromosome 21 can cause a variety of symptoms, the severity of which is not known until birth or even later. Most people with Down syndrome learn to read and write. Others are nonverbal. Some do not have heart defects. Others spend months or even years in and out of the hospital to fix a heart valve. Most have healthy digestive systems. Others lack the nerve endings needed to anticipate bowel movements, necessitating more surgeries, possibly even a stoma bag or diapers. The women who chose abortion feared the worst possible outcomes. Some even grieved the possibility of aborting a child who might have had a mild form of Down syndrome. But in the end, Lou told me, "the uncertainty just becomes too much."

This emphasis on uncertainty came up when I spoke with David Wasserman, a bioethicist at the U.S. National Institutes of Health who, along with his collaborator Adrienne Asch, has written some of the most pointed critiques of selective abortion. (Asch died in 2013.) They argued that prenatal testing has the effect of reducing an unborn child to a single aspect—Down syndrome, for example—and making parents judge the child's life on that alone. Wasserman told me he didn't think that most parents who make these decisions are seeking perfection. Rather, he said, "there's profound risk aversion."

It's hard to know for sure whether the people in Lou's study decided to abort for the reasons they gave or if these were retrospective justifications. But when Lou subsequently interviewed parents who had made the unusual choice to continue a pregnancy after a Down syndrome diagnosis, she found them more willing to embrace uncertainty.

Parents of children with Down syndrome have described to me the initial process of mourning the child they thought they would have: the child whom they were going to walk down the aisle, who was going to graduate from college, who was going to become president. None of this is guaranteed with any kid, of course, but while most parents go through a slow realignment of expectations over the years, prenatal testing was a rapid plummet into disappointment—all those dreams, however unrealistic, evaporating at once. And then the doctors present you with a long list of medical conditions associated with Down syndrome. Think about it this way, Karl Emil's sister, Ann Katrine, said: "If you handed any expecting parent a whole list of everything their child could possibly encounter during their entire life span—illnesses and stuff like that—then anyone would be scared."

"Nobody would have a baby," Grete said.

A peculiar effect of Denmark's universal-screening program and high abortion rate for Down syndrome is that a fair number of babies born with Down syndrome are born to parents who essentially got a false negative. Their first-trimester screening results said their odds were very low—so low that they needed no invasive follow-up testing. They simply went on with what they thought was an ordinary pregnancy. In other words, like the couple Grete once counseled, these are parents who might have chosen to abort, had they known.

The day after I met Grete, I attended a meeting of the local

Copenhagen Down syndrome group. The woman who invited me, Louise Aarsø, had a then-five-year-old daughter with Down syndrome, Elea. Aarsø and her husband had made the unusual choice to opt out of screening. Though they support the right to abortion, they knew they would want to have the baby either way. At the meeting, two of the seven other families told me their pre-natal screening had suggested extremely low odds. At birth, they were surprised. A few others said they had chosen to continue the pregnancy despite a high probability for Down syndrome. Ulla Hartmann, whose son Ditlev was eighteen, noted that he was born before the national screening program began. "We're very thankful we didn't know, because we had two twin boys when I got pregnant with Ditlev and I really don't think we would have been, 'Okay, let's take this challenge when we have these monkeys up in the curtains,'" she told me. "But you grow with the challenge."

Daniel Christensen was one of the parents who had been told the odds of Down syndrome were very low, something like 1 in 1,500. He and his wife didn't have to make a choice, and when he thinks back on it, he said, "what scares me the most is actually how little we knew about Down syndrome." What would the basis of their choice have been? Their son August is four now, with a twin sister, who Christensen half-jokingly said was "almost normal." The other parents laughed. "Nobody's normal," he said.

Then the woman to my right spoke; she asked me not to use her name. She wore a green blouse, and her blond hair was pulled into a ponytail. When we all turned to her, I noticed that she had begun to tear up. "Now I'm moved from all the stories; I'm a lit-tle . . ." She paused to catch her breath. "My answer is not that beautiful." The Down syndrome odds for her son, she said, were 1 in 969.

"You remember the exact number?" I asked.

"Yeah, I do. I went back to the papers." The probability was low enough that she didn't think about it after he was born. "On the one hand I saw the problems. And on the other hand he was per-fect." It took four months for him to get diagnosed with Down syndrome. He is six now, and he cannot speak. It frustrates him, she said. He fights with his brother and sister. He bites because he cannot express himself. "This has just been *so many times,* and you never feel safe." Her experience is not representative of all children with Down syndrome; lack of impulse control is common, but violence is not. Her point, though, was that the image of a happy-go-lucky child so often featured in the media is not always representative either. She wouldn't have chosen this life: "We would have asked for an abortion if we knew."

Another parent chimed in, and the conversation hopscotched to a related topic and then another until it had moved on entirely. At the end of the meeting, as others stood and gathered their coats, I turned to the woman again because I was still shocked that she was willing to say what she'd said. Her admission seemed to violate an unspoken code of motherhood.

Of course, she said, "it's shameful if I say these things." She loves her child, because how can a mother not? "But you love a person that hits you, bites you? If you have a husband that bites you, you can say goodbye . . . but if you have a child that hits you, you can't do anything. You can't just say, 'I don't want to be in a relationship.' Because it's your child." To have a child is to begin a relationship that you cannot sever. It is supposed to be unconditional, which is perhaps what most troubles us about selective abortion—it's an admission that the relationship can in fact be conditional.

Parenting is a plunge into the unknown and the uncontrollable. It is beautiful in this way, but also daunting.

In the cold, scientific realm of biology, reproduction begins with a random genetic shuffling—an act of fate, if you were to be less cold, more poetic. The twenty-three pairs of chromosomes in our cells line up so that the DNA we inherited from our mother and father can be remixed and divided into sets of twenty-three single chromosomes. Each egg or sperm gets one such set. In women, this chromosomal division begins, remarkably, when they themselves are fetuses in *their* mother's womb. The chromosomes freeze in place for twenty, thirty, even forty-plus years as the fetus becomes a baby, a girl, a woman. The cycle finishes only when the egg is fertilized. During the intervening years, the proteins holding chromosomes together can degrade, resulting in eggs with too many or too few chromosomes. This is the biological mechanism behind most cases of Down syndrome—95 percent of people born with an extra copy of chromosome 21 inherited it from their mother. And this is why the syndrome is often, though not always, linked to the age of the mother.

In the interviews I've conducted, and in interviews Lou and researchers across the United States have conducted, the choice of what to do after a prenatal test fell disproportionately on mothers. There were fathers who agonized over the choice too, but mothers usually bore most of the burden. There is a feminist explanation (my body, my choice) and a less feminist one (family is still primarily the domain of women), but it's true either way. And in making these decisions, many of the women seemed to anticipate the judgment they would face.

Lou told me she had wanted to interview women who chose abortion after a Down syndrome diagnosis because they're a silent majority. They are rarely interviewed in the media, and rarely willing to be interviewed. Danes are quite open about abortion —astonishingly so to my American ears—but abortions for a fetal anomaly, and especially Down syndrome, are different. They still carry a stigma. "I think it's because we as a society like to think of ourselves as inclusive," Lou said. "We are a rich society, and we think it's important that different types of people should be here." And for some of the women who end up choosing abortion, "their own self-understanding is a little shaken, because they have to accept they aren't the kind of person like they thought," she said. They were not the type of person who would choose to have a child with a disability.

For the women in Lou's study, ending a pregnancy after a prenatal diagnosis was very different from ending an unwanted pregnancy. These were almost all wanted pregnancies, in some cases very much wanted pregnancies following long struggles with infertility. The decision to abort was not taken lightly. One Danish woman I'll call "L" told me how terrible it was to feel her baby inside her once she'd made the decision to terminate. In the hospital bed, she began sobbing so hard, the staff had difficulty sedating her. The depth of her emotions surprised her, because she was so sure of her decision. The abortion was two years ago, and she doesn't think about it much anymore. But recounting it on the phone, she began crying again.

She was disappointed to find so little in the media about the experiences of women like her. "It felt right for me, and I have no regrets at all," she told me, but it also feels like "you're doing something wrong." L is a filmmaker, and she wanted to make a documentary about choosing abortion after a Down syndrome diagnosis. She even thought she would share her own story. But she hadn't been able to find a couple willing to be in this documentary, and she wasn't ready to put herself out there alone.

When Rayna Rapp, the anthropologist who coined the term *moral pioneers*, interviewed parents undergoing prenatal testing in New York in the 1980s and '90s, she noticed a certain preoccupation among certain women. Her subjects represented a reasonably diverse slice of the city, but middle-class white women especially seemed fixated on the idea of "selfishness." The women she interviewed were among the first in their families to forgo homemaking for paid work; they had not just jobs but *careers* that were central to their identity. With birth control, they were having fewer children and having them later. They had more reproductive autonomy than women had ever had in human history. (Rapp herself came to this research after having an abortion because of Down syndrome when she became pregnant as a thirty-six-year-old professor.) "Medical technology transforms their 'choices' on an individual level, allowing them, like their male partners, to imagine voluntary limits to their commitments to their children," Rapp wrote in her book *Testing Women, Testing the Fetus.*

But exercising those "voluntary limits" on motherhood—choosing not to have a child with a disability out of fear for how it might affect one's career, for example—becomes judged as "selfishness." Medical technology can offer women a choice, but it does not instantly transform the society around them. It does not dismantle the expectation that women are the primary caregivers or erase the ideal of a good mother as one who places no limits on her devotion to her children.

The centrality of choice to feminism also brings it into uncomfortable conflict with the disability rights movement. Anti-abortion-rights activists in the United States have seized on this to introduce bills banning selective abortion for Down syndrome in several states. Feminist disability scholars have attempted to resolve the conflict by arguing that the choice is not a real choice at all. "The decision to abort a fetus with a disability even because it 'just seems too difficult' must be respected," Marsha Saxton, the director of research at the World Institute on Disability, wrote in 1998. But Saxton calls it a choice made "under duress," arguing that a woman faced with this decision is still constrained today—by popular misconceptions that make life with a disability out to be worse than it actually is and by a society that is hostile to people with disabilities.

And when fewer people with disabilities are born, it becomes harder for the ones who *are* born to live a good life, argues Rosemarie Garland-Thomson, a bioethicist and professor emerita at Emory University. Fewer people with disabilities means fewer services, fewer therapies, fewer resources. But she also recognizes how this logic pins the entire weight of an inclusive society on individual women.

No wonder, then, that "choice" can feel like a burden. In one small study of women in the United States who chose abortion after a diagnosis of a fetal anomaly, two-thirds said they'd hoped—or even prayed—for a miscarriage instead. It's not that they wanted their husbands, their doctors, or their lawmakers to tell them what to do, but they recognized that choice comes with responsibility and invites judgment. "I have guilt for not being the kind of person who could parent this particular type of special need," said one woman in the study. "Guilt, guilt, guilt."

The introduction of a choice reshapes the terrain on which we all stand. To opt out of testing is to become someone who *chose* to opt out. To test and end a pregnancy because of Down syndrome is to become someone who *chose* not to have a child with a disability. To test and continue the pregnancy after a Down syndrome diagnosis is to become someone who *chose* to have a child with a disability. Each choice puts you behind one demarcating line or another. There is no neutral ground, except perhaps in hoping that the test comes back negative and you never have to choose what's next.

What kind of choice is this, if what you hope is to not have to choose at all?

Down syndrome is unlikely to ever disappear from the world completely. As women wait longer to have children, the incidence of pregnancies with an extra copy of chromosome 21 is going up. Prenatal testing can also in rare cases be wrong, and some parents will choose not to abort or not to test at all. Others will not have access to abortion.

In the United States—which has no national health care system, no government mandate to offer prenatal screening—the best estimate for the termination rate after a diagnosis of Down syndrome is 67 percent. But that number conceals stark differences within the country. One study found higher rates of termination in the West and Northeast and among mothers who are highly educated. "On the Upper East Side of Manhattan, it's going to be completely different than in Alabama," said Laura Hercher, the genetic counselor.

These differences worry Hercher. If only the wealthy can afford to routinely screen out certain genetic conditions, then those conditions can become proxies of class. They can become, in other words, *other people's problems*. Hercher worries about an empathy gap in a world where the well-off feel insulated from sickness and disability.

For those with the money, the possibilities of genetic selection are expanding. The leading edge is preimplantation genetic testing (PGT) of embryos created through in vitro fertilization, which altogether can cost tens of thousands of dollars. Labs now offer testing for a menu of genetic conditions—most of them rare and severe conditions such as Tay-Sachs disease, cystic fibrosis, and phenylketonuria—allowing parents to select healthy embryos for implantation in the womb. Scientists have also started trying to understand more common conditions that are influenced by hundreds or even thousands of genes: diabetes, heart disease, high cholesterol, cancer, and—much more controversially—mental illness and autism. In late 2018, Genomic Prediction, a company in New Jersey, began offering to screen embryos for risk of hundreds of conditions, including schizophrenia and intellectual disability, though it has since quietly backtracked on the latter. The one test customers keep asking for, the company's chief scientific officer told me, is for autism. The science isn't there yet, but the demand is.

The politics of prenatal testing for Down syndrome and abortion are currently yoked together by necessity: the only intervention offered for a prenatal test that finds Down syndrome is an abortion. But modern reproduction is opening up more ways for parents to choose what kind of child to have. PGT is one example. Sperm banks, too, now offer detailed donor profiles delineating eye color, hair color, education; they also screen donors for genetic disorders. Several parents have sued sperm banks after discovering that their donor may have undesirable genes, in cases where their children developed conditions such as autism or a degenerative nerve disease. In September, the Georgia Supreme Court ruled that one such case, in which a sperm donor had hidden his history of mental illness, could move forward. The "deceptive trade practices" of a sperm bank that misrepresented its donor-screening process, the court ruled, could "essentially amount to ordinary consumer fraud."

Garland-Thomson calls this commercialization of reproduction "velvet eugenics"—*velvet* for the soft, subtle way it encourages the eradication of disability. Like the Velvet Revolution from which she takes the term, it's accomplished without overt violence. But it also takes on another connotation as human reproduction becomes more and more subject to consumer choice: *velvet*, as in quality, high-caliber, premium-tier. Wouldn't you want only the best for your baby—one you're already spending tens of thousands of dollars on IVF to conceive? "It turns people into products," Garland-Thomson says.

None of this suggests that testing should be entirely abandoned. Most parents choosing genetic testing are seeking to spare their children real physical suffering. Tay-Sachs disease, for example, is caused by mutations in the *HEXA* gene, which causes the destruction of neurons in the brain and spinal cord. At about three to six months old, babies begin losing motor skills, then their vision and hearing. They develop seizures and paralysis. Most do not live past childhood. There is no cure.

In the world of genetic testing, Tay-Sachs is a success story. It has been nearly eliminated through a combination of prenatal testing of fetuses; preimplantation testing of embryos; and, in the Ashkenazi Jewish population, where the mutation is especially prevalent, carrier screening to discourage marriages between people who might together pass on the mutation. The flip side of this success is that having a baby with the disease is no longer simple misfortune because nothing could have been done. It can be seen instead as a failure of personal responsibility.

Fertility doctors have spoken to me passionately about expanding access to IVF for parents who are fertile but who might use embryo screening to prevent passing on serious diseases. In a world where IVF becomes less expensive and less hard on a woman's body, this might very well become the responsible thing to do. And if you're already going through all this to screen for one disease, why not avail yourself of the whole menu of tests? The hypothetical that Karl Emil's sister imagined, in which a child's every risk is laid out, feels closer than ever. How do you choose between one embryo with a slightly elevated risk of schizophrenia and another with a moderate risk of breast cancer?

Not surprisingly, those advocating for preimplantation genetic testing prefer to keep the conversation focused on monogenic diseases, where single gene mutations have severe health effects. Talk of minimizing the risk of conditions like diabetes and mental illness—which are also heavily influenced by environment—quickly turns to designer babies. "Why do we want to go there?" says David Sable, a former IVF doctor who is now a venture capitalist specializing in life sciences. "Start with the most scientifically straightforward, the monogenic diseases—cystic fibrosis, sickle cell anemia, hemophilia—where you could define very specifically what the benefit is."

What about Down syndrome, then, I asked, which can be much less severe than those diseases but is routinely screened for anyway? His answer surprised me, considering that he has spent much of his career working with labs that count chromosomes: "The concept of counting chromosomes as a definitive indicator of the truth—I think we're going to look back on that and say, 'Oh my God, we were so misguided.'" Consider the sex chromosomes, he said. "We've locked ourselves into this male-female binary that we enforced with XX and XY." But it's not nearly so neat. Babies born XX can have male reproductive organs; those born XY can have female reproductive organs. And others can be born with an unusual number of sex chromosomes like X, XXY, XYY, XXYY, XXXX, the effects of which range widely in severity. Some might never know there's anything unusual in their chromosomes at all.

When Rayna Rapp was researching prenatal testing back in the '80s and '90s, she came across multiple sets of parents who chose to abort a fetus with a sex-chromosome anomaly out of fear that it could lead to homosexuality—never mind that there is no known link. They also worried that a boy who didn't conform to XY wouldn't be masculine enough. Reading about their anxieties thirty years later, I could sense how much the ground had moved under our feet. Of course, some parents might still have the same fears, but today the boundaries of "normal" for gender and sexuality encompass much more than the narrow band of three decades ago. A child who is neither XX nor XY can fit into today's world much more easily than in a rigidly gender-binary one.

Both sex-chromosome anomalies and Down syndrome were early targets of prenatal testing—not because they are the most dangerous conditions but because they were the easiest to test for. It's just counting chromosomes. As science moves past this relatively rudimentary technique, Sable mused, "the term *Down syndrome* is probably going to go away at some point, because we may find that having that third 21 chromosome maybe does not carry a predictable level of suffering or altered function." Indeed, most pregnancies with a third copy of chromosome 21 end as miscarriages. Only about 20 percent survive to birth, and the people who are born have a wide range of intellectual disabilities and physical ailments. How can an extra chromosome 21 be incompatible with life in some cases and in other cases result in a boy, like one I met, who can read and write and perform wicked juggling tricks with his diabolo? Clearly, something more than just an extra chromosome is going on.

As genetic testing has become more widespread, it has revealed just how many other genetic anomalies many of us live with—not only extra or missing chromosomes, but whole chunks of chromosome getting deleted, chunks duplicated, chunks stuck onto a different chromosome altogether, mutations that should be deadly but that show up in the healthy adult in front of you. Every person carries a set of mutations unique to them. This is why new and rare genetic diseases are so hard to diagnose—if you compare a person's DNA with a reference genome, you come up with hundreds of thousands of differences, most of them utterly irrelevant to the disease. What, then, is normal? Genetic testing, as a medical service, is used to enforce the boundaries of "normal" by screening out the anomalous, but seeing all the anomalies that are compatible with life might actually expand our understanding of normal. "It's expanded mine," Sable told me.

Sable offered this up as a general observation. He didn't think he was qualified to speculate on what this meant for the future of Down syndrome screening, but I found this conversation about genetics unexpectedly resonant with something parents had told me. David Perry, a writer in Minnesota whose thirteen-year-old son has Down syndrome, said he disliked how people with Down syndrome are portrayed as angelic and cute; he found it flattening and dehumanizing. He pointed instead to the way the neurodiversity movement has worked to bring autism and ADHD into the realm of normal neurological variation. "We need more kinds of normal," another father, Johannes Dybkjær Andersson, a musician and creative director in Copenhagen, said. "That's a good thing, when people show up in our lives"—as his daughter, Sally, did six years ago—"and they are just normal in a totally different way." Her brain processes the world differently than his does. She is unfiltered and open. Many parents have told me how this quality can be awkward or disruptive at times, but it can also break the stifling bounds of social propriety.

Stephanie Meredith, the director of the National Center for Prenatal and Postnatal Resources at the University of Kentucky, told me of the time her twenty-year-old son saw his sister collide with another player on the basketball court. She hit the ground so hard that an audible crack went through the gym. Before Meredith could react, her son had already leapt from the bleachers and picked his sister up. "He wasn't worried about the rules; he wasn't worried about decorum. It was just responding and taking care of her," Meredith told me. She had recently been asked a simple but probing question: What was she most proud of about her son that was not an achievement or a milestone? The incident on the basketball court was one that came to mind. "It doesn't have to do with accomplishment," she said. "It has to do with caring about another human being."

That question had stayed with Meredith—and it stayed with me —because of how subtly yet powerfully it reframes what parents should value in their children: not grades or basketball trophies or college-acceptance letters or any of the things parents usually brag about. By doing so, it opens the door to a world less obsessed with achievement. Meredith pointed out that Down syndrome is defined and diagnosed by a medical system made up of people who have to be highly successful to get there, who likely base part of their identity on their intelligence. This is the system giving parents the tools to decide what kind of children to have. Might it be biased on the question of whose lives have value?

When Mary Wasserman gave birth to her son, Michael, in 1961, kids with Down syndrome in America were still routinely sent to state institutions. She remembers the doctor announcing, "It's a mongoloid idiot"—the term used before chromosome counting became common—and telling her "it" should go to the state institution right away. Wasserman had volunteered for a week at such an institution in high school, and she would never forget the sights, the sounds, the *smells*. The children were soiled, uncared for, unnurtured. In defiance of her doctor, she took Michael home.

The early years were not easy for Wasserman, who was a divorced mother for much of Michael's childhood. She worked to support them both. There weren't really any formal day cares then, and the women who ran informal ones out of their homes didn't want Michael. "The other mothers were not comfortable," one of them told her after his first week. Others rejected him outright. She hired private babysitters, but Michael didn't have playmates. It wasn't until he was eight, when a school for kids with disabilities opened nearby, that Michael went to school for the first time.

Michael is fifty-nine now. The life of a child born with Down syndrome today is very different. State institutions closed down after exposés of the unsanitary and cruel conditions that Wasserman had glimpsed as a high school student. After children with disabilities go home from the hospital today, they have access to a bevy of speech, physical, and occupational therapies from the government—usually at no cost to families. Public schools are required to provide equal access to education for kids with disabilities. In 1990, the Americans with Disabilities Act prohibited discrimination in employment, public transportation, day cares, and other businesses. Inclusion has made people with disabilities a visible and normal part of society; instead of being hidden away in institutions, they live among everyone else. Thanks to the activism of parents like Wasserman, all of these changes have taken place in her son's lifetime.

Does she wish Michael had had the opportunities that kids have now? "Well," she says, "I think maybe in some ways it was easier for us." Of course the therapies would have helped Michael. But there's more pressure on kids and parents today. She wasn't shuttling Michael to appointments or fighting with the school to get him included in general classes or helping him apply to the college programs that have now proliferated for students with intellectual disabilities. "It was less stressful for us than it is today," she says. Raising a child with a disability has become a lot more intensive—not unlike raising any child.

I can't count how many times, in the course of reporting this story, people remarked to me, "You know, people with Down syndrome work and go to college now!" This is an important corrective to the low expectations that persist and a poignant reminder of how a transforming society has transformed the lives of people with Down syndrome. But it also does not capture the full range of experiences, especially for people whose disabilities are more serious and those whose families do not have money and connections. Jobs and college are achievements worth celebrating—like any kid's milestones—but I've wondered why we so often need to point to achievements for evidence that the lives of people with Down syndrome are meaningful.

When I had asked Grete Fält-Hansen what it was like to open up her life to parents trying to decide what to do after a prenatal diagnosis of Down syndrome, I suppose I was asking her what it was like to open up her life to the judgment of those parents—and also of me, a journalist, who was here asking the same questions. As she told me, she had worried at first that people might not like her son. But she understands now how different each family's circumstances can be and how difficult the choice can be. "I feel sad about thinking about pregnant women and the fathers, that they are met with this choice. It's almost impossible," she said. "Therefore, I don't judge them."

Karl Emil had grown bored while we talked in English. He tugged on Grete's hair and smiled sheepishly to remind us that he was still there, that the stakes of our conversation were very real and very human.

Contributors' Notes

Nora Caplan-Bricker is a journalist, essayist, and critic in Boston. Her work appears in *The New Yorker*, the *New York Times Magazine*, *Harper's*, and *Ploughshares*, among other places. She is the web editor at *Jewish Currents*.

Julia Craven writes about racism and health disparities for *Slate* magazine. Previously, she spent five years reporting at *HuffPost*, centering her work on racism and politics. She's a born, bred, and educated North Carolina Tar Heel who appreciates good whiskey, good memes, and women rappers. She's a 2021 finalist for a Writers Guild Award, a member of the Center for Health Journalism at the University of Southern California, and a National Association of Black Journalists Salute to Excellence Award nominee.

Meehan Crist is writer-in-residence in biological sciences at Columbia University. Her work has appeared in publications such as the *New York Times*, the *London Review of Books*, *The Atlantic*, *The Nation*, and *Scientific American*. She is co-editor of the nonfiction collection *What Future 2018* (Unnamed Press), a founding member of NeuWrite, and the host of *Convergence: A Show About the Future*. Her nonfiction book on the climate crisis, *Is It OK to Have a Child?*, is forthcoming.

Bathsheba Demuth is author of *Floating Coast*, which was named a best book of 2019 by *Nature* and National Public Radio, among other outlets. An assistant professor of history and environment and society at Brown University, she lives in Rhode Island when not in the Arctic. She is currently writing a book about the Yukon River.

Susan Dominus has worked for the *New York Times* since 2007, first as a Metro columnist and then as a staff writer with the *New York Times Maga-*

zine. She has been a member of two Pulitzer Prize–winning teams: in 2009 her team won the Pulitzer Prize for breaking news, for its coverage of the scandal that resulted in the resignation of Governor Eliot Spitzer, and in 2018 she was part of the team that won the Pulitzer Prize for public service, for reporting on workplace sexual harassment issues. At the *New York Times Magazine,* she is grateful to have worked closely with her superb editor, Ilena Silverman.

Katie Engelhart is a writer and documentary film producer, based in Toronto and New York City, and a National Fellow at New America. Her first book, *The Inevitable: Dispatches on the Right to Die,* was published 2021.

Jiayang Fan is a staff writer at *The New Yorker.* She is working on her first book, *Motherland,* which will be published in 2023.

Latria Graham, a journalist and fifth-generation South Carolina farmer, is a graduate of Dartmouth College and later earned her MFA in creative nonfiction from the New School in New York City. In 2019 she was awarded the Great Smoky Mountain Association's Steve Kemp writer-in-residence position, and for two years she has been in and out of conservation spaces, intent on unearthing long-forgotten Black history that she finds crucial to the narrative we tell about the American South. She holds contributing editor positions at *Garden & Gun* and *Outdoor Retailer.* Her essays, profiles, and reviews have appeared in *The Guardian,* the *New York Times,* the *Los Angeles Times,* espnW, *Southern Living, Bicycling,* and *Backpacker.* You can find more of her work at LatriaGraham.com.

Heather Hogan is a writer and editor who lives in New York City with her wife, Stacy, and their cackle of rescued pets. She's a member of the Television Critics Association and the Gay and Lesbian Entertainment Critics Association and is also a Rotten Tomatoes Tomatometer critic. She's been living with long Covid since March 2020.

Sabrina Imbler is a writer based in Brooklyn. Their essay collection about sea creatures is forthcoming in 2022.

Brooke Jarvis is a contributing writer for the *New York Times Magazine.* A winner of the Livingston Award for National Reporting, she teaches graduate feature writing at New York University. Her work has been anthologized in *The Best American Science and Nature Writing, The Best American Travel Writing, Love and Ruin: Tales of Obsession, Danger, and Heartbreak from "The Atavist Magazine,"* and *New Stories We Tell: True Tales by America's Next Generation of Great Women Journalists.* She lives in Seattle.

Maya L. Kapoor tends to write about the underappreciated organisms, places, and histories of the U.S. West. Her writing can be found in, among other venues, *The Atlantic, Business Insider, Grist, Longreads, Mother Jones, Newsweek, Slate,* and *Undark,* and her work has been anthologized in *How We Speak to One Another* and *The Sonoran Desert: A Literary Field Guide.* She is an associate editor for *High Country News,* where she writes science features and edits reportage, essays, and book reviews.

Katy Kelleher is a writer who lives in the woods of Maine. She's working on a book, *The Ugly History of Beautiful Things,* about the darkness that lies behind some of her favorite pretty objects.

Roxanne Khamsi is an independent science journalist whose articles have appeared in publications such as *The Economist, Wired, Nature, Scientific American,* and the *New York Times.* She has received wide recognition for her work, including the American Medical Writers Association's Walter C. Alvarez Award and multiple first-place awards from the Association of Health Care Journalists. She has taught health reporting and science communication at Stony Brook University's Alan Alda Center and at the Craig Newmark Graduate School of Journalism. She lives in Montreal.

Marina Koren is a staff writer at *The Atlantic.* She covers all things space, from astronaut missions and robotic explorers to the wonders of the solar system and beyond. She has reported from Cape Canaveral in Florida, SpaceX's launch site in South Texas, and NASA's headquarters in Washington, D.C. Before working in science journalism, she was a breaking-news and political reporter.

Amanda Mull is a staff writer at *The Atlantic,* where she covers health and writes Material World, a column on consumerism. A Georgia native and graduate of the University of Georgia, she now resides in Brooklyn, New York, with her chihuahua, Midge.

Susan Orlean has been a staff writer for *The New Yorker* since 1992 and is the author of eight books, including *The Orchid Thief* and *The Library Book.*

Helen Ouyang is an emergency physician, writer, and associate professor at Columbia University. She has written for *The Atlantic, Harper's,* the *Los Angeles Times, New York, The New Yorker,* the *New York Times,* the *New York Times Magazine,* the *Washington Post,* and others. She has also been a National Magazine Award finalist.

Emily Raboteau is the author of *Searching for Zion,* winner of the American

Book Award. She is a regular contributor to the *New York Review of Books* and a contributing editor at *Orion* magazine. Her next book, *Caution: Lessons in Survival,* is forthcoming. The crowd-sourced essay in this volume was inspired by Susan Sontag's experimental short story about the AIDS crisis, "The Way We Live Now," published in 1986.

Julia Rosen is an independent journalist covering science and the environment. She is fascinated by how the world works and how humans are changing it. Her writing has appeared in the *New York Times, The Atlantic, Science, Hakai,* and *High Country News,* among other publications, and she is a former science reporter for the *Los Angeles Times.* She lives in Portland, Oregon, with her family.

Jennifer Senior is a staff writer for *The Atlantic,* where she writes long-form features about a wide variety of subjects. Before joining *The Atlantic,* she was a columnist for the *New York Times* and one of the paper's three daily book critics. She spent almost twenty years as a staff writer for *New York,* writing profiles and cover stories about politics, social science, and mental health. She is also the author of *All Joy and No Fun,* which spent eight weeks on the *New York Times* best-seller list, has been translated into twelve languages, and was named by *Slate* as one of the Top 10 Books of 2014. Awarded many journalism prizes, including a GLAAD Award, the Erikson Prize in Mental Health Media, and two Front Page Awards from the Newswomen's Club of New York, she lives in Brooklyn with her husband and son.

Namwali Serpell is a Zambian writer and a professor of English at Harvard University. She is the author of *Seven Modes of Uncertainty* and *Stranger Faces.* Her first novel, *The Old Drift,* which fictionalized some of the history of the Tonga people she tells here, won a Windham-Campbell Prize for fiction, the Anisfield-Wolf Book Prize for fiction "that confronts racism and explores diversity," the Arthur C. Clarke Award for science fiction, and the *Los Angeles Times* Art Seidenbaum Award for First Fiction in 2020.

Shannon Stirone is a freelance writer based in New York City. She covers space exploration, astronomy, and why humans are enamored with the cosmos. Her work can be found in the *New York Times,* the *Washington Post, Longreads, Wired,* and others.

Zeynep Tufekci is a visiting associate professor at Columbia University and a contributing writer for *The Atlantic* and the *New York Times,* as well as her own newsletter, *Insight.*

Rosanna Xia is an environment reporter for the *Los Angeles Times*. She covers the coast and was a Pulitzer Prize finalist in 2020 for explanatory reporting.

Sarah Zhang is a staff writer at *The Atlantic*, where she covers health and science. She was previously a staff writer at *Wired*, and her writing has appeared in the *New York Times*, *Nature*, and other publications. She has a degree in neurobiology from Harvard.

Other Notable Science and Nature Writing of 2020

THE BEST AMERICAN SERIES®

FIRST, BEST, AND BEST-SELLING

The Best American Essays

The Best American Food Writing

The Best American Mystery and Suspense

The Best American Science and Nature Writing

The Best American Science Fiction and Fantasy

The Best American Short Stories

The Best American Travel Writing

Available in print and e-book wherever books are sold.

Visit our website: MarinerBooks.com/BestAmerican